Environment Engineering and Disaster Management

ENVIRONMENT ENGINEERING AND DISASTER MANAGEMENT

Semester-II, B.Tech. All Branches, RTU, Kota

By

Dr. Sanjay K. Sharma

Professor, Head of Deptt. in Chemistry
Institute of Engineering and Technology
Alwar, Rajasthan

UNIVERSITY SCIENCE PRESS

(An Imprint of Laxmi Publications Pvt. Ltd.)

BANGALORE • CHENNAI • COCHIN • GUWAHATI • HYDERABAD
JALANDHAR • KOLKATA • LUCKNOW • MUMBAI • PATNA
RANCHI • NEW DELHI

Published by :
UNIVERSITY SCIENCE PRESS
(An Imprint of Laxmi Publications Pvt. Ltd.)
113, Golden House, Daryaganj,
New Delhi-110002

Phone : 011-43 53 25 00
Fax : 011-43 53 25 28

www.laxmipublications.com
info@laxmipublications.com

First Edition: 2011

OFFICES

Bangalore	080-26 75 69 30	**Chennai**	044-24 34 47 26
Cochin	0484-237 70 04, 405 13 03	**Guwahati**	0361-251 36 69, 251 38 81
Hyderabad	040-24 65 23 33	**Jalandhar**	0181-222 12 72
Kolkata	033-22 27 43 84	**Lucknow**	0522-220 99 16
Mumbai	022-24 91 54 15, 24 92 78 69	**Patna**	0612-2300097
Ranchi	0651-221 47 64		

UEE-9596-225-ENVIRON ENGG DISAS MGMT-SHA C—

Typesetted at : Shubham Composer, New Delhi. *Printed at* :

CONTENTS

Preface

The thinking behind writing this book and spirit rest much on the following mighty saying of Mahatma Gandhi-

"Nature has enough to satisfy everyone's need but has not enough to satisfy man's greed"

The book Environmental Engineering And Disaster Management is meant for the students of B.Tech. Second Semester and strictly according to the latest and revised syllabus of Rajasthan Technical University, Kota. The syllabus is divided into Five different Units, which presents various facets of the subject containing-Basics of Environmental Science, Air Pollution, Water Pollution, Biodiversity, Noise Pollution, Waste Management and Disaster Management. These five units are further divided into 10 chapters for better understanding of students as well as Teachers.

This subject is first time introduced by RTU, Kota in the present form. So, it was very challenging task for me to prepare a book following syllabus and satisfying the students-teachers at the same time.

Here, for the better understanding of the contents of the book, I am trying to throw light on it, chapterwise-

Chapter-1: Basics of Environment

It covers general introduction of environment, biotic and abiotic environment, adverse effects of environmental pollution and remedial steps to prevent the environment from pollution. To make the students aware about legislature related to environmental pollution, Indian legislation on environment has been discussed at the end of the chapter.

Chapter-2: Ecosystems and Biodiversity

Concept of ecology, basics of species, types of ecosystems, and Biodiversity and energy flow in ecosystems are the main contents of this chapter. Hydrological and chemical cycles have also been discussed systematically.

Chapter-3: Renewable Energy Sources

Solar energy, wind energy, biomass energy, tidal energy, geothermal energy etc. are the main sources of renewable energy. In this chapter we have discussed about them all and about their status in Indian Scenario.

Chapter-4: Water Pollution

Water pollution is the main transporter of fast spreading diseases causing germs, and impurities. In this chapter we have discussed about water pollution, its causes and its control.

Chapter-5 Waste and Wastewater Management

Wastewater can be treated and managed by primary, secondary and tertiary methods. In this chapter we have discussed different treatment processes, onsite sanitation, Environmental Impact Assessment and other wastewater treatment practices.

Chapter-6: Air Pollution

Here we have tried to outline different sources of air pollution, its adverse effects on mankind, air pollution control, greenhouse effect, global warming, acid rain. Ozone depletion etc. all these topics are the burning topics and in discussion everywhere.

Chapter-7: Noise Pollution

Noise pollution causes hearing disabilities, mental and physical disorders as well as different diseases related to digestion system. It can be studied and controlled in a way that we should have recognize that "How loud is too loud?"

Chapter-8: Solid Waste Management

Solid waste and its management is a big problem all over the globe. What are the reasons behind it? How can we control it? What are its adverse effects? These are the main questions , discussed in this chapter.

Chapter-9: Disaster and Disaster Management

In this chapter we have thrown light on many disasters like Earthquakes, Tsunami, Cyclones, Floods, Drought, and Landslides etc. We have also highlighted the Nuclear and Chemical disasters. What Do's and Don'ts are important in this regard have been discussed here in detail.

Chapter-10: Earthquakes and Seismology

Seismology is the study of Earthquakes. What is plate Tectonic Theory? Why and how, it is related to earthquakes is an important question discussed here. Focus, Focal depth, epicenter, seismic zoning and earthquake resistant constructions are the main features of this chapter.

I have spared no effort to make the book useful for the students as well as fellow teachers. I am heartily thankful to all the students and friends, who gave me valuable suggestions during this work.

Constructive suggestions and criticism for further improvement of the book will be gratefully acknowledged in incorporated in the next edition.

At last, I conclude my humble preface with a wise saying of someone-

"We don't inherit the Earth from our parents; we borrow it from our children".

So, I request you all to take care of this Earth.

—Author

SYLLABUS

Environmental Engineering & Disaster Management

Class B. Tech. 2nd Semester	Evaluation
Schedule per week Lectures : 3	Examination Time = Three (3) Hours Maximum Marks = 100 [Mid-term (20) & End-term (80)]

Units	Contents of the Subject
I	Basics of Environment. Adverse effects of environmental pollution and control strategies. Environmental Acts and Regulations. Functional concepts of Ecology. Basics of species. Ecosystem. Hydrological and chemical cycles. Energy flow in ecosystems. Biodiversity, population dynamics. Renewable sources of energy. Potential & present status of renewable sources of energy in India.
II	Quality and quantity of potable water. Surface and ground water sources. Basics of water supply schemes. Treatment of water. Wastewater management, Treatment & disposal of wastewater. Anaerobic digestion, Septic tanks. Reuse and saving in use of water. Onsite sanitation. Environmental Impact Assessment (EIA). Necessity and methodology of EIA.
III	Air Pollution. Harmful effects of Air Pollution. Control of Air Pollution. Noise Pollution. Adverse effects and control of noise pollution. Global warming, Acid rain, Ozone depletion. Solid Waste Management. Classification of solid waste. Collection, transportation, treatment, and disposal of solid waste. Energy recovery. Sanitary landfill.
IV	Type of Disasters: Natural and Manmade (Earthquake, Tsunami/Cyclone, Flood, Drought, Landslides, Nuclear, Chemical, Fire and Environmental Hazards). Disaster Management Cycle and its components. Vulnerability of Indian Continent to different types of Disasters. Do's and Don'ts for safety during these disasters.
V	Introductory : seismology, Occurrence of Earthquakes, Plate Tectonic Theory, types of earthquake. Definitions; Earthquake Magnitude, Intensity and their scales, Focus, Focal Depth, Epicentre, Epicentral Distance, Earthquake Energy. Concept of Seismic Zoning. Basic Concepts of Earthquake Resistant Houses & Construction Practices.

This book is for
Mukul, Nikhil, Guddan, Ashu,Malu, Shalu, Kunal, Kritika, Akku, Shivi,
Ayush,Gunjan,Vishu .... and their friends.
Because-
"We don't inherit the Earth from our parents; we borrow it from our children."

1

BASICS OF ENVIRONMENT

1.1 INTRODUCTION: WHAT IS ENVIRONMENT?

Our Environment is our surrounding. This includes living and non-living things around us. The non-living components of environment are land, water and air. The living components are germs, plants, animals and people.

All plants and animals adjust to the environment in which they are born and live. A change in any component of the environment may cause discomfort and affect normal life. Any unfavorable change or degeneration in the environment is known as '**Environmental Pollution**'. We need to protect our environment to live happily.

How to Maintain a Good Environment?

For better environment, all its components should be protected from pollution and the surroundings should be clean. We need to take good care of our land, water resources, forests and atmosphere. It is also necessary to ensure a balance between these resources and living creatures, to meet our needs.

(*i*) **Protection of Land:** Land is the upper layer of our planet earth. It is made up of soil and rocks. Land stores water and provide a surface to live on. Soil contains nutrients and air to nurse germs and plants. Fertility of the soil has been built up over hundreds of years. Soil is likely to be washed away with rain water or blown away by wind, if it is not protected under the cover of grasses, crops or trees.

(*ii*) **Conservation of Water:** Water is our life. With water, plants grow well, crops yield more and the atmosphere remains cool. Flowing water is also a source of energy and mode of transport. Rain is the main source of water. It is absorbed by soil and stored under the ground. A Part of this water, which is not absorbed, flows into rivers and finally reaches the sea. Being salty, sea water is neither fit for human consumption nor plant growth. It is necessary to keep the water clean for drinking purpose and conserve it for growing crops and trees.

(*iii*) **Protection of Forests:** Forests are an important part of the environment, because trees clean the air and keep the atmosphere cool. We cannot live without plants, because the oxygen need for breathing is produced by plants.

Trees absorb sunlight and reduce the heat. Plants provide fodder for animals, firewood, timber, medicines, honey, wax, gum, lac and food for us. Tree roots penetrate deep into the soil and from cavities in the ground. The dry leaves which fall on ground cover the soil and absorb more rain water, which slowly percolates through the soil. Thus, a large portion of the rain water can be retained in the field, by planting more trees. Flooding or rivers can be prevented by protecting trees in the forest.

Keeping the Air Clean

Clean air is right of everyone. It is MUST for survival of every living being. Air is thick at sea level and as we go above sea level, the air layer becomes thinner with less oxygen. Beyond 12 kilometers above sea level, the composition of the air changes. This layer contains more ozone. Ozone absorbs ultraviolet rays present in sunlight, which otherwise would reach the earth and cause injury to plants and animals. Air is the carrier of sound. In the absence of air, we cannot hear. When the volume of sound increases, it causes noise. This can frighten animals and irritate people. Continuous exposure to noise may lead to loss of hearing. Therefore, quietness is essential to improve the environment.

Clean air is colorless and is composed of the following gases:

Composition of Air	
Gases	*Concentration: Percentage of total volume of air*
Nitrogen	78.00
Oxygen	21.00
Carbon dioxide	0.034
Hydrogen	Traces
Argon	Traces
Methane	Traces
Ozone	Traces
Others Moisture and dust particles vary with location and climate.	

What is Environmental Science?

Environmental Science is that branch of knowledge that involves the study of different components of the environment that interact and result in such phenomena, which can be either beneficial or harmful for the environment and the beings dependent on it for their sustenance. Most of the students who are Environmental Science now study mostly study about pollution, global warming, forest conservation, soil erosion and the like. It also includes observation of changes in the earth's atmosphere, studying the reasons and think about propositions to repair the damages that have been and are being done.

Branches of Environmental Science

Environmental science deals with many areas as is comprehensible as environmental factors work in determining the way life would work in synchronization with the natural factors. For example pollution is related to population, green house effects is related to global warming, pollution is again related to ozone depletion which also contributes to Global warming. There is no definite demarcation between the branches of study of Environmental Science, each overlaps on the scope of the other, yet to make things easier to grasp and to facilitate work upon them, the study of Environment Science has been divided into the following divisions:

- Environmental Chemistry
- Ecology
- Geosciences
- Atmospheric Science
- Environmental Assessment
- Environmental Microbiology.

The last branch of Environmental Science is related with the study of microbes and their communities on the environment. It studies the composition and physiology of the microbes. Environmental Assessment is a branch of study that estimates and assesses the ways and methods to improve and enhance the environment. Atmospheric Science as the name suggests is the study of the gaseous layer that covers the earth like a blanket and has a huge role to play in the earth's environment. This is one branch of Environmental studies that is being worked upon on large scale. Then we come to the next branch, which is called Geosciences, and it studies the demographical part of the environment that would include the study of the erosion, ocean, mountain building and the like. Now we come to the most interesting part of Environmental science, which is related to the impact of environment on the life, living and survival of various species of animals. It studies the reason as to why some species are becoming endangered or the reason behind the behavioral changes in some animals, which are considered as the changes to adjust with nature. Next comes Environmental Chemistry, which involves the study of the chemical part of the Environmental Science and deals with the chemical changes in water, air and soil that are being brought about by the changes in environment.

List of Environmental Problems

There are a number of significant environmental problems today. They are as follows:

- **Loss of forest cover.** Forest is being cut down for timber, construction and for obtaining extra agricultural land.
- **Air pollution.** This is caused by polluting industries and vehicular traffic. Burning incinerators, furnaces and stoves cause it. Wind is responsible for spreading air pollution. The Ozone layer has been damaged because of the use of CFC's (Chlorofluorocarbons) which are used in a number of industrial, commercial, and household applications.
- **Water pollution.** Industries and factories release toxic wastes into the water. Leakage from oil tankers causes pollution of the sea. Pesticides and fertilizers also cause water pollution. Sewage released by cities into water bodies. It is predicted that there will be conflicts among nations because water will become scarce.
- **Overpopulation.** The increase in human population increases demands for the water resources. More land is required for housing. There is increased demand for food for which more agricultural land is needed. More fuel is required. More automobiles cause more pollution.

- **Erosion of soil.** When trees and plants are destroyed that causes increase in soil erosion.
- **Climate change and Global warming.** Human activities cause climate change like increase in temperature. Many animal and plant species may be destroyed by climate change.
- **Introduction of genetically modified species.** Humans are tinkering with the genes of various plants and animals. We do not fully understand how the introduction of new species will affect the environment.
- **Destruction of biodiversity.** Many plant and animal species have become extinct or are approaching extinction because of human activities. These endangered species may have important uses in medicine and science for humans.
- **Devastation of natural habitats.** Species such as tigers and lions are reducing in numbers because humans are steadily destroying the natural habitats which sustain them.
- **Diminishing natural resources.** Mankind is using up natural resources at an alarming rate. Fuel, water, minerals, timber are being consumed rapidly. Oil and natural gas resources are expected to get over by the end of this century.
- **Marine habitats are being polluted.** Over fishing has led to the extinction of many fish species. The oceans are vast but still vulnerable to pollution.

Conserving the Environment

Conserving the environment is a major social and political issue. The following steps can be taken to conserve the environment:

- Reduce wastage of electricity, fuel and food.
- Encourage the use of environment friendly and renewable sources of energy like solar energy, wind energy and tidal energy.
- Establish sanctuaries for endangered wildlife.
- Stop exploitation of areas rich in biodiversity.
- Encourage planting of trees.
- Reduce usage of plastic since it is not biodegradable.
- Stop over fishing.
- Encourage organic agriculture.
- Polluting industries should be closed.
- Discourage harmful genetic engineering.
- Recycle garbage.
- Aggressive family planning to limit human population.
- Elect leaders who work for the environment.
- Promote the development and use of environment friendly technologies.

The environment is a hot topic these days, and there is a lot of information out there about what is wrong and what we can do about it. All this information can be overwhelming, and leave you unsure of where to start. There are simple things that you can do everyday to do your part for the environment, and most of them take nothing more than an extra few seconds of your time.

Five Things You Can Do to Help the Environment Right Now:

5. **Recycle** who cares if I just throw this one can in the garbage instead of the recycle bin? If everyone in the U.S. thought this way, and threw out just one can or plastic bottle a day,

imagine the amount of waste that would be poured into our landfills. Not to mention the amount of packaging that could be recycled instead, and used in a positive way. It is so easy to recycle at home, it's no more difficult than taking out the regular garbage, and you just need to put your recyclables into whatever container your city or town requires for pick up. If you are really ambitious, those soda cans can even be turned into profit if you have a recycling facility nearby that will pay you for them. More and more restaurants, stores and other venues are providing recycling receptacles for you when you are out of the house as well, making it a snap to recycle when you're out and about.

4. **Skip the paper or plastic at the grocery store** there are many reusable shopping bags available, and they are even available in fun colors, patterns, or with fun sayings on them. There are reusable bags that fold up into tiny little packages, making it easier than ever to pop them in your purse or pocket and take them with you wherever you go. Keep a supply in your car so that even when you make an unexpected stop at the store, you don't have to use paper or plastic.
3. **Walk or take a bike ride** is the grocery store less than a mile from your house? Think about walking or biking there. It's good for the environment, great for your health, and can actually be a great stress reliever. If you live within a short distance of your job, you might even think about biking to work. It takes a little bit of effort on your part, but the payoffs can be big. With gas prices as high as they are right now, it will even save you some money!
2. **Use environment friendly cleaning products** take a look at the ingredients in the cleaning products you use in your house. Scary, isn't it? Environmentally friendly cleaning products are becoming more common, more affordable, and easier to find. Not only will it help the environment, but it will also keep toxins and chemicals out of your house, which can have a great affect on your health, and the health of your children or pets.

   And the number one thing you can do to help the environment right now:
1. **Realize that you can make a difference, even though you are just one person.** You often hear people say: "I'm just one person, how can I make a difference?" If everyone in this country thought that way, the end result would be disastrous. You have the power to do your part, and for each person who does their part, this earth will be so much better off for it. We have a responsibility to ourselves, and to our children to do whatever we can to leave this earth a better place than when we got here. Go ahead, take a few small steps, and know that you ARE making a difference.

Our Climate

Intensity of sunlight, cloudiness, temperature moisture content in the air and speed of the wind and collectively referred to as **weather.** These weather conditions including rainfall, length of days and nights of an area at a particular period or season, are collectively known as **climate**. The climate changes in winter, summer and rainy seasons. The imaginary line which divides the earth into two equal halves in south and north directions is known as the **equator.** The region around the equator is warm while Polar Regions which are farther away from the equator, both in North and South directions, are very cold. The plants growing in different regions are adapted to these climate changes in winter, summer and rainy seasons.

Fruits like mango, coconut and banana which are grown in the tropics cannot survive in the cold Himalayas, while apples and plums do not grow well in warm regions.

Animals in Our Environment

People and animals around us are part of our environment. Plants and animals are the sources of our food. Animals also provide manure through dung, needed for plant growth.

It is ideal to have a smaller number or animals and people, because they compete for land, water, food and other necessities. Overpopulation can exhaust these resources in a short period and cause scarcity.

What is Ecology?

Different plants and animals are found in different environments, suitable to them. Any change in that environment may affect their living. Study of plants and animals in relation to one another and to their surroundings is called **ecology**.

In an ideal condition, we can live together with plants and animals, without disturbing each other. This is known as **'ecological balance'**.

Need for Ecological Balance

Plants are basic for all. Some small animals feeding on plants are eaten by larger animals. Plants or animals which serve as food for others become a part of the food chain. Destruction of theses plants or animals may affect the lives of others in the food chain. We can locate several food chains around us. We'll discuss several such food chains later in different chapters.

Existence of different plants and animals in sufficient number to support the food chain is an indication of a good ecological balance. Let us now look at the reasons for the break in food chains. Mosquitoes lay eggs in still water. Frogs eat mosquito larvae and other insects. Frogs are eaten by snakes. Snakes are eaten by mongoose and eagles. If a large number of frogs are collected from paddy fields by people for food, the frog population of mosquitoes and other insects will increase. This may cause damage to the crops and spread diseases like malaria. In the absence of adequate number of frogs, snakes will come to residential areas in search of alternate food like rats, eggs and small birds.

In the same way, when the number of tigers and lions decrease, the population of deer and bosons will increase and these animals that depend on vegetation will destroy forests. When elephants do not get adequate foliage and bamboo shoots in the forest, they will move to nearby villages to graze in agricultural fields.

With increasing population, more and more people have to live on the same piece of land, and this will lead to an ecological imbalance. In the olden days, a large number of people died of diseases caused by germs. With the advancement of science, vaccines were developed to prevent diseases like small pox, cholera, diphtheria, typhoid, tuberculosis, etc. Hence, most of the people now survive deadly diseases and live long. This has resulted in overpopulation. Arrow indicates that insect is the food for frog. With more people, we need additional houses, more food, fuel and other commodities. But, we cannot create additional Land. We need to produce more on the available land itself. In India. In 1951 about 117 people used to live on a square kilometer of land, but now 250 people are living on the same area. A cropping area of 147 million hectares, which was supporting 361 million people in 1951, now has to produce food for 850 million people.

Hence, the agricultural scientists have introduced the following modern practices to increase our food production:

- Development of high yielding varieties of crops. However, new varieties require more manure, assured irrigation, protection from diseases and insect attacks.
- Application of agro-chemicals such as chemical fertilizers, insecticides, fungicides, bactericides and weedicides.
- Application of chemical fertilizers, because a large quantity of cattle manure required for growing high yielding varieties, is not available.
- Spraying of insecticides on crops control insect attack. Fungicides and bactericides control diseases. Weedicides kill the weeds and unwanted plants which compete with crops.

Agro- Chemical help to grow more food, but they can pollute the air, water and soil when used in large quantities. Even with these efforts and risk of health hazards, we may not be able to produce sufficient food to feed everyone, because our population is still increasing.

It is difficult to guess what the hungry people will do, when they cannot get food. They may cut jungles, start eating animals and fight among themselves for survival. These are the worst possible effects of an ecological imbalance. Thus, ecological balance is necessary for our survival! Control oh human and livestock populations and better management of your natural resources would certainly help to achieve this.

What is Environmental Pollution?

Environmental pollution is caused due to over-use of natural resources, presence of a large number of people and livestock in congested areas, use of agro-chemicals, setting up of factories, running of automobiles, burning of fuel, etc.

A change in the environment due to pollution also affects the ecological balance. Environmental pollution is caused both in rural and urban areas. Pollution can cause sickness and discomfort. It also affects the productivity of natural resources, such as land, water, forests and livestock. We can prevent environmental pollution if we understand its causes.

[A] Pollution in Rural Areas

In rural areas, pollution takes place around the houses, colonies and public places.

(*i*) **Pollution around Houses:** In rural areas, most of the houses are poorly ventilated because of small windows and low roofs. Keeping livestock inside the house is an old practice. In the absence of toilets, villagers use open fields. As the families grow, the number of houses increase and the colonies become congested. Thus, pollution occurs in many ways.

(*ii*) **Air Pollution:** In most of the rural houses, traditional chulhas are used for cooking food. Due to a poorly ventilated kitchen and wet fuel wood, the chulha generates a lot of smoke. The smoke inhaled by a rural woman is equivalent to the smoking of 200 cigarettes every day. Imagine what will happen then? Smoke also affects their health and harms their eyes and lungs. The coal used for cooking also releases smoke, with harmful gases. This pollution can be prevented by replacing old chulhas with improved smokeless chulhas. Such chulhas also reduce the requirement of fuel wood. A kitchen can be properly ventilated by installing large windows and fixing a chimney on the roof. A better way of preventing smoke is to install a biogas plant. This would facilitate smokeless cooking at a low cost, besides producing good quality farmyard manure, useful for increasing food production. Keeping livestock inside the house attracts insects like flies, ticks and mosquitoes. Dung and urine release ammonia, methane and other harmful gases having bad odor. Hence, the animals should be kept outside and the dung should be stored properly in manure pits, away from the house.

Dust from barren fields, over-grazed pastures and roads pollute the air. Inhalation of dust can cause sore throat, cough and breathing disorders. We can reduce the dust by planting trees around houses, along roads and in pastures.

(iii) **Water Pollution:** Where do we get drinking water from? It may be from a well, tank canal or river. But it is the rain water which fills these sources. Some portion of the rain water percolates into the ground and reaches the wells and the remaining water flows into rivers. The area near the well should be clean and must not be used for manure pits, toilets and urinals. This will prevent the entry of dirt into the well along with rain water. Washing livestock, clothes and utensils and letting the toilet discharge into ponds should also be avoided, because such practices pollute drinking water with germs and chemicals, thereby causing dangerous diseases. Lack of drainage facilities to remove the discharge from the toilet and cattle shed is another cause of air and water pollutions, Construction of a toilet linked biogas plant can solve this problem. The dirty water from the bathroom and kitchen can be used to irrigate fruit and forestry plants grown around the house, instead of allowing it to stagnate along the road. Such stagnate water promotes the breeding of flies and mosquitoes.

(iv) **Surplus Livestock:** Villagers maintain a large number of cattle, buffaloes, sheep and goats to supplement their income. But they do not have sufficient fodder to feed them. These animals when let out for grazing, damage pastures and forests. Therefore, it is better to have a few high yielding animals which are fed well.

(v) **Chemical Pollution:** Use of agro-chemicals has become a common practice in agriculture of these chemicals. Consumption of fruits and vegetables, harvested soon after the chemical spray is another health hazard.

To avoid this, it is necessary to educate farmers to use agro-chemicals safely and only when needed. Plant products like oil and cake of neem which are non-toxic to us, can be used to control many crop pests. In irrigated fields, when farmers apply large quantities of chemical fertilizers and pesticides, a part of these chemicals may dissolve in water and percolate into the wells. Thus, the water may get polluted. Drinking this water may cause sickness. In such areas, we need to persuade farmers to reduce the doses of agro-chemicals. Effluents from sugar factories and other industries let in the filed may also pollute drinking water in a similar way. Such factories should be told to treat the effluents properly.

(vi) **Sanitation:** Many rural houses do not have toilets and people use backyards, roadsides or community land. Such poor sanitary conditions promote the breeding of flies and germs which spread many diseases. As many of the families cannot build toilets for themselves, it is necessary to build community toilets, and drainages to keep the surroundings clean. This discharge known as sewage can be treated to kill germs and then used to irrigate fruit and forestry plantations.

(vii) **Unpleasant Weather:** In the absence of trees on roadsides, community lands and degrades forests, the temperature of our surroundings increases. Hot winds make the summer worse and field crops wither quickly. To keep the atmosphere cool, we should plat trees along roads, field's bunds and barren lands. To protect trees, livestock should not be let out in the filed. The wildly grown Parthinium weed in fields and along roadsides cases skin and throat irritation for the people living nearby. Therefore, it is better to pull out these plants before the flowering stage.

(viii) **Flood and Drought:** In the absence of tree cover, the land cannot absorb more rain water. So most of the water flows into the river causing floods. This water also carries away fertile soil from the field and deposits it in lakes, reducing the water storage capacity. The rivulets flowing from such forests dry soon. The water level of wells in such an area decreases, causing water scarcity in summer. To prevent floods and droughts, earthen bunds across

the slope, called contour bunds, should be built on hilly areas and agricultural fields. This would retain more water in the field. The surplus water flowing from the field can be stored in percolation tanks developed by constructing bunds across gullies. Trees can be planted along canals. This will enrich the environment and bring prosperity to rural areas.

What Can Rural People do?

People in village can persuade their parents to:

- Use smokeless chulhas or install biogas plants;
- Provide smoke outlets and ventilation in the kitchen;
- Tie livestock outside the house and reduce their number;
- Feed livestock in their sheds without letting them out for grazing;
- Plant trees around the house, on field bunds and along roadsides;
- Develop filed bunds across the slope the retain more water and prevent the soil being washed away. Plough the field across the slope;
- Keep tanks, canals and other water sources clean;
- Use the sewage water for growing trees;
- Keep the surroundings of the house and well clean;
- Prepare compost by using garbage, dung and other wastes;
- Select a suitable site for the toilet, away from water sources and houses. It can also be connected to the biogas plant;
- Use agro-chemicals carefully and try to avoid them. Plant products may be preferred wherever effective to protect crops;
- Protect wildlife like frogs, snakes, mongoose, birds, etc.

[B] Pollution in Urban Areas

Overpopulation, industrial growth and an increasing number of automobiles are the major causes of pollution in cities.

(*i*) **Air Pollution:** When there are fewer plants and more people living in an area, the oxygen supply will decrease and carbon dioxide content in the air will increase, causing air pollution. The tall buildings absorb heat during the day and release it during the night. Thus, the weather remains hot and unpleasant. Automobiles and industries emit smoke containing carbon dust, carbon dioxide, carbon monoxide and other poisonous gases, which are unhealthy. Chemical factories release harmful gases, which can travel long distances. Gases like sulphur dioxide and nitrous oxide come in contact with cloud or rain water and turn into acid and result in **acid rain**. Release of harmful gases like fluorine and nitrous oxide by factories into the atmosphere would destroy the ozone layer and allow the harmful rays of the sun to reach the earth. Gases like carbon dioxide, sulphur dioxide and nitrous oxide, when released from automobiles and industries, absorb heat from sunlight and raise the temperature of our surroundings. Due to deforestation and increasing industrial growth, the average temperature of the atmosphere may rise by 2° C – 5° C during the next fifty years. This is known as **global warming.** The rise in temperature will result in melting of snowy mountains, rise in water level of the sea and sinking of cities and villages located on the sea coast. This is known as the **greenhouse effect.** The best way to purity the polluted air is to grow more trees, wherever we can. If there is no space for tress, then shrubs or creepers which can spread on walls or poles, without occupying any additional space can be planted. Trees direct the gases to move upwards. Some trees absorb harmful gases. Trees take in carbon dioxide and release oxygen. The dust settles on leaves. Trees reduce noise by blocking sound waves and provide shade thereby keeping the atmosphere cool. Automobiles should be avoided whenever possible. Factories should be forced to install pollution control devices.

(*ii*) **Water Pollution:** A large quantity of industrial effluent is generated in cities. This needs to treat to remove harmful substances before letting into the river. However, this is often Not practiced because of the high cost. Such untreated effluent may poison fish, other animals and plants living in water. Sewage containing harmful germs and chemicals can contaminate the sources of drinking water.

The treated effluent and sewage can be used for irrigating fruit and forestry trees.

(*iii*) **Sanitation:** Many villages moving to cities stay in congested huts which lack ventilation, toilet and drainage facilities. Such colonies are called slums often suffer form illness, caused by pollution.

Improper disposal of kitchen wastes known as garbage is another cause of pollution of. Garbage heaps breeds flies and release bad odor. Dropping litter on the roads makes the surroundings dirty and spreads many diseases. Throwing slippery materials like banana skin can cause injury to pedestrians. Garbage and litter should be deposited in dustbins which are collected by the municipal authorities. It is possible to use garbage for producing biogas a source of pollution-free energy. Garbage can also is converted into **compost** by neatly depositing it in a pit. It is possible to recycle waste paper, plastic, tin cans and glass pieces. We must learn ways of making use of these wastes. While keeping our surroundings clean. Remember, wastage leads to pollution. If we waste water, the quantity of sewage will increase. If we waste food, the volume of garbage will increase. The municipal authorities will not have space to dump these wastes and hence pollution will increase. Moreover, food water, electricity will be available to other. In this way, the we can avoid shortages.

How are Our Public Places Polluted?

Maintaining cleanliness in public utilities like buses, bus stands, trains, railway stations, schools, playgrounds. Roads, parks, picnic spots and beaches are necessary. If all of us maintain cleanliness, or surroundings can remain clean. After enjoying good food at a picnic, if we do not throw the garbage into a dustbin, the parks and beaches can turn filthy and others will not be able to enjoy. Let us not expect others to clean our mess. We must do it ourselves.

What Can Urban People do?

Urban People can do a lot to reduce environment pollution.

- Plant trees in school and home premises. Even if you are staying in a flat, plants and creepers can be raises in pots and wooden boxes, in the balcony;
- Protect the trees planted along roadsides;
- Keep public areas clean and avoid littering;
- Plant trees along roadsides, near bus stops, around playgrounds, and in parks to provide shade;
- Avoid dumping garbage on the street. Dispose them in a garbage dump; you can also make a compost pit to convert garbage into manure;
- Waste paper, plastic, glass and metal pieces can be recycled this would reduce the pollution and conserve our resources;

- Avoid using plastic materials such as plates and carry bags which cannot be used again. Moreover, when plastic is thrown away, it does not degrade but remains in the soil, polluting the surroundings;
- Do not make noise in public places; every likes quiet surroundings;
- Request your family members to use automobiles only when necessary; walking or cycling can be a pleasure when the distance is short;
- If someone is causing pollution in your area, inform the authorities through your teachers or parents.

We can protect our environment in many ways. Let us act now and persuade others to join us. This will ensure safety for our future generations.

1.2 POLLUTANTS AND THEIR CLASSIFICATION

Simply, the material which cause pollution of environment are known as **pollutants.** It may be a harmful solid, liquid or gas present in undesirable concentrations.

A pollutant may belong to:

- a chemical substance.
- a biological organism and product.
- a physical property, *e.g.*, noise, sound etc.
- a geochemical material, *e.g.*, dust, etc.

For example, **fertilizers** were used to increase food production to fulfill the demand of food of increasing human population. But, at the same time fertilizers could destroy the environment. They were washed down to water resources like rivers lake, oceans and seas. Resultantly water pollution occurs. Use of insecticides and pesticides is also very alarming, for example- D.D.T.

Pollutants can be classified in number of ways:

(*a*) **On the basis of their existing form:**

(*i*) *Primary pollutants:* Substances emitted directly from an identifiable source. They exist as such into the environmental. For example, CO_2, SO_2, NO_x etc.

(*ii*) *Secondary pollutants:* Substances derived from primary pollutants. They are result of chemical reactions of primary pollutants. For example, nitrogen oxides when react with hydrocarbons in sunlight, give **peroxylacetyl nitrate** (PAN) as a secondary pollutant.

(*b*) **On the basis of the ecosystem point of view:**

(*i*) *Biodegradable pollutants:* Pollutants, which can be decomposed easily in environment, and don't cause (or cause very little) any harmful effect on the environment. Mostly are organic materials. For example, animal and plant debris domestic sewage etc.

(*ii*) *Non biodegradable pollutants:* Pollutants, which do not degrade (ordecompose) or degrade very slowly in the environment. Mostly are, **inorganic material,** like lead salts, iron, D.D.T. mercury etc. These pollutants decrease the rate of photosynthesis and affect adversely the balance exiting between oxygen and carbon dioxide. This situation creates problem for mankind as well as for plants.

It is quite impossible to remove these pollutants from the biosphere. For example, we can not remove lead which has beed dumped into the air we breathe. The major source of lead is vehicles. Motor vehicle emissions are one of the leading cause of air pollution.

Polythene is also a non-biodegradable pollutant and creates great risk for our environment.

Thus pollutant are residues of materials used by us, made by us and thrown away by us which pollute the environment in one way or the other.

1.3 TYPES OF POLLUTION

The pollution may be **natural** (for example-earthquake, forest fires, organic decay etc.) or **artificial,** which is man made pollution.

However it may be classified as:

- ***Air pollution,*** the release of chemicals and particulates into the atmosphere. Common examples included carbon monoxide, sulfur dioxide, chloflurocarbons (CFCs) and nitrogen oxides produced by industries and motor vehicles.
- ***Water pollution*** via surface runoff, leaching to groundwater, liquid spills, waste water discharges, eutrophication and littering.
- ***Soil contamination*** occurs when chemicals are released by Factories or underground storage tank leakage. Among the most significant soil contaminants are hydrocarbons.
- ***Radioactive contamination,*** added in the wake of 20th-century discoveries in atomic physics. It cause very serious and long lasting adversely effects.
- ***Noise pollution,*** which includes roadway noise, aircraft noise, industrial noise as well as high-intensity sound amplifiers.

1.4 ADVERSE EFFECTS OF ENVIRONMENTAL POLLUTION ON HUMAN HEALTH

Defective air quality can destroy many organisms including human beings. It causes many adverse effects on human health, as:

- *Ozone pollution* cause respiratory disease, cardiovascular disease, throat inflammation, chest pain, and congestion.
- *Water pollution* causes approximately 14,000 deaths per day, mostly due to contamination of drinking water by untreated sewage in developing countries.
- *Oil spills* can cause skin irritations and rashes.
- *Noise pollution* causes hearing loss, blood pressure, stress, and sleeping disorders.
- *Sulfur dioxide* and *oxides of nitrogen,* can cause acid rain which reduces the *p*H value of soil. Soil can become infertile and unsuitable for plants. This will affect other organisms ill the food world. Acid rain is also harmful for our historical monuments.
- *Smogs* can reduce the amount of sunlight received by plants to carry out photosynthesis, which reduces the growth of plants.

1.5 AIR POLLUTION

The natural composition of air have nitrogen 78%, oxygen 21%, carbon dioxide 0.03% and inert gases, water vapours in traces. There is life on earth due to presence of oxygen.

Due to excess of industrialization and other human activities the lower layer of atmosphere is undergoing many harmful changes. This is known as air pollution. Different sources, pollute the air are as:

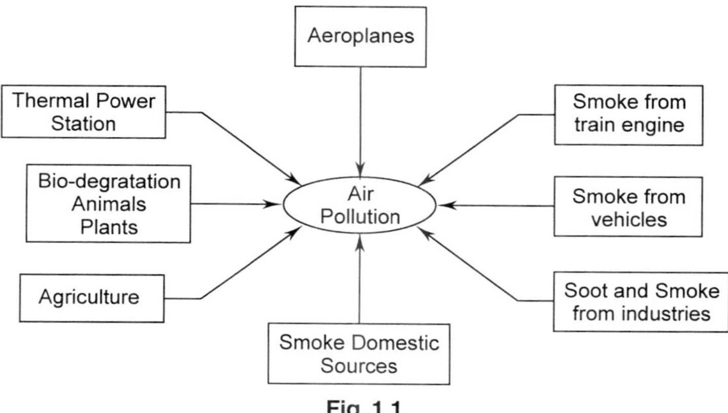

Fig. 1.1

Causes of Air Pollution

Air pollution is caused by natural as well as man-made factors, on this basis the sources, polluting the air can be studied under two types:

(*i*) Natural sources

(*ii*) Human activities (Man made Sources)

(*i*) **Natural sources:** Air is polluted by natural climatic activities, like cyclone, high velocity winds etc., which make the atmosphere full of dust and makes the visibility almost nil and it becomes difficult to inhale this dusty air. Dust cause many types of allergies and mainly affects ear, nose, throat and eyes.

Air pollution is also caused, due to **flying ashes** from volcanos. It becomes difficult for asthmatic patients during the changing seasons.

(*ii*) **Human activities (Man made sources):** This type of air pollution is caused due to various human activities. These are as following:

(*a*) *Combustion:* The process of combustion is one of the main reasons of air pollution. During burning of the fuel many types of gases, soot and small particles are produced. These, mix with the air and pollute it.

The **Suspended Particulate Matter (SPM),** is quite harmful for our respiratory system and eyes. *The following combustion process cause air pollution.*

- **Combustion in domestic activities:** Fire is produced by the combustion of various fuels like, coal, coke, wood, kerosine, etc. This combustion produces gases like carbon monoxide (CO), carbon dioxide (CO_2), sulphur dioxide (SO_2) etc., and mix with the air, in which we breathe, the oxygen needed in combustion of these fuels is also taken from the surroundings. In our country in most of the villages and smaller towns, forest wood, cow-dung, dry farm waste, dry grass and leaves are used as fuels, these fuels produce more harmful gases as well as SPM.
- **Combustion in thermal power stations:** To obtain heat, for producing electricity tons of coal and lignite are burnt. This produces great quantity of SO_2, soot, fly ash, etc., which are just thrown into the air.
- **Combustion in vehicles:** All the means of transport, bus, train, aeroplanes, scooter, car, run on the combustion engines, using petro, diesel as fuel. After combustion, the **vacuum** pipe of the vehicle evolves smoke and soot containing gases like nitrogen dioxide and peroxide,

carbon mono and di-oxide and carbon particles as SPM. If there is lead in the petrol, it is a big pollutant itself.

With the increase in pollution, urbanization and consumerism, number of petrol and diesel driven vehicles is increasing every day the air pollution level of metrocities is given every-day in the news, which is mostly above the-safe limits.

- **Combustion for disposal of garbage:** It is very common practice to burn, all types of garbage, expelled from, house, hotels, hospitals, gardens, farms etc. But all type of bio-degradable or non-biodegradable material produce, smoke and soot polluting the air.
- **Combustion in small and large scale industries:** In most of the industries, bit furances are installed for melting, fusion, evaporation etc., and tons of fuel like, coal, wood, is burnt, the smoke and SPM pollute the air.

(*b*) *The crackers:* It is very common practice to burn crackers for entertainment, many shows are organized, on weddings and Diwali festival and a large amount of solid waste in the form of colloidal particles also get mixed in the air, we breathe. This increase SPM level in the air.

(*c*) *Nuclear explosions:* The nuclear expolsion in an area, pollute the air of a large area at once, inducing many hazardous chemicals, dust particles, in excess,

(*d*) *The waste products coming out of the chimneys of industries:* The second, main, human made factor of air pollution is industrilization. For the allround development of a country, industrialization is must but in absence of the proper treatment plants, all the harmful gases, hydrocarbon wastes and other hazardous chemicals are emitted out, in the atmosphere, polluting it.

Bhopal gas tragedy is worst accident ever happened in the country due to air pollution. The killing gas *methyl isocyanide*, was emitted out from union carbide fertilizer factory on the night of 2nd and 3rd Dec. 1984, which killed thousands of people and left many others handicapped.

(*e*) *Through agricultural activities:* This is also a prominent factor of air pollution. To increase the yield, excess of harmful chemicals are sprayed on the crops, in the form of liquids and powders, these chemicals travel a large distance with air, polluting it. While processing the crops, the small particles in the form of saw dust etc., are also induced in the air.

(*f*) *Due to many appliances, gadgets etc.:* In the devices like fridge, air conditioners etc., *chloroflouro carbons* (CFC) are used. These are the chemicals which diminish the protective ozone layer.

(*g*) *Through smokers:* The people who smoke, cigarettes, bidi, cigar etc., pollute the air with smoke. This harms the non-smokers also as they inhale the smoke.

(*h*) *Through cosmetics:* Most of the synthetic perfumes, deodorants, room fresheners are prepared, using the medium of solvents like ether, alcohol etc., when these sprays are used, they get mixed with the air.

(*i*) *Use of solvents:* In the polish of floor, furniture, etc., and spray painting of vehicles, gudgets, many organic solvents are used, which evaporate at very low temperature. These are low boiling hydrocarbons. These harmful chemicals mix into the air, when used.

(*j*) *Other sources:* Dead animals, rotten garbage, murshy gutter, tanneries, distrillaries, produce gases having very foul smell, this pollution can be felt in every day life in our country.

Table-1.1 gives an account of the pollutant gases, their sources and hazards:

Table. 1.1

S.N.	*Pollutant*	*Main source*	*Effect*	*Remarks*
1.	Carbon di-oxide (CO_2)	Vehicles, burning of fuels for energy, production combustion	Not any direct effet on human beings, but can cause global warming.	Natural component of atmosphere necessary for trees
2.	Carbon monoxide (CO)	Partial combustion of fuel (*e.g.*, vehicles)	Reduces oxygen bearing muscles. Expecially harms asthemtic people.	No contribution of nature. The smokers affected.
3.	Sulphur-di-oxide	Burning of coal, oil, etc., with pungent smell.	More harmful when mixes with smoke. Increases respiratory problems. Chocking, burning of eyes, causes acid rain, when mixed with water vapours, Reduces fertility of soil acidifies soil and water bodies Corrosion of monuments	
4.	Suspended particulate matter (SPM)	Smoke from houses, industries and vehicles. Dust storms, earthquakes volcanos.	Poisonous according to the combination increases effect of SO_2, reduces light. from smog, rusting increases.	Chemical most varried in nature.
5.	Nitrogen Oxides	Burning of fuel in vehicles and forest fires	Respiratory infections in kids, copper coloured fog in the atmosphere of the town. Rust producing.	There are two such oxides, nitrogen oxide(NO_2) and nit-rogen dioxide (NO_2)
6.	Volatile hydrocarbons	Burning of organic fuels, industrial process disposal of solid wastes.	Causes burning in eyes, when mixes with other pollutants (acrolein, aldehyde).ethylene is bad for plants, aerosols reduce visibility.	
7.	Oxides and ozone	Emitted from vehicles, photochemical reaction between nitrogen oxides and hydrocarbon	Burning in eyes damages lungs, corrode buildings and other things, reduces visibility, ozone is hazardous for plants	forms due to mutual reaction between different pollution Ozone is natural and main component of upper parts of atmosphere.

Effect of Pollution

Effect of air pollution is wider than any other type of pollution, because it travels with the wind. Polluted air affects, human and animal health, vegetation, small insects, organisms and environment in general. Some of the harmful effects of air pollution are :

(*i*) **Effect on human health:** Air pollution affects the human health very easily, as every time we breathe we inhale the polluted air. This goes to the lungs, if carbon monoxide (CO) is present in the air, it mixes with the *hemoglobin* of the blood and forms carboxy hemoglobin. CO reacts 200 times more than oxygen (O_2), thus a small amount of carbon monoxide gas (CO) is sufficient to denature the blood, if CO is inhaled in excess it can be fatal.

Even in this 21st century, such accidents can be seen where, during winters a family sleeps in an unventilated room with a slow-burning fire of coke, and the whole family dies, becuase of CO gas poisoning. Due to excess of carbon monoxide (CO) in the inhaled air, oxygen bearing capacity of blood decreases which reduces the activity of lungs and brain. Vehicular smoke contains great amount of CO.

Other gases produced by vehicles and inhaled by people are oxides of sulphur and nitrogen. When the polluted air goes to lungs. It causes many diseases like, *bronchitis, Asthama, sour throat, burning of eyes* and respiratory problems in childern.

If the petrol contains lead, then the vehicular smoke becomes more harmful due to lead poisoning. Lead damages liver and kidney muscles and weakens the bones. Other chemicals like hydrocarbons, benzopyrin move in the form of particulate matter may enter the lungs and cause lung cancer.

(*ii*) **Effect on vegetation:** Plants are damaged due to pollution in the air. The dusty and sooty smoke gets desposited on the leaves and fall. The hydrocarbons, oil etc., close the *stomata*, which affect transpiration and photosynthesis. This can be easily seen, in the plants grown on the divider of a heavy traffic road. They become black and dry.

During the winter and rainy season, the soot and smoke is condensed or runs into the water bodies damaging the vegetation. **Acid rain** affects the colours of flowers and growth of plants.

(*iii*) **Effect on insects and wildlife:** Air pollution also affects insects and animals, adversely. The insects like honey-bee, butterfly, etc., which are helpful in pollination, are destroyed totally in a polluted area, which reduces the production of a crop, production of apples is seduced due to the reduction in the number of honey-bees.

Effects on Environment

(*i*) **Effect on ozone layer:** Ozone layer is a thick layer of ozone gas in *stratosphere*, which shields the earth from harmful U.V. rays of the Sun. This layer has been depleted by pollutant gases like chlorofluoro carbons (CFC) by 2% during last few years, this is known as **ozone hole.** The effect of air pollution by CFCS was first observed at Antarctica, where there is much thinning of this layer.

(*ii*) **Global warming:** According to an estimate the temperature of the earth has increased by 1% during the last fifty years. This is because of the increase in the percentage of carbon dioxide, oxides of nitrogen and chlorofluoro carbons in the atmosphere. These gases are also known as **greenhouse gases.** These gases cause undesirable changes in the atmosphere.

(*iii*) **Photochemical smog:** The colloidal particles present in the vehicular emission, hydrocarbons, reacts in presence of light forming a photochemical smog, this reduces visibility on roads. Due to this smog in some highly industrialized cities of U.S.A., Britain and Japan, traffic has to be stopped, due to visibility.

The **'London-smog'** is well known effects of air pollution, where photochemical smog, encircled this industrial city in 1952, for five days, Thousands of people fell ill due to the smog.

(*iv*) **Effect of monuments:** The black smoke *i.e.*, soot contains high concentration of hydrocarbons in colloidal form. This gets deposited on the monuments of archeological importance and deteriorte their upper layer. **Taj-Mahal** of Agra suffered corrosion due to soot and smoke from Mathura oil refinery, tanneries and other such industries, in the surroundings of Taj. To conserve the beauty of Taj-Mahal, many of the industries have been banned.

Air Pollution Control Strategies and Devices

Pollution control is a term used in environmental science and engineering. It means the control effluents into air, water or soil. Without pollution controls the undersirable waste products from human consumption, industrial and agricultural activities, mining, transportation and other sources will accumulate or disperse and degrade the natural environment. In this regard, pollution prevention and waste minimization are more desirable than pollution control.

(A) **Dust collection systems:**

These are designed to handle dust loads. A dust collector consists of a blower, dust filter, a filter-cleaning system, and a dust removal system.

1. **Cyclones:** *Cyclonic separation* is a method of removing particulates from air without the use of filters through *vortese* separation. Rotational effects and gravity are used to complete the method.

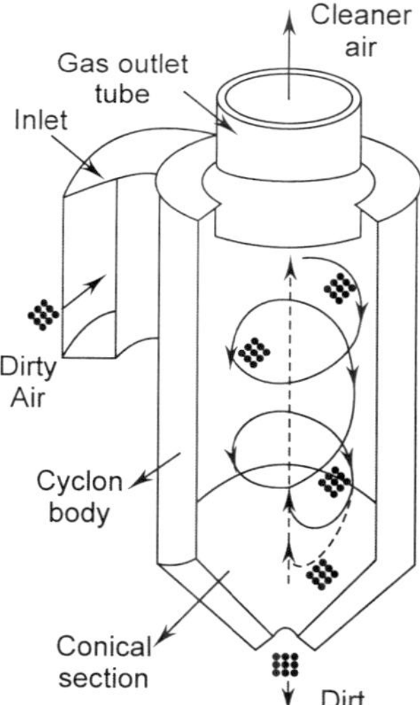

Fig. 1.2 Cyclone Separator

In this method, a high speed rotating air-flow is required within a cylindrical container called a **cyclone.** Air flows in a spiral pattern, beginning at the top (wide end) of the cyclone and ending at the bottom (narrow end) and exit from the top. Larger particles or group of smaller particles in the, rotating air stream have too much inertia to follow the light curve of the air stream and strike the outside wall, falling then to the bottom of the cyclone where they can be collected and removed.

Cyclones are used in saw mills to remove sawdust from extracted air. Cyclones are also used in refineries to separate oils and gases, and in the cement industry as components of kiln preheaters.

Analogous devices for separating particies (or solids) from liquids are called **hydrocyclones.**

2. **Electrostatic precipitators:** An electrostatic precipitator (ESP), or electrostatic air cleaner is a particulate collection device that removes particles form a flowing gas (such as air) using the force of an induced electrostatic charge. Electrostatic precipitators are highly efficient filtration devices that minimally impede the flow of gases through the device, and can easily remove fine particulate matter such as dust and smoke from the air stream.

3. **Baghouses (Fabric Collectors):** Fabric collectors are used to separate dust particulates from dusty gases. They are one of the most efficient and cost effective types of dust collectors available and can achieve a collection efficiently of more than 90% very fine particulates.

   Dust-laden gases enter the baghouse and pass through fabric bags that act as filters. The bags can be of woven cotton, synthetic, or glass-material in a bag shape.

   The dust cake formed on the surfaces of the bags. Because the fabric provides a surface on which dust particulates collected.

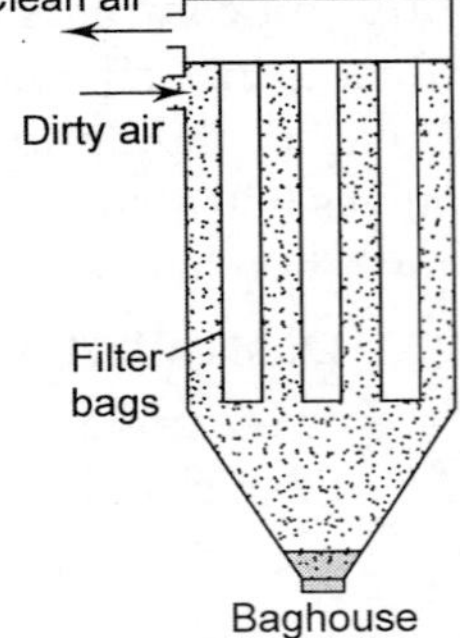

Fig. 1.3 Baghouse

Other Control Measures to Check Air Pollution

(*i*) Instead of using low grade and conventional fuels, use of non-conventional fuels like, Gobar gas, bio-gas compressed natural gas, LPG must be prepared and encouraged.

(*ii*) The vehicular pollution can be controlled by using good quality fuel, by keeping the engine fully effective, so that the fuel is totally burnt, by using unleaded petrol and by minimising the use of motorred vehicles.

(*iii*) The industrial pollution can be reduced by following simple methods :

- Instead of throwing industrial soot and smoke, directly into the air, it should be, first passed through **electrostatic precipitators,** where the **suspended particulate matter** (SPM) and other fine particles of hydrocarbons can be adsorbed and precipitated down, and the smoke becomes clean.
- By adsorption on the surface of activated charcoal. The solid particles present in the industrial, geseous waste are adsorbed when allowed to pass through a bed of activated charcoal. These are used in the chimney's of industries.
- The industrial gaseous waste is passed through **scrubbers** where high pressure water stream is lashed on the gases and they are dissolved in water, from where some useful substances can be recovered.
- Air pollution can also be checked by converting the hazardous chemicals into the less harmful substances like carbon di-oxide, water etc.

(*iv*) The air quality can be improved and maintained by growing and developing green belts and forests.

1.6 WATER POLLUTION

Water is life for all the living beings and activities such as agriculture, domestic and industrial uses on earth. But still the clean and safe drinking water is a commodity, today. Water is being polluted through many factors. Water pollution can be defined as :

"Any change in the physical or chemical properties of water, due to presence of any foreign substance is known as water pollution."

Causes of Water Pollution

Water gets polluted due to natural factors as well as human activities.

The **natural factors** causing pollution of water bodies are:

- The biomass, like dry leaves, twing, remnants of dead animals etc., are taken to the water bodies like river, ponds etc., through rain water.
- The cyclones floods, landslides also change the composition of water, by mixing the different types of material during these natural disastes. The **human activities** causing water pollution are:
- Bathing, washing clothes and excerting in rivers, ponds etc., especially in small towns and villages.
- Mixing of sewage water into the water bodies from, residential and public toilets. Where there is no proper sewage line.
- Disposal of dead human bodies and ashes in the rivers.
- Disposal of dead animals into the water bodies.
- Disposal of industrial effluents directly into the water bodies, without any water treatment.
- During the transporation of oil and from the oil wells at the sea-shores, and due to marine accidents, oil-pollution of the sea water take place.
- Thermal pollution of ponds and rivers etc., takes place when hot water is released from the industrial plants.
- Radio active pollution of water-bodies takes place when there is nuclear explosion : The minute particles spread in the large area and mix with water slowly.

Effects of Water Pollution

As water is essential for human beings, animals and plants the effects of water pollution are also wide spread. Water pollution is harmful for where-ever it is used.

(*i*) **Direct Effect on the human beings:** By drinking polluted water, much damage is caused to human body. The diseases spread by polluted water are known as water born diseases:

- Cholera, typhoid, diarhea, etc., bacterial diseases.
- Hepatitis, jaundice etc., viral diseases.
- Amoebiosis, stomach and intenstinal infection etc., protozoa caused diseases.
- Tape worm, ring worm, water caused diseases.
- Due to contact with polluted water, skin diseases, eye infections and other secondary infections take place.
- When there is excess of a chemical fluoride in water, it causes deformation of bones, staining and weakening of teeth and the disease is known as **fluorosis.**
- Presence of some metal ions like, mercury, arsenic lead, cadmium make the water poisonous and unfit for drinking.

(*ii*) **Effect on aquatic animals:** Pollution in water harms the aquatic life in a great way:

- High level of pollution in water harms fish and aquatic life decreasing their population alarmingly.

- Due to decrease in the amount of oxygen in polluted water, death of fish and other aquatic animals take place.
- The oil pollution in ocean harms aquatic animals in the sea.
- Due to thermal pollution loss of many aquatic lives.

(iii) **Effect on other animals:**

- Other than the aquatic animals, other cattle like cow, buffalows, sheep, goats, etc., drink the polluted water and suffer from intenstinal diseases.
- By drinking the water having excess of fluoride both pet and wild animals suffer from weakening of bones and teeth.

(iv) **Effects on vegetation:** Due to pollution in water vegetation of an area, is either damaged or some species are permanently missing from that area:

- Water becomes, alkaline due to mixing of domestic waste water and detergents etc., in which many aquatic plants can not survive.
- In the agricultural effluents, there is excess of nitrates, phosphaes etc., which enhances the growth of plants like **blue green algae** in a water body and due to excess spreading of this algae, all the other vegetation die.
- Affecting genetic diversity of many water bodies.
- In polluted water light penetration is less which slow-down the process of photosynthesis, diminishing the growth of the plants.
- BOD and COD in polluted water is high.

Control Measures of Water Pollution

Any type of pollution can be controlled in the best way by educating people about the problem and through awarness and public participation, much of the damage to the water can be checked or even stopped. Following are some control measures:

- The domestic, used water, should not be mixed with the water bodies of the area.
- The sewage water must be directed at a particular place and should be cleaned through water treatment plants and then it should be used for agricultural purposes.
- Water bodies should not be used as dumping place for garbage and waste materials.
- The dead animals must never be disposed off in water bodies.
- Human activities like bathing, washing of clothes and utensils, excretion in and near the rivers and ponds, must be stopped.
- Cleaning the cattle in ponds, rivers must be checked.
- Washing of trucks, tractors and other heavy vehicles in the water bodies should not be allowed.
- Use of harmful chemicals as insecticides pesticides and fertilisers must be controlled in farms, so that during rains, these chemicals and their residues do not mix with the near by water bodies.
- The source of drinking water must be protected by some type of natural fencing of trees shrubs etc., or by a wall, so that solid waste does not mix with water.
- Mining areas and industrial areas must be at a distance from residential colonies.
- According to the types of water pollution in an area, education and awareness programs must be launched about cause and **cure** of water pollution.

- Through N.S.S. camps, N.C.C. camps, eco-clubs, etc., maintenance of water bodies and addition of cleaning powders and chemicals may be done.
- The laws of pollution, must be implemened, strictly.
- Awarness compaign must be run in tribal and rural areas about water pollution.

1.7 ROLE OF AN INDIVIDUAL IN PREVENTION OF POLLUTION

It is the duty of a good citizen, to be aware of the national and local problem.

Problem of environmental pollution is related to every human being as one is using the water, air, soil and atmosphere during one's life. As a comon citizen one can contribute for the prevention of pollution in many ways without giving any extra time or money as such:

(*i*) One must not add to any kind of pollution *i.e.,* air, water and land.
(*ii*) By not throwing the household garbage every where on the land.
(*iii*) Separating, wet and dry darbage, and the recyclabe waste from the garbage can reduce the solid waste.
(*iv*) By not using polythene bags, instead using paper and cloth bags.
(*v*) By growing and saving the trees.
(*vi*) By protecting the natural resources.
(*vii*) By keeping the city and monuments clean.
(*viii*) By reducing the use of insecticides and pesticides and chemical fertilizers.
(*ix*) Increasing the use of orgnic fertilizer is good for the crop as well as environment.
(*x*) By reducing the use of spray perfumes and other cosmetics as they contain harmful organic solvents.
(*xi*) By avoiding the equipments containing the chlorofluoro carbons (CFC's).
(*xii*) By not polluting the water bodies.
(*xiii*) By participating and endorsing the environment protection, awareness programmes.
(*xiv*) By following the pollution control laws.

Simple Ways to Save the Environment

- Why not resolve to plant some trees every year? If each of us plants one little tree, it can amount to a great amount of afforestation making the environment healthier. Increasing the use of bicycles or making a habit to walk down short distances can contribute to reduction in air pollution. Try to minimize the use of vehicles. Use CFC-free products. For some destinations, the use of vehicles has no better options. But at least maintain your vehicles; clean their exhaust pipes, keep the pollution they cause under strict control.
- Look at the gadgets you use at home. Are all of them necessary? Do you maintain them well and use them efficiently? Replace the air filters for your air conditioning unit once a month. Turn off the ACs as also the lights and fans in unused rooms. Make sure to switch off the lights, the television or radio systems before you leave the house. Do not keep your computer switched on while you are not using it. Did you know that your refrigerator and water heaters consume a lot of power? A careful use of these gadgets is a good way to save the electricity.
- One of the most important constituents of the environment is water. Preventing the wastage of water and curbing water pollution is one of our primary duties. Turn off the taps; do not

let the buckets overflow! The use of bath showers and heavy-flushes in toilets leads to an excessive usage of water. Do not dump garbage down a storm drain. Do not pollute water bodies.

- Recycling is one of the best measures of saving the environment. Try to use renewable sources of energy. Resort to the use of renewable natural resources. A simple way to do this is to lessen the use of rubber and plastic. Instead, use paper bags and cardboard containers. Even a simple habit of buying in bulk can save a lot of packaging material, thus contributing to saving the environment. Buy the products that you can reuse.
- When in office, print only when it is absolutely necessary. Printing every soft copy leads to a heavy wastage of paper. Remember to switch off your computer when no one is using it. Avoid an excessive use of air conditioners in the office. Use emails instead of paper correspondence. Do not use disposable cups when you have an option of using the ceramic ones.
- Minimize the use of animal products, which involve their killing. Animal fur and ivory are some of the excessively used products that are gradually leading to the extinction of the animals that provide them. Resolve not to hunt animals. Follow the principles of saving the environment and encourage others in taking to environment-saving activities.
- Activities like composting can be of great help in recycling of garbage into useful manures. Avoid an extensive use of chemical fertilizers and synthetic nutrients. Natural manures are a better option for farms and gardens.

These are some of the simple ways in which we can contribute towards saving nature. A small step by each of us can make a huge leap towards saving the environment.

1.8 ENVIRONMENTAL POLLUTION: INDIA SCENARIO

The environmental problems in India are growing rapidly. The increasing economic development and a rapidly growing population that has taken the country from 300 million people in 1947 to more than one billion people today is putting a strain on the environment, infrastructure, and the country's natural resources. Industrial pollution, soil erosion, deforestation, rapid industrialization, urbanization, and land degradation are all worsening problems. Overexploitation of the country's resources is it land or water and the industrialization process has resulted environmental degradation of resources. Environmental pollution is one of the most serious problems facing humanity and other life forms on our planet today.

With India's population at 1.2 billion people and counting, plus internal economic migration to urban areas from the countryside, the country's cities are bursting at the seams. Housing shortages, electricity and water cuts, traffic congestion, pollution and a lack of basic services are the reality for millions. The demographers are predicting that India will add three to four hundred million new people to its population over the next 40 years About 45 per cent of India's land is degraded, air pollution is increasing in all its cities, it is losing its rare plants and animals more rapidly than before and about one-third of its urban population now lives in slums, says the **State of Environment Report India 2009** brought out by the government. The third official report on the state of India's environment published after a gap of eight years and released by Minister of State for Environment and Forests Jairam Ramesh in New Delhi on September 2, 2009.

The cost of environmental damage in India would shave 4 percent off of the country's gross domestic product. Lost productivity from death and disease due to environmental pollution are the primary culprits. The government agency responsible for environmental affairs is the Ministry of Environment and Forests (MoEF). Coping with India's industrial pollution is perhaps the agency's top priority. MoEF recognizes the need to strike a balance between development and protecting the environment in administering and enforcing the country's environmental laws and policies. The government heightened the Ministry's powers with the passage of the 1986 Environment Protection Act. This act built on the 42nd amendment to India's constitution in 1976 that gave the government the right to step in and protect public health, forests, and wildlife. This amendment however had little power as it contained a clause that stated it was not enforceable by any court. India is the first country in the world to pass an amendment to its constitution ostensibly protecting the environment.

Mahatma Gandhi on Environmental pollution. Mahatma Gandhi had said that nature has enough to satisfy everyone's need but has not enough to satisfy man's greed. Sadly our ever-expanding greed has put us in such precarious situation. Will we realize it? The policy of industrialization had helped rich to become richer and poor become poorer. The disparity has widened. It is the democratic system followed in the country which has forced our policy-makers to think of growth for all. That is why we are hearing plans for inclusive growth. Industrialization is not without price. All these have a direct bearing on environmental pollution leading to climatic change. We are all witness to the deleterious effects of climate change. The whole world is now anxious to repair the damage.

Mahatma Gandhi

1.9 BASICS OF ENVIRONMENT

The planet Earth was formed about 4.5 billion years ago, after the collapse of the super massive gaseous body. As time moved on, the Earth cooled down and is still cooling, as of today. As a result of the cooling, denser materials like iron and sulfur sank to the inside of the Earth; whereas, lighter materials like silicates and water floated near the Earth's surface.

Lithosphere

The planet Earth comprises several layers. Starting from the Earth's surface, there are four main layers; namely, the crust, the mantle, the outer core and the inner core. The pressure and temperature increases tremendously from the outer layers to the inner layers. Let's take a look at each of them individually.

The lithosphere, Greek for "rocky sphere," is the outermost shell of the Earth. The term is also used to refer to the outermost rocky shell of other solid planets. It is a relatively thin layer, 50-100 km thick under the oceans, 150 km thick on the continents. The lithosphere is composed of the upper crust, 5 km thick in the oceans and 65 km thick on the continents, and the upper mantle, which makes up the remainder. Separating the crust and the upper mantle is the Mohorovicic discontinuity, the point at which rocks become plastic rather than solid. Beneath the lithosphere is the asthenosphere, which continues the upper mantle, and is approximately the point at which the mantle becomes liquid. In the Earth, the lithosphere includes the crust and the uppermost mantle, which constitute the hard and rigid outer layer of the Earth. The lithosphere is underlain by the asthenosphere, the weaker, hotter, and deeper part of the upper mantle. The boundary between the lithosphere and the underlying asthenosphere is defined by a difference in response to stress: the lithosphere remains rigid for very long periods of geologic time in which it deforms elastically and through brittle failure, while the asthenosphere deforms viscously and accommodates strain through plastic deformation.

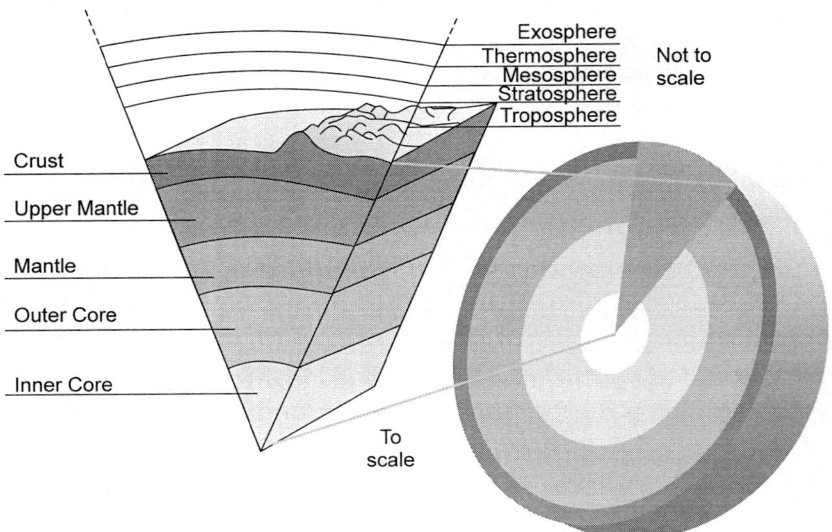

The lithosphere is broken into tectonic plates. It is made of following four layers:

The Crust: The crust or the lithosphere is the outermost layer of the Earth made up of silicate rock materials. The crust makes up only about one percent of the Earth and is the thinnest layer in comparison to the other three layers. Most earthquakes occur in the Earth's crust. The thickness and the composition of the Earth's crust vary in the land and the ocean. For example: the continental crust is about 32 kilometer thick and composed of lighter materials like granite, quartz and feldspar; whereas, the oceanic crust measures about 10 kilometers and is mostly made up of basalt.

The Mantle: The mantle, the largest layer of the Earth, is made up of iron, aluminum, calcium, magnesium, silicone and oxygen. In fact, most of the Earth's mass (about 80 percent) lies in the mantle. The temperature in this layer is estimated to be about 1000 degrees celsius. It is in this layer that volcano magma is present. For better understanding, we can say that the mantle layer is divided into the upper and lower sections. The upper mantle is much cooler than the lower (deeper) section. The overall thickness of the mantle layer is 2900 kilometers.

The Outer Core: The outer core can be regarded as a ball of very hot metals. The outer core is liquid and made up of iron and nickel. The density is very high, but less than pure molten iron. Hence, scientists are of the opinion that sulfur and oxygen may be present in the outer core. (This is due to the fact that they dissolve easily in liquid iron.) The outer core measures 2200 kilometers in thickness. As the Earth rotates, the outer core (consisting of iron) spins over the inner core and generates the Earth's magnetic field, which is the factor responsible for functioning of the magnetic compass.

The Inner Core: The inner core, as the name suggests, is the innermost layer of the Earth, and is characterized by extremely high temperature and pressure conditions. The temperature of the inner core layer is more than the sun's surface. The intense heat reflected from the inner core mobilizes the materials of the outer core and the mantle. It is due to the high pressure that the inner core materials are unable to move, and hence remain solid. The thickness of the inner core is believed to be about 1250 kilometers.

The diameter of the Earth is 12,756 kilometers. The Earth is enveloped by atmosphere, which comprises four major layers, namely, the troposphere, stratosphere, mesosphere and thermosphere. The atmosphere provides air, water, warmth and protects from the harmful rays of the sun. Without atmosphere, life on earth would be impossible!

Atmosphere

Atmosphere is the huge blanket of gas that circles the entire Earth. Without it, life as we know it could not exist.

This blanket of gas starts at ground level and stretches 600 miles into the sky. However, most of this life-supporting shell is squashed down into a layer only six miles thick. The top of Mount Everest barely peeks above the edge of this layer.

The remaining 594 miles cannot support life. However, these layers do protect us from the dangers of the sun's radiation. They also protect us from drifting rocks, big hunks of metal, and other bits and pieces of space junk that collide with our planet from time to time.

Composition of the Atmosphere

The **atmosphere** is made of four different layers. Layer number one is called the **troposphere.** People, plants, animals, and insects live in the troposphere. It is the layer where all weather occurs.

Scientists who study weather are called meteorologists. They refer to the troposphere at "The Weather Zone." The troposphere begins at ground level and extends six miles up into the sky where it meets with the second layer.

Layer number two is called the **stratosphere.** The stratosphere begins at the six-mile point and reaches nearly 31 miles into the sky.

An important layer of atmosphere containing ozone is located inside the stratosphere. Ozone is a special form of oxygen, and the ozone layer is very important to all life on Earth. Ozone blocks large amounts of solar ultraviolet radiation from entering the troposphere. Too much solar radiation can harm living things, including people.

The third major layer is the **mesosphere.** The mesosphere begins 32 miles above the Earth's surface. Temperatures are warmest at the lowest level of the mesosphere and become colder at its highest level.

The **thermosphere** is the fourth layer and begins 50 miles above the Earth. Temperatures in the thermosphere become hotter and hotter when moving farther away from ground level. **How can this be?**

Together, the troposphere, stratosphere, mesosphere, and thermosphere act as a giant safety blanket. They keep the temperature on the Earth's surface from dipping to extreme icy cold that would freeze everything solid, or from soaring to blazing heat that would burn up all life.

Layers of the Atmosphere

The atmosphere of the Earth may be divided into several distinct layers, as the following figure indicates.

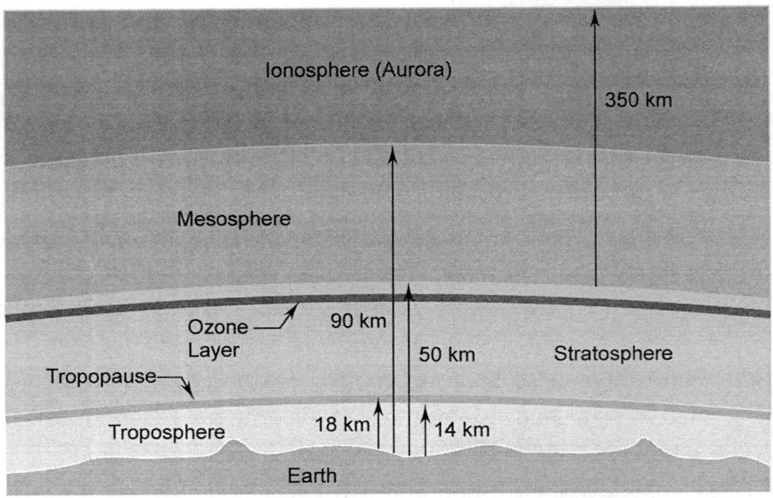

Layers of the Earth's Atmosphere

The Troposphere

The *troposphere* is where all weather takes place; it is the region of rising and falling packets of air. The air pressure at the top of the troposphere is only 10% of that at sea level (0.1 atmospheres). There is a thin buffer zone between the troposphere and the next layer called the *tropopause*.

The Stratosphere and Ozone Layer

Above the troposphere is the *stratosphere*, where air flow is mostly horizontal. The thin ozone layer in the upper stratosphere has a high concentration of ozone, a particularly reactive form of oxygen. This layer is primarily responsible for absorbing the ultraviolet radiation from the Sun. The formation of this layer is a delicate matter, since only when oxygen is produced in the atmosphere can an ozone layer form and prevents an intense flux of ultraviolet radiation from reaching the surface, where it is quite hazardous to the evolution of life. There is considerable recent concern that manmade fluorocarbon compounds may be depleting the ozone layer, with dire future consequences for life on the Earth.

The Mesosphere and Ionosphere

Above the stratosphere is the mesosphere and above that is the ionosphere (or *thermosphere*), where many atoms are ionized (have gained or lost electrons so they have a net electrical charge). The ionosphere is very thin, but it is where aurora takes place, and is also responsible for absorbing the most energetic photons from the Sun, and for reflecting radio waves, thereby making long-distance radio communication possible.

The structure of the ionosphere is strongly influenced by the charged particle wind from the Sun (solar wind), which is in turn governed by the level of solar activity. One measure of the structure of the ionosphere is the free electron density, which is an indicator of the degree of ionization.

Hydrosphere

The hydrosphere is the liquid water component of the Earth. It includes the oceans, seas, lakes, ponds, rivers and streams. The hydrosphere covers about 70% of the surface of the Earth and is the home for many plants and animals.

The hydrosphere, like the atmosphere, is always in motion. The motion of rivers and streams can be easily seen, while the motion of the water within lakes and ponds is less obvious. Some of the motion of the oceans and seas can be easily seen while the large scale motions that move water great distances such as between the tropics and poles or between continents are more difficult to see. These types of motions are in the form of currents that move the warm waters in the tropics toward the poles, and colder water from the Polar Regions toward the tropics. These currents exist on the surface of the ocean and at great depths in the ocean (up to about 4km).

The characteristics of the ocean which affects its motion are its temperature and salinity. Warm water is less dense or lighter and therefore tends to move up toward the surface, while colder water is more dense or heavier and therefore tends to sink toward the bottom. Salty water is also more dense or heavier and thus tends to sink, while fresh or less salty water is less dense or lighter and thus tends to rise toward the surface. The combination of the water's temperature and salinity determines whether it rises to the surface, sinks to the bottom or stays at some intermediate depth.

The oceans currents are also affected by the motion of the atmosphere, or winds, above it. The energy in the wind gets transferred to the ocean at the ocean surface affecting the motion of the water there. The effect of wind is largest at the ocean surface.

The ocean serves two main purposes in the climate system. First, it is a large reservoir of chemicals that can contribute to the greenhouse effect in the atmosphere and energy absorbing 90% of the solar radiation which hits the surface. This reservoir changes very slowly limiting how fast the climate can change. Second, it works with the atmosphere to redistribute the energy received from the sun such that the heat in the topics, where a lot of energy is received from the sun, is transferred toward the poles, where heat is generally lost to space.

Biosphere

The biosphere is made up of all that is living on earth, from the smallest bacterium to the largest whale. It includes between 3 and 30 million species of plants, animals, and fungi, and also two kinds of single-celled living things: those without nuclei ("prokaryotes") such as bacteria, and those with nuclei ("eukaryotes") such as protozoans. The biosphere extends over the Earth's surface in a thin layer from a few kilometers into the atmosphere, in very cold environments, to the deep-sea vents of the ocean depths, in very hot environments.

The biosphere is a life-supporting global ecosystem, where each living things depends on each other and the environment. The ecosystem includes all living organisms and the abiotic or nonliving environment on which they depend for their energy and the nutrients they need to live.

1.10 ENVIRONMENTAL ACTS AND REGULATIONS FOR ENVIRONMENTAL PROTECTION

In the Constitution of India it is clearly stated that it is the duty of the state to 'protect and improve the environment and to safeguard the forests and wildlife of the country'. It imposes a duty on every citizen 'to protect and improve the natural environment including forests, lakes, rivers, and wildlife'. Reference to the environment has also been made in the Directive Principles of State Policy as well as the Fundamental Rights. The Department of Environment was established in India in 1980 to ensure a healthy environment for the country. This later became the Ministry of Environment and Forests in 1985.

The constitutional provisions are backed by a number of laws - acts, rules, and notifications. The EPA (Environment Protection Act), 1986 came into force soon after the Bhopal Gas Tragedy and is considered an umbrella legislation as it fills many gaps in the existing laws. Thereafter a large number of laws came into existence as the problems began arising, for example, Handling and Management of Hazardous Waste Rules in 1989.

Following is a list of the environmental legislations that have come into effect:

1. **General**
2. **Forest and wildlife**
3. **Water**
4. **Air**

1. **General Legislations**

1986 - The Environment (Protection) Act authorizes the central government to protect and improve environmental quality, control and reduce pollution from all sources, and prohibit or restrict the setting and /or operation of any industrial facility on environmental grounds.

1986 - The Environment (Protection) Rules lay down procedures for setting standards of emission or discharge of environmental pollutants.

1989 - The Objective of Hazardous Waste (Management and Handling) Rules is to control the generation, collection, treatment, import, storage, and handling of hazardous waste.

1989 - The Manufacture, Storage, and Import of Hazardous Rules define the terms used in this context, and sets up an authority to inspect, once a year, the industrial activity connected with hazardous chemicals and isolated storage facilities.

1989 - The Manufacture, Use, Import, Export, and Storage of Hazardous Micro-organisms/ Genetically Engineered Organisms or Cells Rules were introduced with a view to protect the environment, nature, and health, in connection with the application of gene technology and microorganisms.

1991 - The Public Liability Insurance Act and Rules and Amendment, 1992 was drawn up to provide for public liability insurance for the purpose of providing immediate relief to the persons affected by accident while handling any hazardous substance.

1995 - The National Environmental Tribunal Act has been created to award compensation for damages to persons, property, and the environment arising from any activity involving hazardous substances.

1997 - The National Environment Appellate Authority Act has been created to hear appeals with respect to restrictions of areas in which classes of industries etc., are carried out or prescribed subject to certain safeguards under the EPA.

1998 - The Biomedical Waste (Management and Handling) Rules is a legal binding on the health care institutions to streamline the process of proper handling of hospital waste such as segregation, disposal, collection, and treatment.

1999 - The Environment (Sitting for Industrial Projects) Rules, 1999 lay down detailed provisions relating to areas to be avoided for sitting of industries, precautionary measures to be taken for site selecting as also the aspects of environmental protection which should have been incorporated during the implementation of the industrial development projects.

2000 - The Municipal Solid Wastes (Management and Handling) Rules, 2000 apply to every municipal authority responsible for the collection, segregation, storage, transportation, processing, and disposal of municipal solid wastes.

2000 - The Ozone Depleting Substances (Regulation and Control) Rules have been laid down for the regulation of production and consumption of ozone depleting substances.

2001 - The Batteries (Management and Handling) Rules, 2001 rules shall apply to every manufacturer, importer, re-conditioner, assembler, dealer, auctioneer, consumer, and bulk consumer involved in the manufacture, processing, sale, purchase, and use of batteries or components so as to regulate and ensure the environmentally safe disposal of used batteries.

2002 - The Noise Pollution (Regulation and Control) (Amendment) Rules lay down such terms and conditions as are necessary to reduce noise pollution, permit use of loud speakers or public address systems during night hours (between 10:00 p.m. to 12:00 midnight) on or during any cultural or religious festive occasion.

2002 - The Biological Diversity Act is an act to provide for the conservation of biological diversity, sustainable use of its components, and fair and equitable sharing of the benefits arising out of the use of biological resources and knowledge associated with it.

2. **Forest and Wildlife Legislations**

1927 - The Indian Forest Act and Amendment, 1984, is one of the many surviving colonial statutes. It was enacted to 'consolidate the law related to forest, the transit of forest produce, and the duty livable on timber and other forest produce'.

1972 - The Wildlife Protection Act, Rules 1973 and Amendment 1991 provides for the protection of birds and animals and for all matters that are connected to it whether it be their habitat or the waterhole or the forests that sustain them.

1980 - The Forest (Conservation) Act and Rules, 1981, provides for the protection of and the conservation of the forests.

3. **Water Legislations**

1882 - The Easement Act allows private rights to use a resource that is, groundwater, by viewing it as an attachment to the land. It also states that all surface water belongs to the state and is a state property.

1897 - The Indian Fisheries Act establishes two sets of penal offences whereby the government can sue any person who uses dynamite or other explosive substance in any way (whether coastal or inland) with intent to catch or destroy any fish or poisonous fish in order to kill.

1956 - The River Boards Act enables the states to enroll the central government in setting up an Advisory River Board to resolve issues in inter-state cooperation.

1970 - The Merchant Shipping Act aims to deal with waste arising from ships along the coastal areas within a specified radius.

1974 - The Water (Prevention and Control of Pollution) Act establishes an institutional structure for preventing and abating water pollution. It establishes standards for water quality and effluent. Polluting industries must seek permission to discharge waste into effluent bodies.

The CPCB (Central Pollution Control Board) was constituted under this act.

1977 - The Water (Prevention and Control of Pollution) Cess Act provides for the levy and collection of cess or fees on water consuming industries and local authorities.

1978 - The Water (Prevention and Control of Pollution) Cess Rules contains the standard definitions and indicate the kind of and location of meters that every consumer of water is required to affix.

1991 - The Coastal Regulation Zone Notification puts regulations on various activities, including construction, are regulated. It gives some protection to the backwaters and estuaries.

4. **Air Legislations**

1948 - The Factories Act and Amendment in 1987 was the first to express concern for the working environment of the workers. The amendment of 1987 has sharpened its environmental focus and expanded its application to hazardous processes.

1981 - The Air (Prevention and Control of Pollution) Act provides for the control and abatement of air pollution. It entrusts the power of enforcing this act to the CPCB.

1982 - The Air (Prevention and Control of Pollution) Rules defines the procedures of the meetings of the Boards and the powers entrusted to them.

1982 - The Atomic Energy Act deals with the radioactive waste.

1987 - The Air (Prevention and Control of Pollution) Amendment Act empowers the central and state pollution control boards to meet with grave emergencies of air pollution.

1988 - The Motor Vehicles Act states that all hazardous waste is to be properly packaged, labeled, and transported.

1.11 CENTRAL POLLUTION CONTROL BOARD *(COURTESY: WWW.CPCB.NIC.IN)*

The **Central Pollution Control Board (CPCB)**, statutory organization, was constituted in September, 1974 under the Water (Prevention and Control of Pollution) Act, 1974. Further, CPCB was entrusted with the powers and functions under the Air (Prevention and Control of Pollution) Act, 1981.

It serves as a field formation and also provides technical services to the Ministry of Environment and Forests of the provisions of the Environment (Protection) Act, 1986. Principal **Functions** of the CPCB, as spelt out in the Water (Prevention and Control of Pollution) Act, 1974, and the Air (Prevention and Control of Pollution) Act, 1981, (*i*) to promote cleanliness of streams and wells in different areas of the States by prevention, control and abatement of water pollution, and (*ii*) to improve the quality of air and to prevent, control or abate air pollution in the country.

Air Quality Monitoring is an important part of the air quality management. The **National Air Monitoring Programmed** (NAMP) has been established with objectives to determine the present air quality status and trends and to control and regulate pollution from industries and other source to meet the air quality standards. It also provides background air quality data needed for industrial sitting and towns planning.

Besides this, CPCB has an automatic monitoring station at ITO Intersection in New Delhi. At this station Resirable Suspended Particulate Matter (RSPM), Carbon Monoxide (CO), Ozone (O_3), Sulphur Dioxide (SO_2), Nitrogen Dioxide (NO_2) and Suspended Particulate Matter (SPM) are being monitored regularly. This information on Air Quality at ITO is updated every week.

Fresh water is a finite resource essential for use in agriculture, industry, propagation of wildlife & fisheries and for human existence. India is a riverine country. It has 14 major rivers, 44 medium rivers and 55 minor rivers besides numerous lakes, ponds and wells which are used as primary source of drinking water even without treatment. Most of the rivers being fed by monsoon rains, which are limited to only three months of the year, run dry throughout the rest of the year often carrying wastewater discharges from industries or cities/towns endangering the quality of our scarce water resources. The parliament of India in its wisdom enacted the Water (Prevention and Control of Pollution) Act, 1974 with a view to maintaining and restoring wholesomeness of our water bodies. One of the mandates of CPCB is to collect, collate and disseminate technical and statistical data relating to water pollution. Hence, **Water Quality Monitoring (WQM)** and Surveillance are of utmost importance.

Functions of the Central Board

[A] Functions of the Central Board at the National Level

1. Advise the Central Government on any matter concerning prevention and control of water and air pollution and improvement of the quality of air.
2. Plan and cause to be executed a nation-wide program for the prevention, control or abatement of water and air pollution;
3. Co-ordinate the activities of the State Board and resolve disputes among them;
4. Provide technical assistance and guidance to the State Boards, carry out and sponsor investigation and research relating to problems of water and air pollution, and for their prevention, control or abatement;
5. Plan and organize training of persons engaged in programmed on the prevention, control or abatement of water and air pollution;
6. Organize through mass media, a comprehensive mass awareness programme on the prevention, control or abatement of water and air pollution;
7. Collect, compile and publish technical and statistical data relating to water and air pollution and the measures devised for their effective prevention, control or abatement;
8. Prepare manuals, codes and guidelines relating to treatment and disposal of sewage and trade effluents as well as for stack gas cleaning devices, stacks and ducts;
9. Disseminate information in respect of matters relating to water and air pollution and their prevention and control;
10. Lay down, modify or annul, in consultation with the State Governments concerned, the standards for stream or well, and lay down standards for the quality of air;

[B] Functions of the Central Board as State Boards for the Union Territories

Perform such other function as may be prescribed by the Government of India.

Advise the Governments of Union Territories with respect to the suitability of any premises or location for carrying on any industry which is likely to pollute a stream or well or cause air pollutions; Lay down standards for treatment of sewage and trade effluents and for emissions from automobiles, industrial plants, and any other polluting source; Evolve efficient methods for disposal of sewage and trade effluents on land; develop reliable and economically viable methods of treatment of sewage, trade effluent and air pollution control equipment; Identify any area or areas within Union Territories as air pollution control area or areas to be notified under the Air (Prevention and Control of Pollution) Act, 1981; Assess the quality of ambient water and air, and inspect wastewater treatment installations, air pollution control equipment, industrial plants or manufacturing process to evaluate their performance and to take steps for the prevention, control and abatement of air and water pollution.

As per the policy decision of the Government of India, the CPCB has delegated its powers and functions under the Water (Prevention and Control of Pollution) Act, 1974, the Water (Prevention and Control of Pollution) Cess Act, 1977 and the Air (Prevention and Control of Pollution) Act, 1981 with respect to Union Territories to respective local administrations. CPCB along with its counterparts State Pollution Control Boards (SPCBs) are responsible for implementation of legislations relating to prevention and control of environmental pollution.

EXERCISE

1. *Environmental Science is a multidisciplinary Study*. Explain it.
2. What is environmental pollution? Write an essay on components of Environment.
3. Write short notes on:
   (*i*) Lithosphere
   (*ii*) Atmosphere
   (*iii*) Hydrosphere
   (*iv*) Biosphere
4. Discuss pollutants and their types.
5. What do you understand by environmental engineering? Give your specific comments on the objectives of Environmental Education. [*Raj. Tech. Univ., Sem.II, 2009*]
6. (*a*) What are the adverse effects of Air Pollution, noise Pollution and Water Pollution?
   (*b*) What are the functions and powers of Central Pollution Control Board? What are the aims and objectives of Laws made for protection of environment in our country?
   [*Raj. Tech. Univ., Sem.II, 2009*]

2

ECOSYSTEMS AND BIODIVERSITY

2.1 INTRODUCTION: WHAT IS ECOLOGY?

Ecology is the study of the interactions between organisms and their environment. Ecology is a science that studies relationships between living organisms and the environment they live in. In 1869, the german biologist Ernst Haeckel coined the term ecology, refering to the Greek origin of the word (oikos, house; logo, science, study, treaty). According to Haeckel, ecology had study species in its biological relations with environment. Other scientists took care later of the surroundings in which each species lives and of its symbiotic and antagonistic relations with others.

Towards 1925, August Thienemann, Charles Elton and others stimulated the ecology of the communities. They worked with concepts as the one of food web, or the one of pyramid of species, in which the number of individuals diminishes progressively from the base to the peak, the plants to the herbivorous animals and the carnivores.

The alive beings are in permanent interaction between each other and with the environment in which they live. The ecology analyzes how each element of an ecosystem affects the other components and how this is affected too. It is a synthesis science, because to understand the complex plot of relations that exist in an ecosystem it takes knowledge from botany, zoology, physiology, genetics and other disciplines like the physics, chemistry and geology.

Its branches focus on studying specific kinds of these organisms and are called respectively. For example: ecology of river beaver or ecology of sturgeon etc. The object can be researched in light of its biotope, places and conditions of its reproduction, its mating patterns and other interaction with representative of the same species, with competitors and enemies (predators, parasites etc.) or just neighbors.

Human being can also be an object of such research. He is studied mostly in his working environment (sea, coal mines, space flight etc.)

A separate sub-branch of ecology of human being is protection of his environment. In many countries there are special ecological services as well as voluntary environmentalist organizations, like Greenpeace. The environmentalists struggle with pollution of water reservoirs, of air and soil, as well as for preserving natural forests and fauna. They protest against nuclear weapons testing, fight for banning of chemical, biological and nuclear weapons and antipersonnel mines that cripple civilians and animals. They also call for banning of the usage of traps in fur trade, which cause incredible sufferings to animals.

The environmentalist movement deserves highest praise and every possible kind of support. The applied ecological studies of professional activities of man are also very important. But in this book we will attempt to see that this is not everything that the ecology of human being may study.

Various different species living in the same place, interacting amongst themselves and with their environment together form an ecosystem.

Ecologists are people that study the interactions between organisms and their environment within food webs or other ecological relationships. Fieldwork is an essential component of this study. Laboratory experiments are also applied, under field conditions. Most of the time ecologists are involved in studying the natural environment and communities, but some are involved in applied ecology, using ecological knowledge in ecosystems directed by humans, commonly known as agro-ecosystems.

After all, man is not only his body, but also, and primarily, consciousness. And he not only lives in the close contact with air, water, soil, other people, animals, plants, viruses and bacteria. He also interacts with non-embodied individual consciousness (spirits of people and animals). And also — which is the most important of all — with God.

In realty we live in a multidimensional environment. And the material world that we can perceive with our senses is just a small portion of it. Other layers (eons or lokas) of the multidimensional universe usually remain completely out of our perception, although they are filled with life that can see us and which influences us.

That is what we will talk about now. And we will see that ecological science has a much broader field of study, than one ordinarily thinks.

While many people confuse the work of an "ecologist" with that of an "environmentalist," there are significant differences between the two occupations. Environmentalists seek to preserve natural systems. Ecologists, while they may share the ideals of environmentalists, are mainly involved with gathering information about the communities of animals and plants, as well as the physical elements (such as rocks and soil) present in a given geographical area.

What is Ecosystem?

An ecosystem is simply an easy way to refer to all the lifeforms (plant and animal) in a cohesive, relatively independent area, and their relationship to one another. In a perfect world, an ecosystem is in balance. That is, the predator species keep the prey species' population in check without completely destroying them, and plantlife remains varied without one type predominating and crowding out the rest.

Ecosystems are rarely in balance, and this is often cause for alarm. For instance, pollution runoff from a factory can effect marine life in a lake, causing the fish population to plummet. This destabilizes the entire surrounding ecosystem, and birds and animals that prey on the fish either die off or migrate to areas with more food.

Attempting to counteract damage to an ecosystem requires a complete understanding of all the interrelationships between plants, animals and atmosphere in the system. Since such perfect understanding is rarely possible, unintended consequences are often the result.

In nature, no ecosystem can be said to be independent of neighboring ecosystems or indeed, the global ecosystem, since ecosystems are open and plants and animals can move between them. All ecosystems are effected by planet-wide trends that effect the atmosphere, such as pollution and global warming.

A key idea in ecology is that of the ecosystem. An ecosystem consists of all the organisms and creatures in a biotope, or particular environment type, that in turn is made up of biotic and abiotic (non-living) components. Related to the ecosystem is the notion of the biome, but this unit is defined by the makeup of its vegetation, such as tundra, desert, or savanna.

In every ecosystem there is a food chain, though it is more of a circle than a chain. At the bottom of the chain are the plants that (for the most part) are capable of photosynthesis. These are normally called autotrophs, because they generate their own energy from sunlight and by uptake of nutrients from soils. Creatures that feed on autotrophs are known as consumers. Primary consumers are the herbivores that eat plants. Secondary consumers eat the animals that eat the plants. The final link is the decomposers. These are the organisms such as bacteria that process the organic matter produced by other life forms in the food chain. As a result of their activities, nutrients return to the soil for plants to take up, and the cycle continues.

In this sense, ecology studies flows of energy throughout systems. Everything, including ourselves, belongs to one or more systems. Humans have a unique capacity to intervene in those systems and alter their functioning.

2.2 CLASSIFICATION OF ECOLOGY

1. On the basis of nature of study ecology can be subdivided as (*a*) Autecology and (*b*) Synecology.

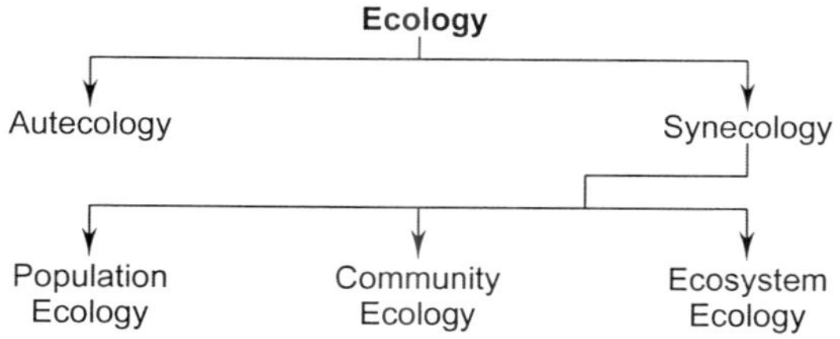

2. On the basis of the kind of environment or habitat, ecology has been sub divided into the following branches:

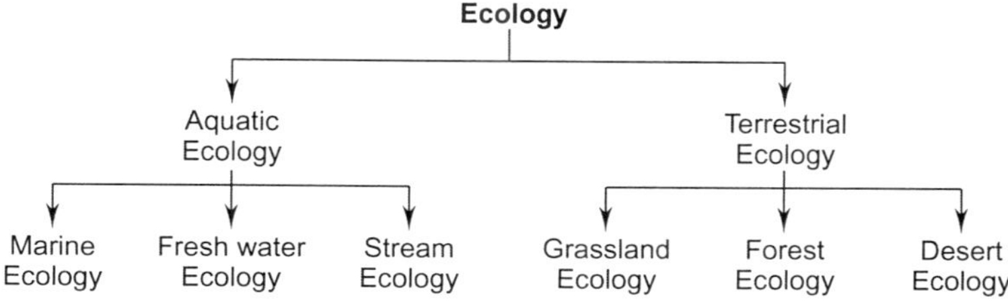

3. With advancing trends in the fields of ecology present day ecologists decide ecology into the following branches:

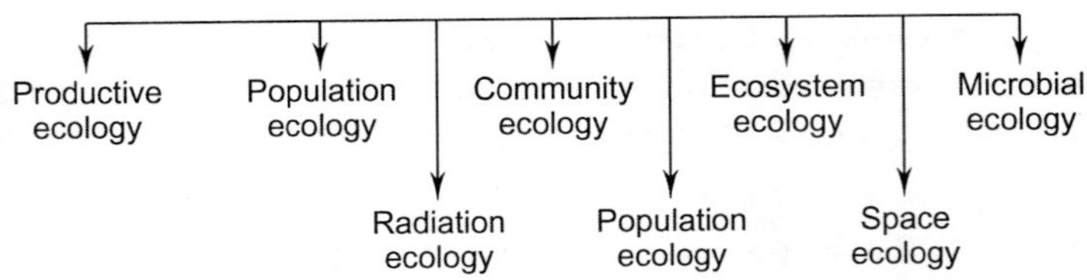

Autecology

Autecology deals with the study of the individual organism or an individual species and its population.

Here, the focus on life-histories and behaviour, as means of adaptation to the environment, is made.

Autecology is also called **"species ecology."**

Synecology

Synecology deals with the study of groups of organisms which are associated together as a unit. In other words it is study of communities, their composition, their behaviour and relation to the environment.

Synecology is also called **"Ecology of communities."**

Synecology is further divided into three fields.

Synecology is further sub divided into :

(*i*) **Population ecology:** It relates the organism with various groups of organisms and their inter-relationship.

(*ii*) **Gene ecology:** It deals with the genetic make up of species in relation to the environment.

(*iii*) **Taxonomic ecology:** It includes ecology of taxonomic groups as microbial ecology. Vertebrate or invertebrate ecology.

For *e.g.*, if a study is carried out of the relation of a Neem tree with the environment, then the approach is said to be autecological in nature. But if the study is carried out of the forest (of which Neem tree is a part), then the approach is said to be synecological in nature.

2.3 CONCEPT OF ECOSYSTEM

A biotic community lives in an environment, which provides material, energy requirement and other living conditions to it. "An ecological system in which the community and the non-living environment function together is called Ecosystem."

The term Ecosystem was introduced by Sir **Arthur Tansely** (1935). *An ecosystem is a group of biotic communities of species interacting with one another and with their non living environment exchanging energy and matter. Now ecology is often defined as* ***"the study of ecosystems"****. In other words,* an ecosystem can be defined *i.e.*, as a structural and functional unit of biosphere or segment of nature consisting of community of living beings and the physical environment both interacting and exchanging materials between them.

In an ecosystem the biotic communities and abiotic environment influence each other. This relationship is called **"holocenosis".**

An ecosystem is a natural grouping of nutrients, minerals, plants animals and their wastes linked together by flow of food, nutrients and energy from on part of the system to another part.

Thus, the ecosystem is the smallest units of biosphere that has all the characteristic to sustain life. Pond stream, seas, deserts, grasslands etc. are all examples of ecosystem.

2.4 TYPES OF ECOSYSTEM

An ecosystem may be natural (*e.g.*, forest, grassland, lake etc) or man made (*e.g.*, on aquaria, crop field etc.) temparary (rainfed pond) or permanent (*e.g.*, lake, forest etc.) or terrestrial (*e.g.*, grassland, forest etc.). The aquatic ecosystem can be either fresh water (*e.g.*, ponds, lakes, streams) or salt water (*e.g.*, marine, estuaries type) Fig. 2.1.

1. **Natural Ecosystem:** These systems operate by themselves under natural conditions without any major interferance by man. These are further divided into:
   (*i*) **Terrestrial ecosystem:** which include forest, grassland and desert etc.
   (*ii*) **Aquatic ecosystem:** may be further distinguished.
      (*a*) *Fresh water:* as which may be **lotic** (running water as spring, stream or rivers) or **lentic** (standing water as lake, pond, pools, ditch, puddles, swamp etc.).
      (*b*) *Marine water:* such as ocean (deep bodies) or sea (shallow ones).
2. **Artificial (Man-made) Ecosystem:** These are maintained artificial by man where, by addition of energy and planned manipulations, natural balance is destributed regularly. For example, croplands like maize, wheat, rice fields etc. Where man tries to control the biotic community as well as the physio-chemical environment, are man engineered ecosystem.

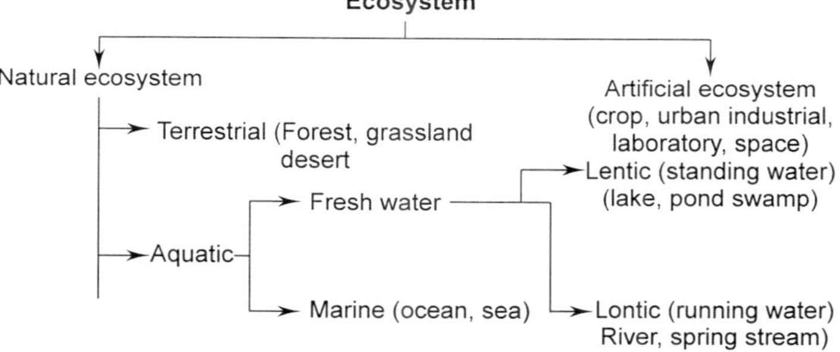

Fig. 2.1 Types of Ecosystem

2.5 ECOSYSTEM—STRUCTURE AND FUNCTION

The structure of an ecosystem is characterised by the physical organisation of biotic and abiotic components. The major structural features of ecosystem are: species composition, stratification, trophic organisation and nutrients.

1. **Species composition.** Each ecosytem has its own type of species composition. Different ecosystems have different species composition. A great variety of species is found in forest ecosystem, whereas a few species occurs in a desert ecosystem.

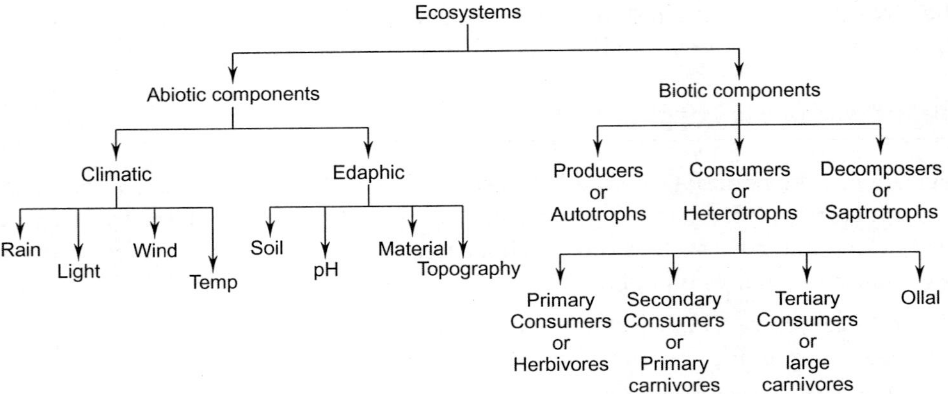

Schematic representation of the structure of an Ecosystem

2. **Stratification.** The organisms in each ecosystem form one or more layers or strata, each comprising the population kind of species. In some ecosystems, such as tropical rain forest, the crown of trees, bushes and ground vegetation form different strata and are occupied by different species. On the other hand, the desert ecosystem shows a low discontinuous herb layer consisting of fewer and extensive bare patches of soil.
3. **Trophic organisation.** Food relationship of producers and consumers is another way to depict ecosystem structure. Trophic (food) structure of ecosystem is based on the existence of several trophic levels in the ecosystem. The producers (autotrophs) form the first trophic level or T_1, herbivores the second or T_2 and carnivores constitute the third or T_3.

   There may be 2-3 levels of carnivores. The top carnivore belong to T_4 or T_5 trophic levels. Decomposers form ultimate or detritus trophic level (Fig. 2.2). Parasites feed upon the organisms of all trophic levels. Therefore, they do not have a fixed trophic level. Similarly there is no fixed trophic level for omnivorous organism like man.

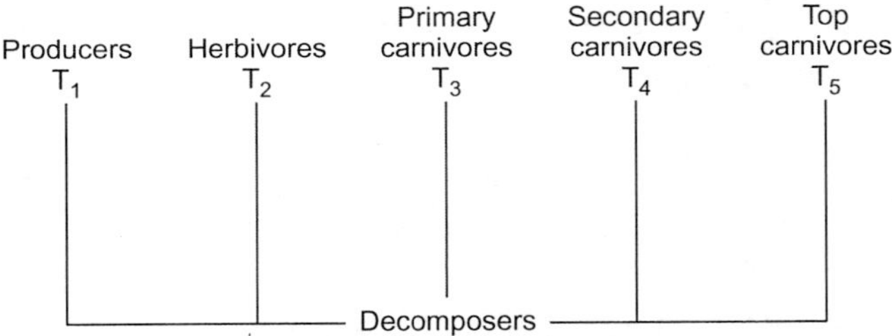

Fig. 2.2 Trophic Levels in an Ecosystem

In an ecosystem there can be only 4–5 successive trophic levels, because:

(*i*) All the food available at one trophic level is not eaten by the animals of the next trophic level. Some of the food energy is lost in this manner.

(*ii*) All the food eaten by an animals is not useful, thus a good deal of energy containing food is passed out as waste.

(*iii*) A large amount of energy is lost in respiration to derive organism's metabolism, and thus, there is not much energy left to support higher trophic levels.

Trophic structure may also be described in terms of the amount of living material present in different trophic levels. The amount of the living material presv different trophic levels at a given time is called **standing crop.** It is commonly expressed as the number or biomass of organisms per unit area. The biomass of a species may be expressed in terms of either fresh or dry weight. But, generally dry weight is preferred so that the variations in weight due to seasonal moisture differences in biomass can be avoided.

4. **Nutrients.** In each ecosystem, the nutrients necessary for the growth of living organism are accumulated in the biomass and the abiotic components like soil. The amount of nutrients such as nitrogen phosphorus and calcium present in the soil at any given time is termed as **standing state.** Different ecosystems have different standing state of nutrients. The standing state of nutrients may vary at different time even in the same ecosystem.

2.6 MAJOR ECOSYSTEMS

Different ecosystems like a pond, a lake, a river, a stream, a spring, an estuary, the sea, a forest, grassland, a desert, and a cropland are operating as self-sufficient interacting systems in the biosphere. The major ecosystems with their groups of climax plants and associated animals are called biomes. These ecosystem have a more of less similar fundamental plan of their gross structure and function. However, they differ in respect of their species composition and productivity rates. The ecological characteristics of the major ecosystems are described below:

1. **Forest Ecosystem.** Forests are natural plant communities with dominance of trees. In india, the forest occupies roughly 19% of the total land area. The major forest biomes found in India are: (*i*) Tropical rain forest, (*ii*) Tropical deciduous forest (*iii*) temperate brad leaf forest (*iv*) Temperate needle leaf or coniferous forest.

   The different components of forest ecosystem are as follows:

   Abiotic components: These include inorganic and organic substances present in the soil and atmosphere. The climate (temperature, light rainfall etc.) and soil (minerals) vary from forest to forest. In addition to minerals the occurrence of litter is characteristic feature of majority of forests.

   Biotic components:

   (*i*) ***Producers:*** These are mainly trees show much species diversity and greater degree of stratification especially in tropical rain and tropical deciduous forests. Besides trees, there are also present shurbs and ground vegetation. In these forests, the producers include the dominant tree species such as *Dipterocarpus. Tectona grandis (Teak), Butea frondosa (Dhak) and shorea robusta (sal).* In temperature broad leaf forests several species of Quercus (Oak), occur ad dominant species, where in temperature coniferous forests the dominant trees are *Pinus, Cedrus (Deodar), Picea (Spruce) and Abies (Silver fir), etc.*

   (*ii*) ***Consumers:*** Following types of consumers occur in the forest biomes.

   (*a*) *Primary consumers:* These are the herbivores that include smaller animals feeding on tree leaves as ants, flies, beetls, leaf-hoppers, bugs, spides, etc., and larger animals grazing on shoot and or fruits of producers as elephants, neelgai, deer, moles, squirrels, shrews flying foxes, mangooses etc.

   (*b*) *Secondary consumers:* These are the carnivores like snakes, birds, lizards, fox, etc., feeding on the herbivores.

   (*c*) *Tertiary consumers:* These are the top carnivores like lion, tiger, etc. that eat upon carnivores of secondary consumer's level.

(*iii*) ***Decomposers:*** These are wide variety of microorganisms including fungi (species of *aspergillus, Polyporus, Alternaria, Fusarium, Trichoderma, etc.), bacteria (species of Bacillus, Pseudomonas, Clostridium* etc.), and actionmycetes (species of Streptomyces). Rate of decomposition in tropical and subtropical forests is more rapid than in the temperate ones.

2. **Grassland Ecosystem.** The grassland biomass occur in the regions, where the climate is cool to cold during winters and hot in summers. The different components of a grassland ecosystem are as follows.

Abiotic components: These include nutrients present in soil and the atmosphere. Thus the elements like C, H, O, N, P, S, etc., are supplied by carbon dioxide, water, nitrates, phosphates and sulphates present in air and soil of the area.

Biotic components:

(*i*) ***Producers:*** The grasslands are characterised by treeless herbaceous plant cover dominant by a wide variety of grass species. The main grasses are species of *Dichanthium, Cynodon, Phragmites, cenchrus, Imparata and Saccharum,* etc. Besides them a few herbs and shurbs also occur as producers.

(*ii*) ***Consumers:*** The various types of consumers occuring in a grassland are as follows:

(*a*) *Primary consumers:* The herbivores feeding on grasses the mainly such grazing animals as cows, buffaloes, deer, rabbit, mouse, etc. Besides them, there are also present some insect species as *Leptocorisa, Dysdercus, Oxyrhachis, Cicindella, Cocinella,* and some termites millipedes, etc. that feed on the leaves of grasses.

(*b*) *Secondary consumers:* Snakes, lizards, birds, jackals, fox, etc. are common secondary consumers which feed on herbivores.

(*c*) *Tertiary consumers:* These include hawks, which fed on secondary consumers.

(iii) ***Decomposers:*** Several fungi *(Mucor, Aspergillus, Penicillium, Ladosporium, Fusarisum etc),* actinomycetes and bacteria decay the dead organic matter of different forms of higher life. They bring about minerals back to the soil, thus making them available to the producers.

3. **Desert Ecosystem.** The desert biomes are characterised by extremely low rainfall (less than 25 cm). They occupy about 17% of the land. Due to scarcity of water and high temperature, the biota is poorly represented. The various components of desert biomes are as follows.

Abiotic components: In desert ecosystem temperature is found to be very high and rainfall is very low. The nutrient cycling is poor is due to scanty biota.

Biotic components:

(*i*) ***Producers:*** These are shrubs, especially bushes, some grasses, and a few trees, the shrubs have widespread-branched root system with their leaves, branches and stems variously modified. Sometimes a few succulents like cacti are also present. Some lower plants like lichens and xerophytic mosses may also be present.

(*ii*) ***Consumers:*** Insects, reptiles, nocturnal rodents, birds camels, etc. are the main consumers.

(*iii*) ***Decomposers:*** These are very few, as due to poor vegetation the amount of dead organic matter is correspondingly less. These are some fungi and bacteria, most of them are thermophilic.

4. **Aquatic Ecosystem.** There are many kinds of aquatic ecosystems renging from small temporary puddles to large ocean. They differ widely with regard to abiotic factors and living organisms. The important aquatic ecosystems are described below.

A. Pond or Lake Ecosystem

A pond or lake is a good example of a self-sufficient and self regulating ecosystem (Fig.2.3). Location size, depth and substratum of a pond or lake constitute the biology of the ecosystem. The various components of the ecosystem are as follows :

Abiotic components: Temperature, light, water, and several inorganic and organic substances like, C, H, O, N, P, Ca, S and carbohydrates, proteins, and lipids make abiotic components. Some proportions of nutrients are in solution state but most of them are present stored in particulate matter as well as in living organisms. The amount of minerals present at any time in the physical environment of the pond is called **standing state.**

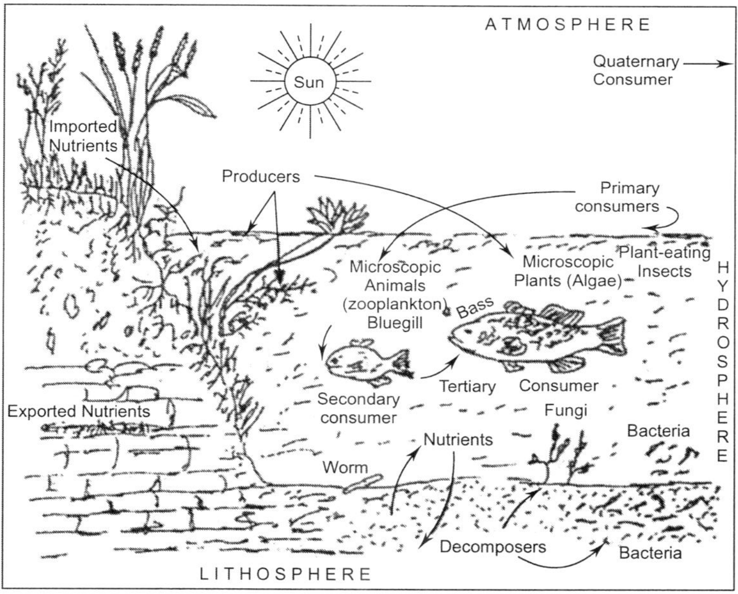

Fig. 2.3 Pond Ecosystem

Biotic components:

(*i*) ***Producers:*** These include green photosynthetic organisms and are of two types: *Phytoplanktons* and *Macrophytes*

(*a*) *Phytoplanktons:* These are minute floating or suspended lower plants belong to some algae are flagellates. *Ulothrix, Spirogyra, Oedogonium, Chalamydomonas, Zygnema, Volvox, Pandorina, Cosmarium, Scenedesmus, Closterium, Anabaena, Pediastrum, Microcystis, diatoms* etc. are common algal phytoplanktons.

(*b*) *Macrophytes: Ceratophyllum, Hydrilla, Utricularia, Vallisneria, Jussiaea, Nitella, Wolfia, Lemna, Spirodella, Pistia, Eichhornia, Azolla, Salvinia, Trapa, Typha, Marsilea,* etc. are included in this category. This may be classified further into submerged, free floating and amphibious plants.

(*ii*) ***Consumers***

(*a*) *Primary consumers:* These include zooplanktons and benthos forms. Zooplankton comprises ciliates, flagellates, other protozoans, small crustacean like Copepods and Daphnia, etc. These animals drift ith water current and are found along with phytoplankton upon which they feed. Benthos or bottom forms comprise the bottom dwelling animals, *e.g.*, annelids and molluscs, which feed on plants directly or on plant remains at the bottom.

(*b*) *Secondary consumers:* These are the carnivores, which feed on the herbivores, *e.g.*, insetcs and fish.

(*c*) *Tertiary consumers:* These are some large fish as game fish that feed on the smaller fish.

(*iii*) ***Decomposers (or microconsumers):*** Several bacteria fungi, (*Aspergillus, Cephadosporium, Phythium, etc.*) and actinomycetes represents the group.

B. River or Stream Ecosystem

The running water of a stream or a river is usually well oxygenated because it has a large surface area to absorb oxygen from the air. The various components of river or stream ecosystem are as follows:

Abiotic components: The river or stream water has lesser mineral content and greater penetration of light. In the lower reaches water is usually muddy cutting down the light at the river bed.

Biotic components:

(*i*) ***Producers:*** In slow moving water of banks, phytoplanktons, attached algae, water grasses and other amphibious plants are the producers. In swift water floating populations of plankton are generally absent.

(*ii*) ***Consumers:*** These include flatworms, leeches, water insects, snails, fishes, and crocodiles. Many birds and mammals also get their food from rivers and streams.

(*iii*) ***Decomposers:*** Several bacteria and fungi represents this group.

C. Ocean or Marine Ecosystem

Oceans cover more than two third of the earth's surface. The marine environment is characterised by its high concentration of salts and mineral ions. The ocean represents a very large and stable ecosystem. The main components of the ocean ecosystem are as follows :

Abiotic components: Marine environment as compared with fresh water, appears to be more stable in chemical composition due to being saline. The other physio-chemical factors such as dissolved oxygen, light and temperature are also different waves of various kinds and tides prevail there. Ocean also show distinct horizontal and vertical zones (Fig. 2.4).

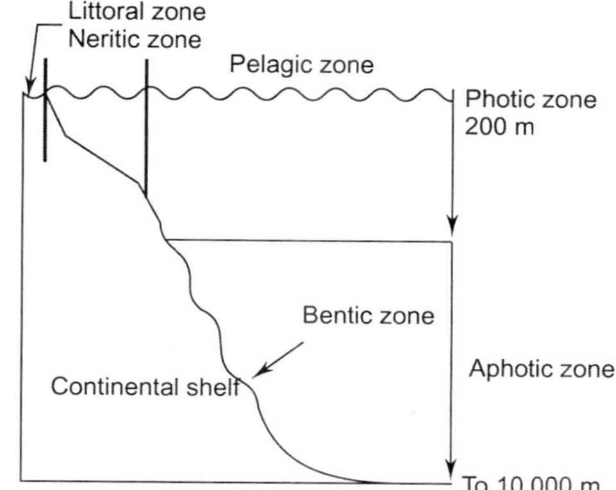

Fig. 2.4 Horizontal and Vertical Zonation in the Sea

Biotic components:

(*i*) ***Producers:*** This category includes phytoplanktons and larger marine plants. The former group includes diatoms and dinoflagelates. The later group includes sea weeds (algae) belonging to chlorophyceae, phaeophyceae and rhodophyceae and angiosperms. *Ruppia, Zostera, Posidonia, Halophila, Enhalus,* etc. are true marine angiosperm while various species of *Rhizophora.*

Ecological Balance

Especially for those who embrace philosophical ideas about a holistic universe, the notion of ecological balance is an important one. But not all ecologists view the situation in such holistic terms. If we bring in the idea of natural evolution, the idea of ecological balance becomes more complicated. It is true that an ecosystem (or other unit) can maintain a balanced state, sometimes for very long time spans, and that there are sometimes 'self-correcting' mechanisms.

In this sense, ecosystems may be resilient. They may adapt to a toxin, or climate change, or the arrival of a new species. But we also know that in the past, events have occurred that have caused irreversible damage to ecosystems. We know too that ecosystems are constantly changing. Some ecologists therefore dispute the idea that there is some 'normal' ecological equilibrium to which ecosystems and environments naturally return.

Ecological Awareness

In the past, people had little or no concept of ecological wholes, evolutionary rhythms or the ripple effects that human activity can have on the natural world. Not only is awareness of ecological principles vital in a world threatened by ecological catastrophe resulting from global warning (*sic*), but such principles can be the basis for our behaviour. Ecology demands more harmonious relations between people and nature, and even between nations, who are sometimes able to bury their differences in the cause of a larger and greater good.

Food Chains and Food Web

Food chains, in ecology, are the sequence of transfers of matter and energy from organism to organism in the form of food. Food chains do not normally encompass more than five trophic levels because energy, in the form of heat, is lost at each step. Food chains combine into highly complex food webs because most organisms consume more than one type of animal or plant.

A **food chain** shows how each living thing gets its food. Some animals eat plants and some animals eat other animals. For example, a simple food chain links the trees & shrubs, the giraffes (that eat trees & shrubs), and the lions (that eat the giraffes). Each link in this chain is food for the next link. *A food chain always starts with plant life and ends with an animal.*

1. Plants are called **producers** because they are able to use light energy from the Sun to produce food (sugar) from carbon dioxide and water.
2. Animals cannot make their own food so they must eat plants and/or other animals. They are called **consumers.** There are three groups of consumers.
   (*a*) Animals that eat ONLY PLANTS are called **herbivores** (or primary consumers).
   (*b*) Animals that eat OTHER ANIMALS are called **carnivores.**
   - carnivores that eat herbivores are called secondary consumers
   - carnivores that eat other carnivores are called tertiary consumers
3. Animals and people who eat BOTH animals and plants are called **omnivores.**
4. Then there are **decomposers** (bacteria and fungi) which feed on decaying matter.

These decomposers speed up the decaying process that releases mineral salts back into the food chain for absorption by plants as nutrients.

Do you know why there are more herbivores than carnivores?

In a food chain, energy is passed from one link to another. When a herbivore eats, only a fraction of the energy (that it gets from the plant food) becomes new body mass; the rest of the energy is lost as waste or used up by the herbivore to carry out its life processes (*e.g.*, movement, digestion, reproduction). Therefore, when the herbivore is eaten by a carnivore, it passes only a small amount of total energy (that it has received) to the carnivore. Of the energy transferred from the herbivore to the carnivore, some energy will be "wasted" or "used up" by the carnivore. The carnivore then has to eat many herbivores to get enough energy to grow.

Because of the large amount of energy that is lost at each link, the amount of energy that is transferred gets lesser and lesser.

1. **The further along the food chain you go, the less food (and hence energy) remains available.**

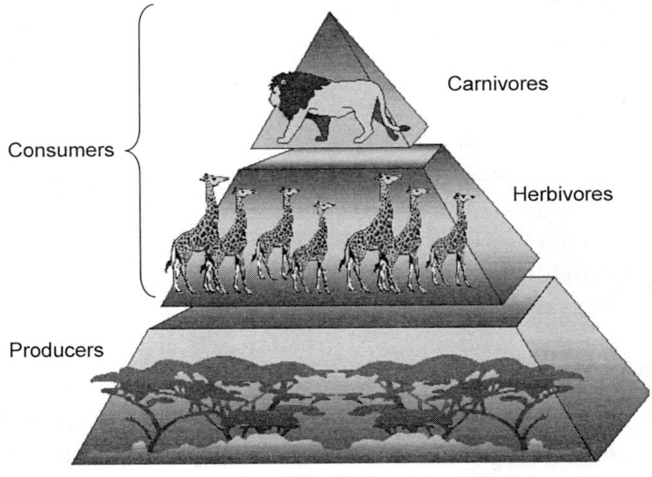

Fig. 2.5

The above energy pyramid shows many trees and shrubs providing food and energy to giraffes. Note that as we go up, there are fewer giraffes than trees and shrubs and even fewer lions than giraffes. as we go further along a food chain, there are fewer and fewer consumers. In other words, a large mass of living things at the base is required to support a few at the top many herbivores are needed to support a few carnivores.

2. **Most food chains have no more than four or five links.**

   There cannot be too many links in a single food chain because the animals at the end of the chain would not get enough food (and hence energy) to stay alive.

Most animals are part of more than one food chain and eat more than one kind of food in order to meet their food and energy requirements. These interconnected food chains form a **food web**.

A change in the size of one population in a food chain will affect other populations.

This interdependence of the populations within a food chain helps to maintain the balance of plant and animal populations within a community. For example, when there are too many giraffes; there will be insufficient trees and shrubs for all of them to eat. Many giraffes will starve and die. Fewer giraffes means more time for the trees and shrubs to grow to maturity and multiply. Fewer giraffes also means less food is available for the lions to eat and some lions will starve to death. When there are fewer lions, the giraffe population will increase.

Grazing Food Chain: This food chain starts from producers (living green plant base) and extends towards herbivores and then carnivores. Some common examples of food chains (Fig. 2.6) are given below:

(*a*) Producer → Grasshopper → Shrew → Hawk
(*b*) Producer → Rabbit → Fox → Wolf → Tiger
(*c*) Producer → Frog → Snake → Peacock
} *Terrestrial Food Chain*

(*d*) Phytoplankton → Zooplankton → Small Fish → Large Fish → Shark
(*e*) Phytoplankton → Zooplankton → Fish → Crave → Hawk
} *Aquatic Food Chain*

Most of the ecosystem (such as grassland, forest ect.) in nature follow grazing food chain. It is directly dependent on influx of solar energy.

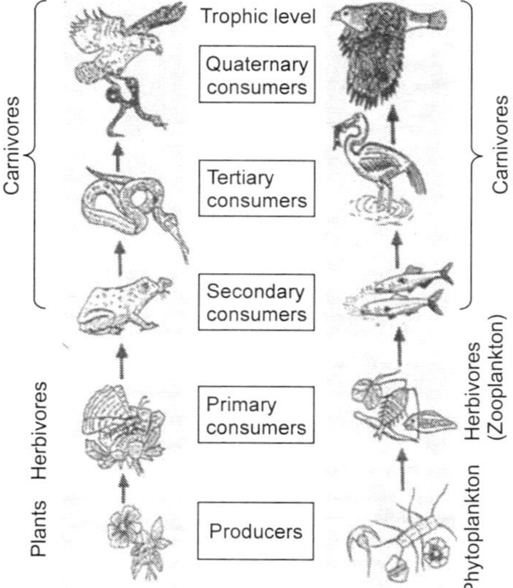

Fig. 2.6 Terrestrial and Aquatic Food Chains

Important Terms Involved:

- *Food Chain:* A food chain is the series of organisms showing feeding relationships. A food chain almost always begins with a green plant (producer) which is eaten by an animal (consumer). The arrow means 'is eaten by', and shows the **flow of matter and energy** along the food chain. There are no decomposers in a food chain.

  Example of a Food Chain

  Grass(Producer) → Grasshopper(1st order Consumer) → Kookaburra(2nd order Consumer)
- *Producer:* usually a green plant that produces its own food by photosynthesis
- *First-order Consumer:* the organism that eats the producer
- *Second-order Consumer:* the organism that eats or derives nutrients from the first-order consumer
- *Herbivore:* a plant eater
- *Carnivore:* an organism that obtains nutrients from the blood or flesh of an animal
- *Omnivore:* an organism which eats both plant and animal matter
- *Scavenger:* an consumer that eats dead animals (*e.g.*, crab)
- *Detritivore:* a consumer that obtains its nutrients from detritus
- *Decomposer:* an organism such as bacteria and fungi that breaks down dead organisms and their wastes
- *Trophic Level:* A trophic level is each level in a food chain. Matter is always 'lost' as heat energy at each trophic level.
- *Basal Energy Requirement (B.E.R.):* the amount of energy used by an organism's body just to keep alive, when no food is being digested and no muscular work is being done
- *Food Web:* a network of interrelated food chains in a given area.

Ecological Pyramids

A rational study of a various trophic levels in ecosystem is termed as '*ecological pyramid*'. This study is based on various factors *viz.* number of individuals, biomass or energy in a particular trophic level. Generally producers form the base of pyramid due to large constituent. After that it decreases in size as go to the successive trophic level and become taper at top.

Ecological pyramids are of three types:

(*i*) Pyramid of Number

(*ii*) Pyramid of Biomass

(*iii*) Pyramid of Energy

(*i*) **Pyramid of Number:** This pyramid is based on the number of individuals per unit area at each trophic level. In this pyramid generally, number of individuals decreases successively and it show relationship between producers, herbivores and carnivores and it is known as upright pyramid. But, number of pyramid may be inverted and spindle shaped. Inverted pyramid show increasing number of individuals at successive levels and spindle shaped pyramid show both sides tapered (Fig. 2.7).

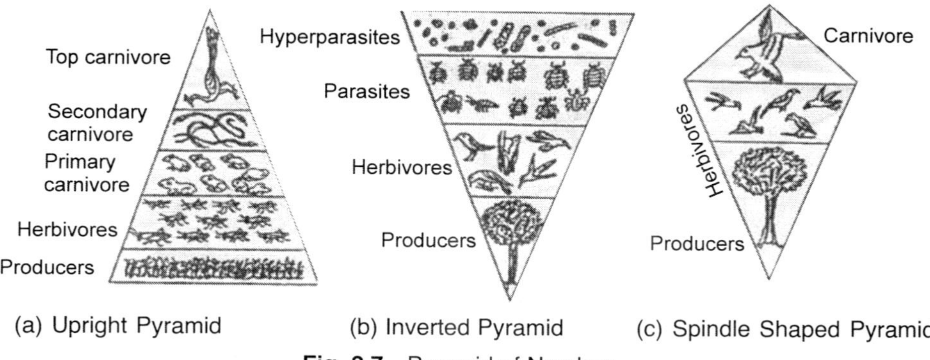

Fig. 2.7 Pyramid of Number

(*ii*) **Pyramid of Biomass:** Total amount of living material in any organism is known as biomass. It is measured in terms of dry weight of organisms per unit area. In terrestrial ecosystem, there is gradual decrease in biomass at different trophic levels. But, in aquatic ecosystem, inverted biomass pyramid are also seen (Fig. 2.8).

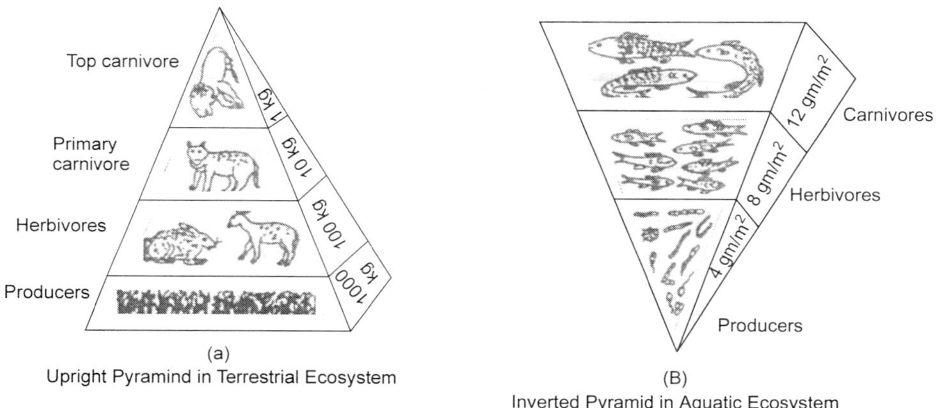

Fig. 2.8 Pyramid of Biomass

(*iii*) **Pyramid of Energy:** This pyramid always found in upright pattern because ~ 10% energy is lost at each successive trophic level. Producers have higher energy than herbivores and carnivores. Energy is represented in terms of calories (Fig. 2.9).

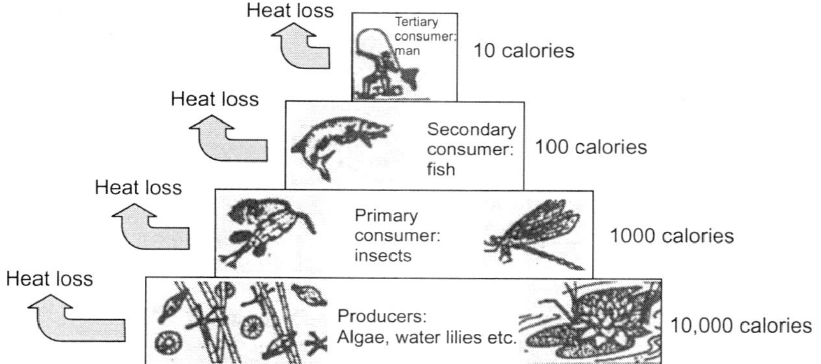

Fig. 2.9 Pyramid of Energy in Terms of Calories

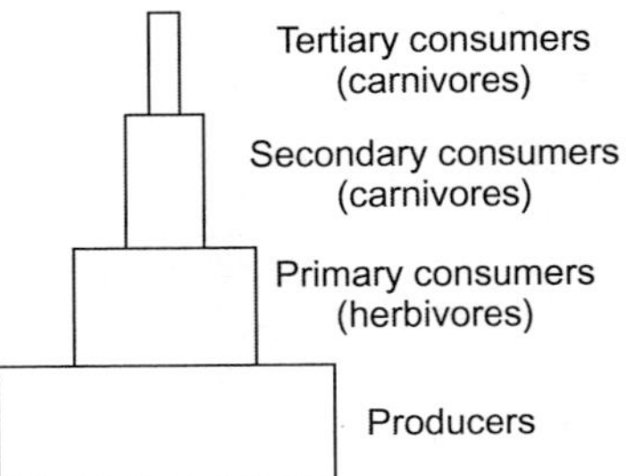

Fig. 2.10 Pyramid of Energy (Kcal per Unit Area within Unit Time, Season or Years) in any Ecosystem

Ecological Food Webs

Ecological food webs are based on the productivity of green plants (or photoautotrophs), which are the only organisms capable of utilizing diffuse solar **radiation** to synthesize simple organic compounds from **carbon dioxide** and **water**. The fixed energy of the simple organic compounds, plus inorganic **nutrients**, are then used by plants in more complex metabolic reactions to synthesize a vast diversity of biochemicals. Plants utilize the fixed energy of their biochemicals to achieve growth and reproduction. On average, **plant photosynthesis** utilizes less than 1% of the solar radiation that is received at the surface of the **earth**. Higher efficiencies are impossible for a number of reasons, including the second law of thermodynamics, but also other constraining factors such as the availability of nutrients and moisture, appropriate temperatures for growth, and other environmental limitations. *Ecological pyramids are based on the productivity of organisms. Plants account for 90% of the total productivity of the food web, and herbivores account for most of the rest. Carnivores are responsible for less than 1% of ecological productivity.* Courtesy of Gale Research. However, even relatively fertile plant communities can only achieve conversion efficiencies of 10% or so, and only for relatively short periods of **time**.

The solar energy fixed by green plants in photosynthesis is, of course, the energetic basis of the productivity of all heterotrophic organisms that can only feed upon living or dead biomass, such as animals and **microorganisms**. Some of the biomass of plants is consumed as food by animals in the next trophic level, that of herbivores. However, herbivores cannot convert all of the energy of the vegetation that they eat into their own biomass. Depending on the digestibility of the food being consumed, the efficiency of this process is about 1–20%. The rest of the fixed energy of the plant foods is not assimilated by herbivores, or is converted into heat. Similarly, when carnivores eat other animals, only some of the fixed energy of the **prey** is converted into biomass of the **predator**. The rest is ultimately excreted, or is converted into heat, in accordance with the requirement for entropy to increase during any energy transformation.

What is Biodiversity?

Simply put, it's the variety of life. More broadly, biodiversity is our collective life support system.

Ontario's Biodiversity Strategy (OBS) adopts the definition of biodiversity that is used in the *Canadian Biodiversity Strategy* and the *United Nations Convention on Biological Diversity*:

Biodiversity is the variability among living organisms from all sources, including inter alia [among other things], terrestrial, marine and other aquatic ecosystems and the ecological complexes of which they are a part; this includes diversity within species, between species and of ecosystems.

The key points of the above definition are:

biodiversity is scalable — that is it exists on many levels, *e.g.*, the genes within populations, the populations within species, the species within ecosystems, the ecosystems within landscapes, the landscapes within bioregions, and so on.

biodiversity is a key element of ecosystems — without the connections biodiversity creates, ecosystems fall apart.

The first challenge to protecting biodiversity is understanding what it means and then, why you should care about it.

Biodiversity is the variety of species, their genetic make-up, and the natural communities in which they occur. It includes all of the native plants and animals in Pennsylvania and the processes that sustain life on Earth. Pennsylvania is home to over 25,000 different species of organisms, and of this total, over 800 are considered to be rare, threatened, or endangered. For many groups of organisms, such as insects, fungi, and algae, very little is known about them — not even what species occur in Pennsylvania! The need to understand the state's rich natural resources has never been more critical.

The term ecosystem is defined as a community of living organisms combined with their associated physical environment. It is our "home system" that makes life possible. Ecosystems are the full tapestry of nature that support life and they also provide valuable services.

- Wetland ecosystems filter out toxins, clean the water, and control floods.
- Estuaries act as marine-life nurseries.
- Forest ecosystems supply fresh water, provide oxygen, control erosion, and remove carbon from the atmosphere.

Many species, working together, are needed to provide these critical services. The loss of biodiversity reduces nature's ability to perform these functions. As greater fluctuations occur, ecosystems as a whole become less stable. Instability causes ecosystems to be more vulnerable to extreme conditions and may also decrease productivity.

Why is Biodiversity Important?

1. **You are a part of biodiversity.** The same things that affect bugs, trees and fish have an impact on you too — the quantity of clean, fresh water, or the quality of the air we breathe.
2. **Everything is connected.** The health of biodiversity affects you. And what you do affects biodiversity. Everything we do either uses natural resources or returns them as waste. The amount of land and resources that a population or a person uses is called an ecological footprint. We all can do things to make our personal footprints smaller.
3. **Natural systems based on healthy biodiversity provide all kinds of services for you... for free!** Things like cooling and filtering air, controlling floods, pollinating plants, controlling pests, aerating soil, and filtering and storing water. These ecosystem services would cost a lot if we had to (or even could) use technology to provide them. These services are called ecosystem services.
4. **You have to live with what's left.** A wise saying states "We don't inherit the earth from our parents, we borrow it from our children." If you're a parent, what kind of environment will you be leaving your children? If you're one of the children, you might want to know what you'll be getting in the future. What we do or don't do now, matters greatly to our future quality of life.

While the term "biodiversity" may not be well known or understood, the ecological services provided by biodiversity are vital to everyday life. Not a day, hour, or even second goes by that we do not depend on biodiversity for survival.

- The air we breathe is a product of photosynthesis by green plants.
- Insects, worms, bacteria, and other tiny organisms break down wastes and aid in the decomposition of dead plants and animals to enrich soils.
- More than 90 percent of the calories consumed by people worldwide are produced from 80 plant species.
- Almost 30 percent of medicines are developed from plants and animals, and many more are derived from these sources.

Biodiversity is the variety and differences among living organisms from all sources, including terrestrial, marine, and other aquatic ecosystems and the ecological complexes of which they are a part. This includes genetic diversity within and between species and of ecosystems. Thus, in essence, biodiversity represents all life. India is one of the mega biodiversity centres in the world and has two of the world's 18 'biodiversity hotspots' located in the Western Ghats and in the Eastern Himalayas. The forest cover in these areas is very dense and diverse and of pristine beauty, and incredible biodiversity.

According to an MoEF Report (1996), the country is estimated to have over 45, 000 plant species and 81, 000 animal species representing 7% of the world's flora and 6.5% of its fauna. The 1999 figures are 49, 219 plant species representing 12.5% and 81, 251 animal species representing 6.6%.

The sacred groves of India are some of the areas in the country where the richness of biodiversity has been well preserved. The Thar desert and the Himalayas are two regions rich in biodiversity in India. There are 89 national parks and 504 wildlife sanctuaries in the country, the Chilika Lake being one of them. This lake is also an important wetland area. Learn more through map on biodiversity in India.

Over the last century, a great deal of damage has been done to the biodiversity existing on the earth. Increasing human population, increasing consumption levels, and decreasing efficiency of use of our resources are some of the causes that have led to overexploitation and manipulation of ecosystems. Trade in wildlife, such as rhino horn, has led to the extinction of species. Consequences of biodiversity loss can be great as any disturbance to one species gives rise to imbalance in others. In this the exotic species have a role to play.

To prevent such loss, the Government of India is setting up biosphere reserves in different parts of the country. These are multipurpose protected areas to preserve the genetic diversity in different ecosystems. Till 1999, ten biosphere reserves had been set up, namely Nilgiri, Nandadevi, Nakrek, Great Nicobar, Gulf of Mannar, Manas, Sunderbans, Similipal, and Dibru Saikhowa. A number of NGOs are being involved in the programme to create awareness. But legal protection is provided only to national parks and sanctuaries, which cover about 4.5% of India's land area.

Different Levels of Biodiversity

1. **Genetic diversity** is:
   - A building block of life
   - Responsible for the variability among individuals within any species, based on variations in genes

Genetic variability increases the chance that a species will adapt to changing environmental conditions or impacts, since some individuals will be able to handle the change better than others. The more individuals there are, the greater the chance of genetic variation. Species with a small population of individuals have limited variability and therefore have limited ability to respond to change. This is why populations of "species at risk" can be so difficult to recover. Once you get below a certain number of individuals, it is virtually impossible based on reproductive potential. *Genetic variation is the cornerstone of all biodiversity.*

Population diversity: While we often hear about species, what we generally see and interact with are populations — distinct groups of members of a particular species that have a limited exchange of genetic material among the groups. They can reproduce together but they don't often do so.

As a result, the genetic differences between populations tend to increase, even though the variability within any one population may be less than across the species as a whole. Also, because of the isolation, local impacts on one population may not be felt by another. A conservative first estimate indicates that about 220 populations per species puts the total number of populations world-wide into at least the low billions .

Extreme population variability can be a double-edged sword. For example, lake trout in Ontario's Great Lakes were once very diverse. There were at least 15 to 20 different forms of lake trout recognized by commercial fishermen before the sea lamprey appeared. The lake trout differed in where they were found, when they spawned, and in their appearance. They were given such names as blacks, redfins, yellowfins, paper bellies, fats, humpers and sand trout. Undoubtedly, the number of genetically distinct populations was much higher.

However, even all this diversity could not withstand over-harvest, sea lamprey predation and loss of habitat, particularly inshore rubble shoals required for spawning. The catches of lake trout plunged to 10% of the original yield in Lake Superior and down to almost nothing in the other Great Lakes. When conditions improved and it came time to try and reintroduce lake trout, the results were disappointing in all but Lake Superior where enough wild populations survived to make a decent comeback.

Fig. 2.11

All those discrete lake trout stocks had evolved for a reason: reproductive success of lake trout in each area. The fish were in effect "tailor-made" for the area. Now many of those stocks have disappeared forever. It will take a lot of time and effort to find stocks that might be reasonable replacements.

2. **Species diversity:** Is all of the different kinds of living things found in a certain habitat or ecosystem. World-wide more than 1.4 million species have been identified but estimates of the actual

number vary from 5 million up to 100 million. Fourteen million appears to be an estimate that is commonly quoted in the literature (Global Biodiversity Assessment, 2001 Summary).

In Ontario, over 30,000 species have been identified including:

- over 20,000 macro-invertebrates (insects, spiders, etc.)
- more than 4,800 plants
- more than 150 fish
- more than 1000 fungi
- 53 reptiles and amphibians
- 479 birds
- 81 mammals

Globally the estimated numbers of species are:

- 35,000 micro-organisms
- 70,000 fungi
- 273,000 plants
- 875,000 invertebrates (insects, spiders, etc.)
- 19,000 fish
- 10,500 reptiles and amphibians
- 9,000 birds
- 4,000 mammals
- 105,000 other animals

Species diversity, however, is more than just the number of species in a given area, habitat or ecosystem. Some species' importance can be out of line with their numbers, for example *keystone species*. There can also be great differences in species composition over time. Species diversity can also be greatly affected by physical conditions in the ecosystems where they live, such as differences in temperature, light, structure and chemical composition.

The point is, biodiversity cannot be reduced to a single number. There are dimensions to diversity, many of them.

3. **Ecosystem diversity:** Is the variety of ecosystems within a landscape or region including wetlands, prairies or savannahs, lakes and rivers, forests and agricultural landscapes. The basic principles of biodiversity apply here as well but the scope is much larger. It is at this level that the interactions and links among species and the consequences of those links are evident. Less diverse ecosystems, such as coldwater streams or small lake trout lakes, contribute to the functioning and productivity of larger areas such as bioregions.

Energy Flow

In ecology,**energy flow** (**calorific flow**) refers to the flow of energy through a food chain. In an ecosystem, ecologists seek to quantify the relative importance of different component species and feeding relationships.

A general energy flow scenario follows:

- Solar energy is fixed by the photoautotrophs, the so called *primary producers*, like green plants. They absorb most of the stored energy in the plant through digestion, and transform it into the form of energy they need, adenosine triphosphate, through respiration. A part of the energy received by the herbivore is converted to bodily heat (an effect of respiration), which is radiated away and lost from the system. Energy loss also occurs in the expulsion of egesta, which contains undigested energy compounds.

- Secondary Consumers then consume the primary consumers. Energy that had been used by the primary consumers for growth and storage is thus absorbed into the secondary consumers through the process of digestion. As with primary consumers, secondary consumers convert this energy into a more suitable form (ATP) during respiration. Again some energy is lost from the system, since energy which the primary consumers had used for respiration cannot be utilised by the secondary consumers.
- Tertiary consumers then consume the secondary consumers, and most of the energy is passed along, while some is again lost in the ways described above and below.
- A final link in the food chain is decomposers which break down the organic matter of the tertiary consumers (or whichever consumer is at the top of the chain) and release nutrients into the soil. Decop Saprotrophic bacteria and fungi are decomposers, and play a pivotal role in the nitrogen and carbon cycles.

The energy is passed on from trophic level to trophic level and each time most (about 90%) of the energy is lost, with some being lost as heat into the environment (an effect of respiration) and some being lost as egesta. This means the top consumer of a food chain receives the least energy, as a lot of the food chain's energy has been lost between trophic levels. This loss of energy at each level limits typical food chains to only 4–6 links.

Energy Flow in Ecosystems

Energy capture and use: At the base of an ecosystem, primary producers are actively converting solar energy into stored chemical energy. **Photosynthesis** is the process of converting solar energy, water, and carbon dioxide into carbohydrates and oxygen. The process occurs in two steps: first light energy is absorbed by chlorophyll to split a molecule of water releasing hydrogen and oxygen. The second step uses the energy to convert carbon dioxide to carbohydrates.

$$6CO_2 + 12H_2O \xrightarrow[\text{(chlorophyll)}]{\text{Solar Energy}} C_6H_{12}O_6 + 6O_2 + 6H_2O$$

The carbohydrate ($C_6H_{12}O_6$) can be converted into starch and stored by the plant. Carbohydrate can be combined with other sugar molecules to make cellulose, the basic structural material of a plant.

Oddly enough, of all the solar radiation striking a plant, only about 1 percent is used in photosynthesis. The rate of photosynthesis is dependent on several things, especially the amount of light received ... up to a point. As solar radiation increases the rate of photosynthesis increases. For many plants there is an upper limit to the rate of photosynthesis. In some plants as incident solar radiation increases the rate of photosynthesis levels off, or may decrease. The increasing solar energy load causes the plant to be too hot and the need to cool the plant increases. As a result, transpiration takes over as the dominate plant process. **Transpiration**, the loss of water from plants, acts to cool the plant by releasing latent energy. Adequate supplies of water, carbon dioxide and the availability of nutrients in the soil affect photosynthesis.

Respiration: While photosynthesis builds stored chemical energy in a plant, **respiration** is the process of "burning" stored chemical energy, basically through oxidation, for maintaining plant metabolism. During plant respiration, carbohydrates combine with oxygen and is reduced to carbon dioxide, water, and heat.

$$C_6H_{12}O_6 + 6O_2 \rightarrow 6CO_2 + 6H_2O + 2830 \text{ kj}$$

While photosynthesis operates only during day when sunshine is available, respiration goes on both night and day. Plant growth occurs so long as photosynthesis exceeds respiration.

The Hydrological and Chemical Cycles

Cycles of Nature:

- *Law of Conservation of Matter:* Matter cannot be created nor destroyed in any chemical reaction. It can only be transformed from one form to another.

Cycle - *A cycle shows the reusing of certain elements and compounds (e.g. water, carbon, oxygen, nitrogen, phosphorus) in different forms in ecosystems.*

1. **Water Cycle**
   1. **Precipitation:** Transported through the circulating atmosphere the clouds move themselves inland, as a result of gravity, and lose their water as it falls back unto the ground. This phenomenon is called rain or precipitation.
   2. **Infiltration:** Rainwater infiltrates into the ground and sinks to the saturated zone, where it becomes groundwater. Groundwater slowly moves from places of high elevation and pressure to places with low elevation and pressure. It moves from the area of infiltration through an aquifer and out to a discharge area, which can be either a sea or an ocean.
   3. **Transpiration:** Plants and other forms of vegetation take up water from the soil and excrete it again as water vapour. About 10% of the precipitation that falls on the ground vapourizes again through transpiration of plants, the rest evaporates from seas and oceans.
   4. **Surface run-off:** The rainwater that does not infiltrate into the soil will directly reach the surface water, as it will run-off to rivers and lakes. After that it will be transported back to the seas and oceans. This water is called surface run-off.
   5. **Evaporation:** Due to the influence of sunlight the water in oceans and lakes will warm up. As a result of that it will evaporate and rise up into the atmosphere. There it will form clouds that will eventually cause rainwater to fall back on earth. The evaporation of oceans is the most important kind of evaporation.
   6. **Condensation:** In contact with the atmosphere the water vapour will transform back to liquid, so that it will be visible in the air. These accumulations of water in the air are what we call clouds.

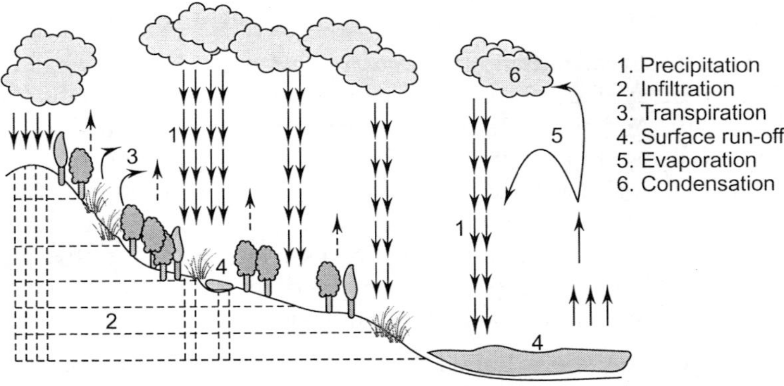

Fig. 2.12 The Hydrological Cycle

A Water Balance

A considerable portion of river flow does not reach the ocean, having evaporated those areas with no natural surface run-off channels. On the other hand, some groundwater bypasses river systems altogether and goes directly to the ocean or evaporates.

Every year, the turnover of water on Earth involves 577, 000 km^3 of water. This is water that evaporates from the ocean surface (502, 800 km^3) and from land (74, 200 km^3). The same amount of water falls as atmospheric precipitation, 458,000 km^3 on the ocean and 119,000 km^3 on land. The difference between precipitation and evaporation from the land surface (119, 000 – 74, 200 = 44, 800 km^3/year) represents the total run-off of the Earth's rivers (42, 700 km^3/year) and direct groundwater run-off.

How does Water Supply and Sewage Disposal fit into the Hydrological Cycle?

We have seen that water flows into rivers, lakes and into groundwater storage. Most importantly, for our daily water needs, it also flows into our homes and taps. A network of underground pipes, pumping stations and treatment works ensures that clean, fresh drinking water is delivered by our local water utility to our homes every day of the week. After water has been used, the water utility collects, transports and then cleans this dirty water and returns it safely back into rivers where it can continue its journey downstream to the sea. The water utility's responsibility begins at the precipitation stage of the hydrological cycle. Utilities in some water-scarce countries encourage the collection of rainwater from rooftops (**rainwater harvesting**) but in most of Europe the hydrological cycle begins with surface waters.

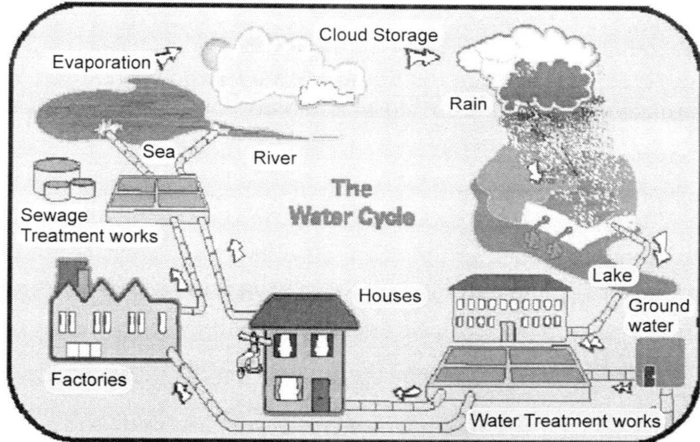

Fig. 2.13 The Water Utility's Role in the Hydrological (Water) Cycle

Surface Waters

Rain water searches for the quickest route to the sea and flows into rivers, streams, lakes and underground stores. The water in the surface waters is clean enough to support a variety of wildlife. Although it is clean, it is not safe to drink and needs to be treated in a water treatment works to remove any substances that may be harmful.

Water Treatment Works

Water is abstracted from underground sources via boreholes or alternatively is pumped from rivers and stored in reservoirs before being passed through sand filter beds which trap any dirt and organisms. It is then treated using the most up to date advanced water treatment (AWT) technology such as ozonation and carbon filtration (granular activated carbon) which remove the substances that we cannot see.

Water Distribution

Clean, fresh drinking water is pumped into an underground network of pipes and storage reservoirs and is not seen again until it reaches your tap. This guarantees that the water you drink remains clean and fresh.

Water Use

On average, in European countries, each person uses around 55,000 litres of water every year. Baths, showers, washing up, washing clothes and toilet flushing all use large amounts of water. These activities transform clean tap water into dirty wastewater. The water utility not only supplies clean drinking water but also collects, transports and disposes of the dirty water after it has been used.

Sewerage

Dirty water or sewage is collected firstly in drains and then in underground sewers and is transported via a sewerage system (a network of pipes and tunnels) to a sewage treatment works.

Sewage Treatment Works

These works use natural micro-organisms to remove harmful substances from dirty water. The solid material (sludge) is separated from the liquid (effluent) and both are treated to produce clean effluent that can be released back to the river and bio-solids that can be used in agriculture as a fertiliser or incinerated to produce energy.

Completing the Cycle

The river continues its journey back to the sea where the cycle starts again. Water evaporates to form clouds, condenses to droplets and eventually falls as rain on to the ground.

2.7 CARBON AND OXYGEN CYCLES

- **Carbon** occurs in all living organisms in the forms of carbon dioxide, carbohydrates (sugars and starches), proteins and fats.
- **Oxygen** occurs in all living organisms in the forms of oxygen gas, water, carbon dioxide, carbohydrates, proteins and fats.
- **Photosynthesis:** Photosynthesis is the process used by plants containing chlorophyll to utilise sunlight, carbon dioxide and water to form sugar (as glucose) and oxygen.

$$\text{Carbon Dioxide} + \text{Water} \xrightarrow[\text{Chlorophyll}]{\text{Sunlight \&}} \text{Glucose Sugar} + \text{Oxygen}$$

$$CO_2 + H_2O \xrightarrow[\text{Chlorophyll}]{\text{Sunlight \&}} C_6H_{12}O_6 + O_2$$

- **Respiration:** Respiration is the process that most living organisms (including animals, plants, fungi and micro-organisms) use to obtain energy from glucose sugar and oxygen. The wastes formed are carbon dioxide and water.

$$\text{Glucose} + \text{Oxygen} \rightarrow \text{Carbon Dioxide} + \text{Water} + \text{Energy}$$

$$C_6H_{12}O_6 + O_2 \rightarrow CO_2 + H_2O + \text{Energy}$$

1. **Carbon Cycle:** Balance is essential to the earth. Chemical cycles keep the amount of elements on the earth in a perfect balance.

   Carbon is essential for the chemical processes that support life. It plays such an important role in life that sometimes we say that life is "carbon-based". But there is only a limited amount of carbon on the earth. So carbon is constantly cycling around the earth, turning up in a lot of different forms and places. The reactions that move carbon around make up a giant web called the carbon cycle.

   Plants get carbon by taking carbon dioxide from the air. They use the carbon dioxide and the energy from sunlight to make food. Animals get their carbon by eating those plants or by eating animals that have eaten those plants. When organisms breathe, they take oxygen from the air. During respiration, the oxygen reacts with food to provide energy. Respiration produces carbon dioxide which is released to the air.

   Volcanic eruptions are a source of carbon. When a volcano erupts, it releases huge amounts of carbon dioxide. But remember—the earth needs it elements to stay in balance. So the effect of volcanoes is balanced by weathering which is a chemical reaction between rainwater and rocks that absorbs carbon dioxide from the air to create rock carbonate minerals. Left to themselves, these natural processes are in perfect balance. But human activities can disturb the cycle and increase the amount of carbon dioxide in the atmosphere. This could cause problems in the future because carbon dioxide is vital for controlling the world's climate.

   The Carbon Cycle describes the flow of carbon between living organisms and the non-living environment. Carbon cycles through the environment in the form of a gas, carbon dioxide (CO_2). The atmosphere of Earth contains .04 percent CO_2.

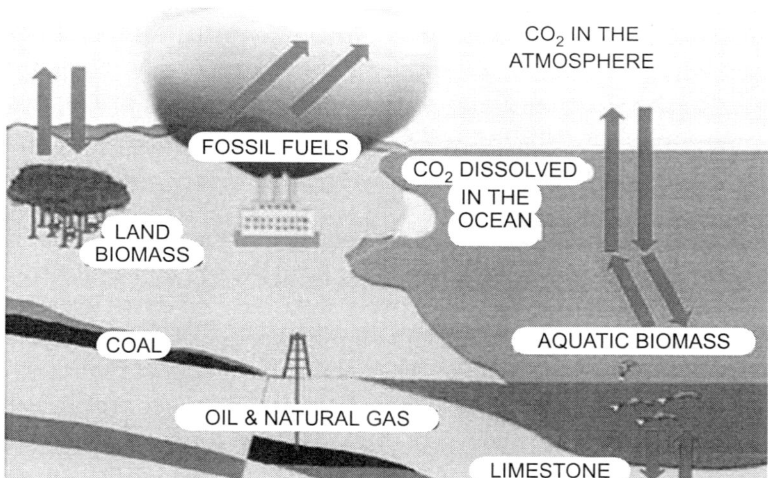

Fig. 2.14

Living organisms provide two important steps in the carbon cycle:

- Plants absorb CO_2 from the atmosphere to use during photosynthesis.
- Other organisms release CO_2 into the air during respiration.
- In addition, there are several important non-organic storage areas of carbon in the environment:

- A large portion of the Carbon on the Earth is stored in rocks.
- The Earth's oceans hold a large amount of CO_2 because it easily dissolves in water.
- Coal, oil, and limestone store carbon that once formed ancient organisms. Burning fossil fuels, like coal and oil, will release CO_2 into the atmosphere.

2. **Nitrogen Cycle:** Nitrogen atoms are constantly moving in a giant circle from the air, through the soil, into the bodies of plants and animals, and eventually back to the air. This whole process is called the nitrogen cycle. All living things need nitrogen to develop and grow. Even though the earth's atmosphere is made up of 78% nitrogen, plants and animals cannot use it in this form because the nitrogen atoms are too firmly bound together in molecules. So plants must draw their nitrogen from nitrogen compounds dissolved in the soil, and animals get their nitrogen by eating plants or by eating other animals that eat plants.

The nitrogen gets into the soil in a couple of different ways. A small quantity of the nitrogen found the soil by way of lightning. Lightning changes atmospheric nitrogen into nitrogen dioxide which is soluble in water. The nitrogen oxides dissolve in rainwater to form nitric acid which is absorbed by soil. The rest of the nitrogen in soil comes from bacteria. Bacteria are the only living things capable of getting nitrogen directly from the air. This is called "fixing". The process is started by certain kinds of bacteria in the soil that can extract nitrogen from the air. Then other bacteria convert the nitrogen into nitrogen compounds called nitrates, This process is called nitrification. Plants absorb the nitrates and turn them into more complex nitrogen compounds. Bacteria also help return nitrogen to the air. Bacteria in the soil decompose animal waste and the remains of dead animals and plants and produce ammonia. Nitrifying bacteria turn the ammonia into nitrates. Other bacteria, called denitrifying bacteria, convert some of the nitrates back into nitrogen gas, which is released into the air. All these different steps form a massive cycle. The effect is that, over time, bacteria in the soil return almost the same amount of nitrogen to the air as other bacteria take from the air. This keeps the nitrogen content of the earth and its atmosphere in a perfect balance.

Unfortunately, humans are interfering with the natural balance when they overuse artificially produced nitrates as agricultural fertilizers. Before these nitrates can be converted into atmospheric nitrogen, they are often carried off from the soil by rain or irrigation. These dissolved nitrates are carried to streams and rivers and even seep down to groundwater. In some parts of the world, water for humans and animals contains such high concentrations of nitrates that it is unsafe for consumption. These excessive amount of nitrates, when they reach rivers and lakes, cause too much algae to grow. This over-abundance of algae uses up too much of the oxygen in the water. When oxygen levels fall, other forms of life in the water die off.

Organisms require nitrogen to produce amino acids. Nitrogen makes up seventy-eight percent of the atmosphere, but most organisms can not use this form of nitrogen, and must have the fixed form. The nitrogen cycle produces the fixed form of nitrogen these organisms need.

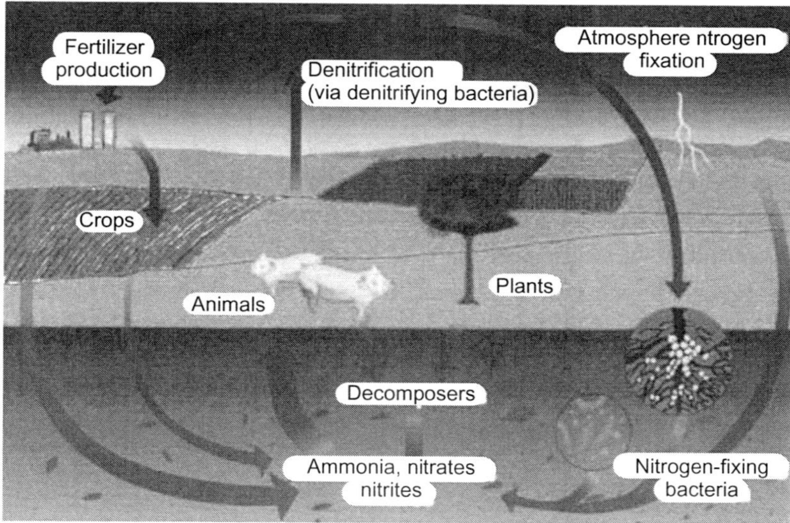

Fig. 2.15

Step 1: A special type of bacteria called nitrogen fixing bacteria take in atmospheric nitrogen and produce ammonia (NH_3).

Step 2: Other bacteria use this ammonia to produce nitrates and nitrites, which are nitrogen and oxygen containing compounds.

Step 3: The nitrates and nitrites are used by plants to make amino acids which are then used to make plant proteins.

Step 4: Plants are consumed by other organisms which use the plant amino acids to make their own.

Step 5: Decomposers convert the nitrogen found in other organisms into ammonia and return it to the soil. A few of these type of bacteria return nitrogen to the atmosphere by a process called denitrification, however this amount is small.

3. **Oxygen Cycle**: The **oxygen cycle** is the biogeochemical cycle that describes the movement of oxygen within and between its three main reservoirs: the atmosphere (air), the biosphere (living things), and the lithosphere (Earth's crust). The main driving factor of the oxygen cycle is photosynthesis, which is responsible for the modern Earth's atmosphere and life.

   The amount of oxygen in and around the earth is fixed. But this oxygen is fed again and again through the world's living systems in a never-ending circle called the oxygen cycle. Our needs are just part of this cycle. The cycle involves a continual exchange of gases between the air and animals and plants. In a process called respiration, animals and plants take oxygen from air and give back carbon dioxide. In a process called photosynthesis, plants take carbon dioxide from air and water and give back oxygen. Respiration and photosynthesis are effectively opposite processes. Respiration is an oxidation reaction, which takes oxygen from the air or from water. Photosynthesis is a reduction reaction. It adds oxygen gas to the air. Enormous quantities of oxygen are taken in by plants and animals every day, and huge quantities of oxygen are returned to the air by plants. These amount exactly balance so that overall the amount of oxygen in the air stays the same.

   Oxygen, like carbon and hydrogen, is a basic element of life. In addition, in the form of O_3, ozone, it provides protection of life by filtering out the sun's UV rays as they enter the stratosphere. In addition to constituting about 20% of the atmosphere, oxygen is ubiquitous. It also occurs in combination as oxides in the Earth's crust and mantle, and as water in the oceans.

Early in the evolution of the Earth, oxygen is believed to have been released from water vapor by UV radiation and accumulated in the atmosphere as the hydrogen escaped into the earth's gravity. Later, photosynthesis became a source of oxygen. Oxygen is also released as organic carbon in CHO, and gets buried in sediments. The role of oxygen in life is describe in the unit on Biological Systems.

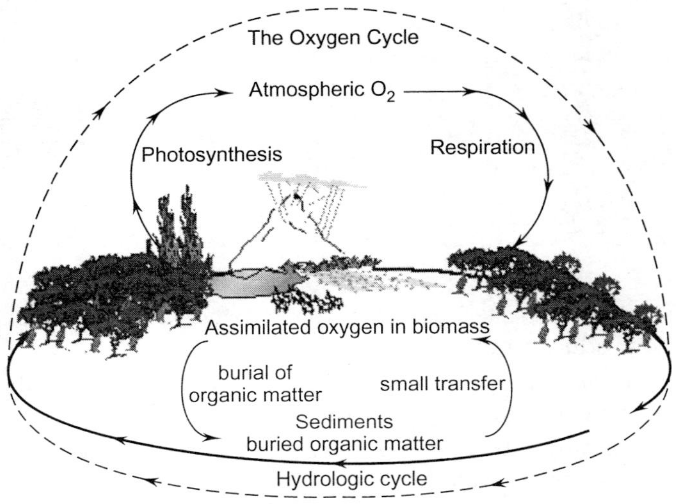

Fig. 2.16 The Oxygen Cycle

Oxygen is highly reactive. A colorless, odorless gas at ordinary temperatures, it turns to a bluish liquid at – 183° C. Burning or combustion is essentially oxidation, or combination with atmospheric oxygen. Fig. 2.16 shows a very broad overview of oxygen cycling in nature. The environment of oxygen in numerous reactions make it hard to present a complete picture.

Oxygen is vital to us in many ways (beside the most obvious—for breathing). Water can dissolve oxygen and it is this dissolved oxygen that supports aquatic life. Oxygen is also needed for the decomposition of organic waste. Wastes from living organisms are "biodegradable" because there are aerobic bacteria that convert organic waste materials into stable inorganic materials. If enough oxygen is not available for these bacteria, for example, because of enormous quantities of wastes in a body of water, they die and anaerobic bacteria that do not need oxygen take over. These bacteria change waste material into H_2S and other poisonous and foul-smelling substances. For this reason, the content of biodegradable substances in waste waters is expressed by a special index called "biological oxygen demand" (BOD), representing the amount of oxygen needed by aerobic bacteria to decompose the waste.

4. **Phosphorus Cycle:** Phosphorus is an essential nutrient for plants and animals in the form of ions PO_4^{3-} and HPO_4^{2-}. It is a part of DNA-molecules, of molecules that store energy (ATP and ADP) and of fats of cell membranes. Phosphorus is also a building block of certain parts of the human and animal body, such as the bones and teeth.

Phosphorus can be found on earth in water, soil and sediments. Unlike the compounds of other matter cycles phosphorus cannot be found in air in the gaseous state. This is because phosphorus is usually liquid at normal temperatures and pressures. It is mainly cycling through water, soil and sediments. In the atmosphere phosphorus can mainly be found as very small dust particles.Phosphorus moves slowly from deposits on land and in sediments, to living organisms, and than much more slowly back into the soil and water sediment. The phosphorus cycle is the slowest one of the matter cycles that are described here.

Phosphorus is most commonly found in rock formations and ocean sediments as phosphate salts. Phosphate salts that are released from rocks through weathering usually dissolve in soil water and will be absorbed by plants. Because the quantities of phosphorus in soil are generally small, it is often the limiting factor for plant growth. That is why humans often apply phosphate fertilizers on farmland. Phosphates are also limiting factors for plant-growth in marine ecosystems, because they are not very water-soluble. Animals absorb phosphates by eating plants or plant-eating animals.

Phosphorus cycles through plants and animals much faster than it does through rocks and sediments. When animals and plants die, phosphates will return to the soils or oceans again during decay. After that, phosphorus will end up in sediments or rock formations again, remaining there for millions of years. Eventually, phosphorus is released again through weathering and the cycle starts over. A schematic representation of the phosphorus cycle:

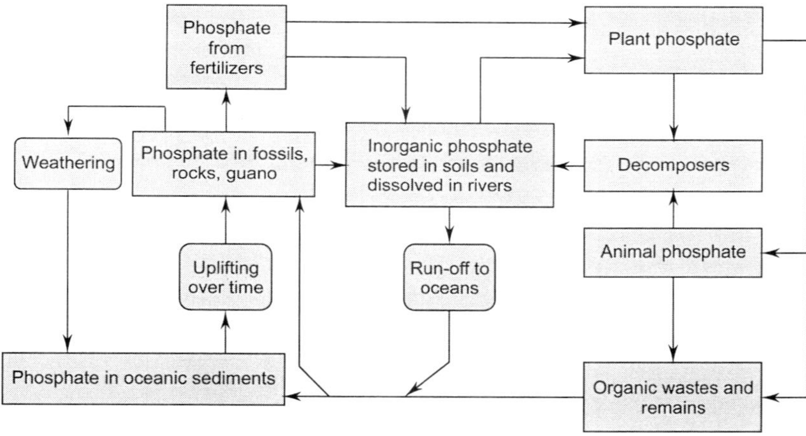

Fig. 2.17

5. **Sulphur Cycle:** Sulphur is one of the components that make up proteins and vitamins. Proteins consist of amino acids that contain sulphur atoms. Sulphur is important for the functioning of proteins and enzymes in plants, and in animals that depend upon plants for sulphur. Plants absorb sulphur when it is dissolved in water. Animals consume these plants, so that they take up enough sulphur to maintain their health. Most of the earth's sulphur is tied up in rocks and salts or buried deep in the ocean in oceanic sediments. Sulphur can also be found in the atmosphere. It enters the atmosphere through both natural and human sources. Natural resources can be for instance volcanic eruptions, bacterial processes, evaporation from water, or decaying organisms. When sulphur enters the atmosphere through human activity, this is mainly a consequence of industrial processes where sulphur dioxide (SO_2) and hydrogen sulphide (H_2S) gases are emitted on a wide scale.

When sulphur dioxide enters the atmosphere it will react with oxygen to produce sulphur trioxide gas (SO_3), or with other chemicals in the atmosphere, to produce sulphur salts. Sulphur dioxide may also react with water to produce sulphuric acid (H_2SO_4). Sulphuric acid may also be produced from demethylsulphide, which is emitted to the atmosphere by plankton species.

All these particles will settle back onto earth, or react with rain and fall back onto earth as acid deposition. The particles will than be absorbed by plants again and are released back into the atmosphere, so that the sulphur cycle will start over again.

A schematic representation of the sulphur cycle:

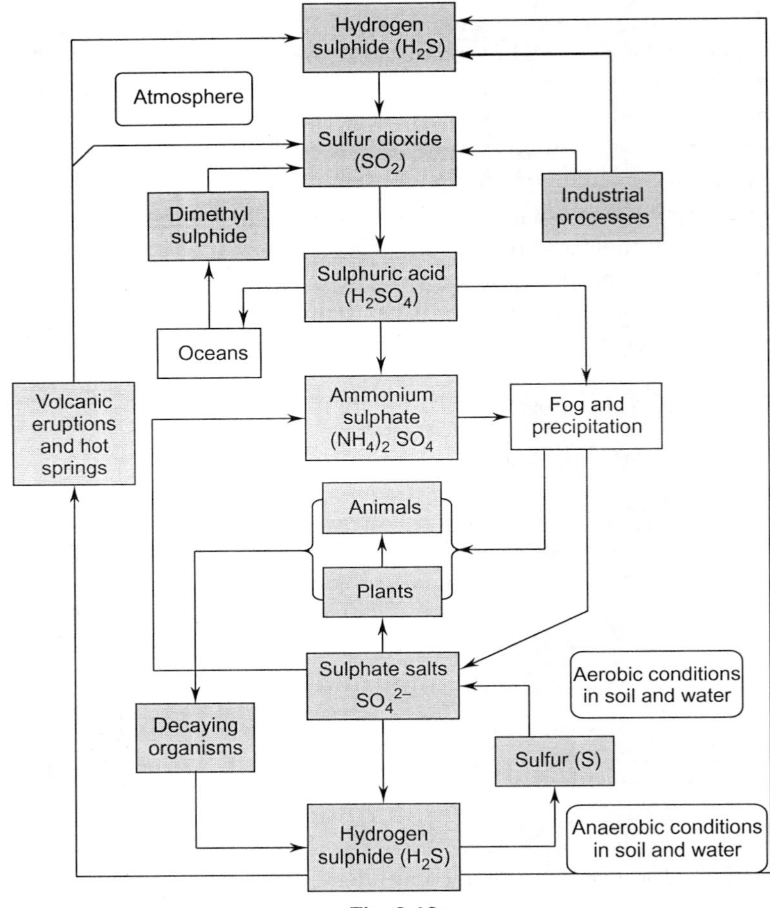

Fig. 2.18

Population Dynamics

Population is the number of individuals of a particular species in an area. Animal and plant populations depend on many things for survival. Limiting factors like the availability of food, water and shelter can impact an organism's population.

In nature, populations of animals and plants are linked together like a puzzle. The pieces of the plant and animal puzzle depend on each other to form a complete picture. If something is wrong with any of the pieces, it affects the whole puzzle.

A fishery is an area with an associated fish or aquatic population which is harvested for its commercial or recreational value. Fisheries can be wild or farmed. Population dynamics describes the ways in which a given population grows and shrinks over time, as controlled by birth, death, and emigration or immigration. It is the basis for understanding changing fishery patterns and issues such as habitat destruction, predation and optimal harvesting rates. The population dynamics of fisheries is used by fisheries scientists to determine sustainable yields. The basic accounting relation for population dynamics is the BIDE model:

$$N_1 = N_0 + B - D + I - E$$

where N_1 is the number of individuals at time 1, N_0 is the number of individuals at time 0, B is the number of individuals born, D the number that died, I the number that immigrated, and E the number that emigrated between time 0 and time 1. While immigration and emigration can be present in wild fisheries, they are usually not measured.

A fishery population is affected by three dynamic rate functions:

- **Birth rate or recruitment.** Recruitment means reaching a certain size or reproductive stage. With fisheries, recruitment usually refers to the age a fish can be caught and counted in nets.
- **Growth rate.** This measures the growth of individuals in size and length. This is important in fisheries where the population is often measured in terms of biomass.
- **Mortality.** This includes harvest mortality and natural mortality. Natural mortality includes non-human predation, disease and old age.

If these rates are measured over different time intervals, the **harvestable surplus** of a fishery can be determined. The harvestable surplus is the number of individuals that can be harvested from the population without affecting long term stability (average population size). The harvest within the harvestable surplus is called **compensatory mortality**, where the harvest deaths are substituting for the deaths that would otherwise occur naturally. Harvest beyond that is **additive mortality**, harvest in addition to all the animals that would have died naturally.

EXERCISE

1. Draw a neat sketch of Hydrological Cycle and discuss the various sources of water. *[Raj. Tech. Uni. , Sem II, 2009]*
2. What are the objectives of ecological study? How do you classify the various aspects of ecology? *[Raj. Tech. Uni. , Sem II, 2009]*
3. What is biodiversity? What are the basic laws of biodiversity? Describe the population characteristics analysis of ecology. *[Raj. Tech. Uni. , Sem II, 2009]*
4. What is Ecology? Explain different Ecosystems.
5. Write an essay on food chains and food web.
6. How many types of biodiversity you know?
7. What is energy flow? Explain 'Energy flow in ecosystem'.
8. Write short notes on-
   (*i*) Water cycle
   (*ii*) Carbon cycle
   (*iii*) Nitrogen cycle
   (*iv*) Phosphorus cycle
   (*v*) Sulphur cycle
   (*vi*) Population Dynamics.

3

Renewable Energy Sources

3.1 INTRODUCTION: WHAT IS RENEWABLE ENERGY?

The United States currently relies heavily on coal, oil, and natural gas for its energy. Fossil fuels are non-renewable, that is, they draw on finite resources that will eventually dwindle, becoming too expensive or too environmentally damaging to retrieve. In contrast, renewable energy resources-such as wind and solar energy-are constantly replenished and will never run out.

Most renewable energy comes either directly or indirectly from the sun. Sunlight, or solar energy, can be used directly for heating and lighting homes and other buildings, for generating electricity, and for hot water heating, solar cooling, and a variety of commercial and industrial uses.

The sun's heat also drives the winds, whose energy, is captured with wind turbines. Then, the winds and the sun's heat cause watèr to evaporate. When this water vapor turns into rain or snow and flows downhill into rivers or streams, its energy can be captured using hydroelectric power.

Along with the rain and snow, sunlight causes plants to grow. The organic matter that makes up those plants is known as biomass. Biomass can be used to produce electricity, transportation fuels, or chemicals. The use of biomass for any of these purposes is called bioenergy.

Hydrogen also can be found in many organic compounds, as well as water. It's the most abundant element on the Earth. But it doesn't occur naturally as a gas. It's always combined with other elements, such as with oxygen to make water. Once separated from another element, hydrogen can be burned as a fuel or converted into electricity.

Not all renewable energy resources come from the sun. Geothermal energy taps the Earth's internal heat for a variety of uses, including electric power production, and the heating and cooling of buildings. And the energy of the ocean's tides comes from the gravitational pull of the moon and the sun upon the Earth.

In fact, ocean energy comes from a number of sources. In addition to tidal energy, there's the energy of the ocean's waves, which are driven by both the tides and the winds. The sun also warms the surface of the ocean more than the ocean depths, creating a temperature difference that can be used as an energy source. All these forms of ocean energy can be used to produce electricity.

Do you know what renewable sources of energy are and why we should think of these alternative energy sources?

In the past century, it has been seen that the consumption of non-renewable sources of energy has caused more environmental damage than any other human activity. Electricity generated from fossil fuels such as coal and crude oil has led to high concentrations of harmful gases in the atmosphere. This has in turn led to many problems being faced today such as ozone depletion and global warming. Vehicular pollution has also been a major problem.

Therefore, alternative sources of energy have become very important and relevant to today's world. These sources, such as the sun and wind, can never be exhausted and therefore are called renewable. They cause less emissions and are available locally. Their use can, to a large extent, reduce chemical, radioactive, and thermal pollution. They stand out as a viable source of clean and limitless energy. These are also known as non-conventional sources of energy. Most of the renewable sources of energy are fairly non-polluting and considered clean though biomass, a renewable source, is a major polluter indoors.

Why is Renewable Energy Important?

Renewable energy is important because of the benefits it provides. The key benefits are:

Environmental Benefits

Renewable energy technologies are clean sources of energy that have a much lower environmental impact than conventional energy technologies.

Energy for Our Children's Children's Children

Renewable energy will not run out. Ever. Other sources of energy are finite and will some day be depleted.

Jobs and the Economy

Most renewable energy investments are spent on materials and workmanship to build and maintain the facilities, rather than on costly energy imports. Renewable energy investments are usually spent within the United States, frequently in the same state, and often in the same town. This means your energy dollars stay home to create jobs and fuel local economies, rather than going overseas.

Meanwhile, renewable energy technologies developed and built in the United States are being sold overseas, providing a boost to the U.S. trade deficit.

Energy Security

After the oil supply disruptions of the early 1970s, our nation has increased its dependence on foreign oil supplies instead of decreasing it. This increased dependence impacts more than just our national energy policy.

What are these Alternative Sources of Energy?

Under the category of renewable energy or non-conventional energy are such sources as the sun, wind, water, agricultural residue, firewood, and animal dung. The non-renewable sources are the fossil fuels such as coal, crude oil, and natural gas. Energy generated from the sun is known as solar energy. Hydel is the energy derived from water. Biomass —firewood, animal dung, biodegradable waste from cities and crop residues— is a source of energy when it is burnt. Geothermal energy is derived from hot dry rocks, magma, hot water springs, natural geysers, etc. Ocean thermal is energy derived from waves and also from tidal waves.

When you burn a piece of wood it turns into ash. Can you use this ash to again light a fire? No, You cannot do this. This is exactly what happens to the non renewable sources of energy such as coal, natural gas and oil. Once you burn them they cannot be reused. Other than this it also causes extensive damage to the environment.

1. **Solar Energy:** Solar energy is the most readily available source of energy. It does not belong to anybody and is, therefore, free. It is also the most important of the non-conventional sources of energy because it is non-polluting and, therefore, helps in lessening the greenhouse effect.

   Solar energy has been used since prehistoric times, but in a most primitive manner. Before 1970, some research and development was carried out in a few countries to exploit solar energy more efficiently, but most of this work remained mainly academic. After the dramatic rise in oil prices in the 1970s, several countries began to formulate extensive research and development programmes to exploit solar energy.

   When we hang out our clothes to dry in the sun, we use the energy of the sun. In the same way, solar panels absorb the energy of the sun to provide heat for cooking and for heating water. Such systems are available in the market and are being used in homes and factories.

   In the next few years it is expected that millions of households in the world will be using solar energy as the trends in USA and Japan show. In India too, the Indian Renewable Energy Development Agency and the Ministry of Non-Conventional Energy Sources are formulating a programme to have solar energy in more than a million households in the next few years. However, the people's initiative is essential if the programme is to be successful.

India is one of the few countries with long days and plenty of sunshine, especially in the Thar desert region. This zone, having abundant solar energy available, is suitable for harnessing solar energy for a number of applications. In areas with similar intensity of solar radiation, solar energy could be easily harnessed. Solar thermal energy is being used in India for heating water for both industrial and domestic purposes. A 140 MW integrated solar power plant is to be set up in Jodhpur but the initial expense incurred is still very high.

Form of Energy: Thermal energy. **This Energy is used for:** Cooking/Heating, Drying/Timber seasoning, Distillation, Electricity/Power generation, Cooling, Refrigeration, Cold storage. **Some of the Gadgets and other Devices:** Solar cooker, Flat plate solar cookers, Concentrating collectors, Solar hot water systems (Domestic and Industrial), Solar pond, Solar hot air systems, Solar Dryers, Solar timber kilns, solar stills, Solar photovoltaic systems, Solar pond, Concentrating collectors, Power Tower, Air conditioning, Solar collectors, coupled to absorption, Refrigeration systems.

Solar energy can also be used to meet our electricity requirements. Through Solar Photovoltaic (SPV) cells, solar radiation gets converted into DC electricity directly. This electricity can either be used as it is or can be stored in the battery. This stored electrical energy then can be used at night. SPV can be used for a number of applications such as:

(*a*) Domestic lighting
(*b*) Street lighting
(*c*) Village electrification
(*d*) Water pumping
(*e*) Desalination of salty water
(*f*) Powering of remote telecommunication repeater stations and
(*g*) Railway signals.

If the means to make efficient use of solar energy could be found, it would reduce our dependence on non-renewable sources of energy and make our environment cleaner.

2. **Wind Energy:** We have been harnessing the wind's energy for hundreds of years. From old Holland to farms in the United States, windmills have been used for pumping water or grinding grain. Today, the windmill's modern equivalent — a wind turbine — can use the wind's energy to generate electricity.

Wind turbines, like windmills, are mounted on a tower to capture the most energy. At 100 feet (30 meters) or more aboveground, they can take advantage of the faster and less turbulent wind. Turbines catch the wind's energy with their propeller-like blades. Usually, two or three blades are mounted on a shaft to form a rotor.

A blade acts much like an airplane wing. When the wind blows, a pocket of low-pressure air forms on the downwind side of the blade. The low-pressure air pocket then pulls the blade toward it, causing the rotor to turn. This is called lift. The force of the lift is actually much stronger than the wind's force against the front side of the blade, which is called drag. The combination of lift and drag causes the rotor to spin like a propeller, and the turning shaft spins a generator to make electricity.

Wind turbines can be used as stand-alone applications, or they can be connected to a utility power grid or even combined with a photovoltaic (solar cell) system. For utility-scale sources of wind energy, a large number of wind turbines are usually built close together to form awind plant. Several electricity providers today use wind plants to supply power to their customers.

Stand-alone wind turbines are typically used for water pumping or communications. However, homeowners, farmers, and ranchers in windy areas can also use wind turbines as a way to cut their electric bills.

Small wind systems also have potential as distributed energy resources. Distributed energy resources refer to a variety of small, modular power-generating technologies that can be combined to improve the operation of the electricity delivery system.

Wind energy is the kinetic energy associated with the movement of atmospheric air. It has been used for hundreds of years for sailing, grinding grain, and for irrigation. Wind energy systems convert this kinetic energy to more useful forms of power. Wind energy systems for irrigation and milling have been in use since ancient times and since the beginning of the 20th century it is being used to generate electric power. Windmills for water pumping have been installed in many countries particularly in the rural areas.

Wind turbines transform the energy in the wind into mechanical power, which can then be used directly for grinding etc., or further converting to electric power to generate electricity. Wind turbines can be used singly or in clusters called 'wind farms'. Small wind turbines called aero-generators can be used to charge large batteries.

Five nations — Germany, USA, Denmark, Spain and India — account for 80% of the world's installed wind energy capacity. Wind energy continues to be the fastest growing renewable energy source with worldwide wind power installed capacity reaching 14,000 MW.

Realizing the growing importance of wind energy, manufacturers have steadily been increasing the unit size of the wind electric generators since the late 1980s. Another important development has been the offshore (*i.e.,* in the sea) wind farms in some regions of Europe, which have several advantages over the on-shore ones. The third major development has been the use of new techniques to assess the wind resource for techno-commercial viability.

Form of Energy: Kinetic energy **This Energy is used for**: Sailing ships, Pumping water/ Irrigation, Grinding Grains, Power generation **Some of the Gadgets and other Devices:** Sails, Windmills, Wind turbines.

In India the states of Tamilnadu and Gujarat lead in the field of wind energy. At the end of March 2000 India had 1080-MWs capacity wind farms, of which Tamilnadu contributed 770-MW capacity. Gujarat has 167MW followed by Andhra Pradesh, which has 88 MW installed wind farms.There are about a dozen wind pumps of various designs providing water for agriculture, afforestation, and domestic purposes, all scattered over the country.

The design of the Auroville multi-blade windmill has evolved from the practical experience gained in operating these mills over a period of 20 years or so. It has a high tripod tower and its double-action pump increases water output by about 60% compared to the conventional single-action pumps.

3. **Hydel Energy (Hydropower):** Flowing water creates energy that can be captured and turned into electricity. This is called *hydroelectric power* or *hydropower*.

   The most common type of hydroelectric power plant uses a dam on a river to store water in a reservoir. Water released from the reservoir flows through a turbine, spinning it, which in turn activates a generator to produce electricity. But hydroelectric power doesn't necessarily require a large dam. Some hydroelectric power plants just use a small canal to channel the river water through a turbine.

   Another type of hydroelectric power plant - called a *pumped storage plant* — can even store power. The power is sent from a power grid into the electric generators. The generators then spin the turbines backward, which causes the turbines to pump water from a river or lower reservoir to an upper reservoir, where the power is stored. To use the power, the water is released from the upper reservoir back down into the river or lower reservoir. This spins the turbines forward, activating the generators to produce electricity.

   A small or micro-hydroelectric power system can produce enough electricity for a home, farm, or ranch.

Energy from Water Sources

The energy in the flowing water can be used to produce electricity. Waves result from the interaction of the wind with the surface of the sea and represent a transfer of energy from the wind to the sea. Energy can be extracted from tides by creating a reservoir or basin behind a barrage and then passing tidal waters through turbines in the barrage to generate electricity.

Mini or Micro Hydropower

Hydropower is one of the best, cheapest, and cleanest source of energy, although, with big dams, there are many environmental and social problems as has been seen in the case of the Tehri and the Narmada Projects. Small dams are, however, free from these problems. This is in fact one of the earliest known renewable energy sources, in the country (since the beginning of the 20th century).

In fact, for the last few hundred years, people living in the hills of the Himalayas have been using water mills, or *chakki*, to grind wheat. The 130 KW small hydropower plant in Darjeeling set up in 1897, was the first in India. Besides being free from the problem of pollution, such plants are also free from issues and controversies that are associated with the bigger projects, namely affecting the lives of thousands of people living along the banks of the rivers, destruction of large areas under forest, and seismological threats.

New environmental laws affected by the danger of global warming have made energy from small hydropower plants more relevant. These small hydropower plants can serve the energy needs of remote rural areas independently. The real challenge in a remote area lies in successful marketing of the energy and recovering the dues. Local industries should be encouraged to use this electricity for sustainable development.

It is a technology with enormous potential, which could exploit the water resources to supply energy to remote rural areas with little access to conventional energy sources. It also eliminates most of the negative environmental effects associated with large hydro projects.

Energy from the Sea — Ocean Thermal, Tidal and Wave Energy

Large amounts of solar energy is stored in the oceans and seas. On an average, the 60 million square kilometre of the tropical seas absorb solar radiation equivalent to the heat content of 245 billion barrels of oil. Scientists feel that if this energy can be tapped a large source of energy will be available to the tropical countries and to other countries as well. The process of harnessing this energy is called OTEC (ocean thermal energy conversion). It uses the temperature differences between the surface of the ocean and the depths of about 1000m to operate a heat engine, which produces electric power.

Form of Energy: Potential/Kinetic energy. **This Energy is being used for:** Power generation. **Some of the Gadgets and other Devices:** Turbine generators.

Energy is also obtained from waves and tides. The first wave energy, project with a capacity of 150MW, has been set up at Vizhinjam near Trivandrum. A major tidal wave power project costing of Rs.5000 crores, is proposed to be set up in the Hanthal Creek in the Gulf of Kutch in Gujarat.

In some countries such as Japan small scale power generators run by energy from waves or the ocean, have been used as power sources for channel marking buoys.

4. **Geothermal Energy:** We live between two great sources of energy, the hot rocks beneath the surface of the earth and the sun in the sky. Our ancestors knew the value of geothermal energy; they bathed and cooked in hot springs. Today we have recognized that this resource has potential for much broader application.

   Geothermal energy is the heat from the Earth. It's clean and sustainable. Resources of geothermal energy range from the shallow ground to hot water and hot rock found a few miles beneath the Earth's surface, and down even deeper to the extremely high temperatures of molten rock called magma.

   Almost everywhere, the shallow ground or upper 10 feet of the Earth's surface maintains a nearly constant temperature between 50° and 60° F (10° and 16° C). Geothermal heat pumps can tap into this resource to heat and cool buildings. A geothermal heat pump system consists of a heat pump, an air delivery system (ductwork), and a heat exchanger—a system of pipes buried in the shallow ground near the building. In the winter, the heat pump removes heat from the heat exchanger and pumps it into the indoor air delivery system. In the summer, the process is reversed, and the heat pump moves heat from the indoor air into the heat exchanger. The heat removed from the indoor air during the summer can also be used to provide a free source of hot water.

In the United States, most geothermal reservoirs of hot water are located in the western states, Alaska, and Hawaii. Wells can be drilled into underground reservoirs for the generation of electricity. Some geothermal power plants use the steam from a reservoir to power a turbine/generator, while others use the hot water to boil a working fluid that vaporizes and then turns a turbine. Hot water near the surface of Earth can be used directly for heat. Direct-use applications include heating buildings, growing plants in greenhouses, drying crops, heating water at fish farms, and several industrial processes such as pasteurizing milk.

Hot dry rock resources occur at depths of 3 to 5 miles everywhere beneath the Earth's surface and at lesser depths in certain areas. Access to these resources involves injecting cold water down one well, circulating it through hot fractured rock, and drawing off the heated water from another well. Currently, there are no commercial applications of this technology. Existing technology also does not yet allow recovery of heat directly from magma, the very deep and most powerful resource of geothermal energy.

The core of the earth is very hot and it is possible to make use of this geothermal energy (in Greek it means heat from the earth). These are areas where there are volcanoes, hot springs, and geysers, and methane under the water in the oceans and seas. In some countries, such as in the USA water is pumped from underground hot water deposits and used to heat people's houses.

The utilization of geothermal energy for the production of electricity dates back to the early part of the twentieth century. For 50 years the generation of electricity from geothermal energy was confined to Italy and interest in this technology was slow to spread elsewhere. In 1943 the use of geothermal hot water was pioneered in Iceland.

Form of Energy: Thermal energy. **This Energy is being used for**: Heating/Power Generation. **Some of the Gadgets and Other Devices**: Heat exchanger, Steam turbines.

In India, Northwestern Himalayas and the western coast are considered geothermal areas. The Geological Survey of India has already identified more than 350 hot spring sites, which can be explored as areas to tap geothermal energy. Satellites like the IRS-1 have played an important role, through infrared photographs of the ground, in locating geothermal areas. The Puga valley in the Ladakh region has the most promising geothermal field. An experimental 1-kW generator is already in operation in this area. It is being used mainly for poultry farming, mushroom cultivation, and pashmina-wool processing, all of which need higher temperature.

5. **Bio-Energy (Biomass Energy):** We have used biomass energy or bioenergy — the energy from organic matter — for thousands of years, ever since people started burning wood to cook food or to keep warm.

   And today, wood is still our largest biomass energy resource. But many other sources of biomass can now be used, including plants, residues from agriculture or forestry, and the organic component of municipal and industrial wastes. Even the fumes from landfills can be used as a biomass energy source.

   The use of biomass energy has the potential to greatly reduce our greenhouse gas emissions. Biomass generates about the same amount of carbon dioxide as fossil fuels, but every time a new plant grows, carbon dioxide is actually removed from the atmosphere. The net emission of carbon dioxide will be zero as long as plants continue to be replenished for biomass energy purposes. These energy crops, such as fast-growing trees and grasses, are called *biomass feed-stocks*. The use of biomass feedstocks can also help increase profits for the agricultural industry.

   Biomass is a renewable energy resource derived from the carbonaceous waste of various human and natural activities. It is derived from numerous sources, including the by-products from the timber industry, agricultural crops, raw material from the forest, major parts of household waste and wood.

   Biomass does not add carbon dioxide to the atmosphere as it absorbs the same amount of carbon in growing as it releases when consumed as a fuel. Its advantage is that it can be used to generate electricity with the same equipment or power plants that are now burning fossil fuels. Biomass is an important source of energy and the most important fuel worldwide after coal, oil and natural gas.

   Traditional use of biomass is more than its use in modern application. In the developed world biomass is again becoming important for applications such as combined heat and power generation. In addition, biomass energy is gaining significance as a source of clean heat for domestic heating and community heating applications. In fact in countries like Finland, USA and Sweden the per capita biomass energy used is higher than it is in India, China or in Asia.

   Biomass fuels used in India account for about one third of the total fuel used in the country, being the most important fuel used in over 90% of the rural households and about 15% of the urban households.

Instead of burning the loose biomass fuel directly, it is more practical to compress it into briquettes (compressing them through a process to form blocks of different shapes) and thereby improve its utility and convenience of use. Such biomass in the dense briquetted form can either be used directly as fuel instead of coal in the traditional chulhas and furnaces or in the gasifier. Gasifier converts solid fuel into a more convenient-to-use gaseous form of fuel called producer gas.

Form of Energy: Chemical energy. **This Energy is being used for**: Cooking, Mechanical, Applications/Pumping, Power generation, Transportation. **Some of the Gadgets and other Devices:** Biogas plant/Gasifier/Burner, Gasifier engine pump sets, Stirling engine pump sets, Producer gas/ Biogas based engine generator sets.

Scientists are trying to explore the advantages of biomass energy as an alternative energy source as it is renewable and free from net CO_2 (carbon dioxide) emissions, and is abundantly available on earth in the form of agricultural residue, city garbage, cattle dung, firewood, etc. Bio-energy, in the form of biogas, which is derived from biomass, is expected to become one of the key energy resources for global sustainable development.

At present, biogas technology provides an alternative source of energy in rural India for cooking. It is particularly useful for village households that have their own cattle. Through a simple process cattle dung is used to produce a gas, which serves as fuel for cooking. The residual dung is used as manure.

Biogas plants have been set up in many areas and are becoming very popular. Using local resources, namely cattle waste and other organic wastes, energy and manure are derived. A mini biogas digester has recently been designed and developed, and is being in-field tested for domestic lighting.

Indian sugar mills are rapidly turning to bagasse, the leftover of cane after it is crushed and its juice extracted, to generate electricity. This is mainly being done to clean up the environment, cut down power costs and earn additional revenue. According to current estimates, about 3500 MW of power can be generated from bagasse in the existing 430 sugar mills in the country. Around 270 MW of power has already been commissioned and more is under construction.

6. **Fuel Cells (Hydrogen Energy):** Hydrogen is the simplest element. An atom of hydrogen consists of only one proton and one electron. It's also the most plentiful element in the universe. Despite its simplicity and abundance, hydrogen doesn't occur naturally as a gas on the Earth - it's always combined with other elements. Water, for example, is a combination of hydrogen and oxygen (H_2O).

   Hydrogen is also found in many organic compounds, notably the *hydrocarbons* that make up many of our fuels, such as gasoline, natural gas, methanol, and propane. Hydrogen can be separated from hydrocarbons through the application of heat - a process known as *reforming*. Currently, most hydrogen is made this way from natural gas. An electrical current can also be used to separate water into its components of oxygen and hydrogen. This process is known as *electrolysis*. Some algae and bacteria, using sunlight as their energy source, even give off hydrogen under certain conditions.

   Hydrogen is high in energy, yet an engine that burns pure hydrogen produces almost no pollution. NASA has used liquid hydrogen since the 1970s to propel the space shuttle and other rockets into orbit. Hydrogen fuel cells power the shuttle's electrical systems, producing a clean byproduct - pure water, which the crew drinks.

A fuel cell combines hydrogen and oxygen to produce electricity, heat, and water. Fuel cells are often compared to batteries. Both convert the energy produced by a chemical reaction into usable electric power. However, the fuel cell will produce electricity as long as fuel (hydrogen) is supplied, never losing its charge.

Fuel cells are a promising technology for use as a source of heat and electricity for buildings, and as an electrical power source for electric motors propelling vehicles. Fuel cells operate best on pure hydrogen. But fuels like natural gas, methanol, or even gasoline can be reformed to produce the hydrogen required for fuel cells. Some fuel cells even can be fueled directly with methanol, without using a reformer.

In the future, hydrogen could also join electricity as an important energy carrier. An energy carrier moves and delivers energy in a usable form to consumers. Renewable energy sources, like the sun and wind, can't produce energy all the time. But they could, for example, produce electric energy and hydrogen, which can be stored until it's needed. Hydrogen can also be transported (like electricity) to locations where it is needed.

What are fuel cells? Fuel cells are electrochemical devices that convert the chemical energy of a fuel directly and very efficiently into electricity (DC) and heat, thus doing away with combustion. The most suitable fuel for such cells is hydrogen or a mixture of compounds containing hydrogen. A fuel cell consists of an electrolyte sandwiched between two electrodes. Oxygen passes over one electrode and hydrogen over the other, and they react electrochemically to generate electricity, water, and heat.

Though fuel cells have been used in space flights and combined supplies of heat and power, electric vehicles are the best option available to dramatically reduce urban air pollution. Compared to vehicles powered by the internal combustion engine, fuel-cell powered vehicles have very high energy conversion efficiency, (almost double that of currently used engines) and near-zero pollution, CO_2 and water vapour being the only emissions. Fuel-cell-powered EV's (electric vehicles) score over battery operated EV's in terms of increased efficiency and easier and faster refuelling.

Canada's Ballad's Power Systems and Germany's Dailmer-Benz are world leaders in the application of fuel cell technology for meeting transportation needs. In India, diesel-run buses are a major means of transport and these emit significant quantities of SPM and SO_2. Thus, fuel-cell powered buses could be introduced with relative ease and yet make a positive impact on urban air quality. Such buses are already in operation in Vancouver in Canada and in Illinois and California in the USA. Though rapid progress has been made, high initial cost is still the biggest hurdle in the widespread commercialization of fuel cells.

Fuel cells for power generation — India has a large gap between the demand for and supply of power. Conventional large-scale power plants use non-renewable fuels with significant adverse ecological and environmental impacts. Fuel cell systems are excellent candidates for small-scale decentralized power generation.

Fuel cells can supply combined heat and power to commercial buildings, hospitals, airports and military installation at remote locations. Fuel cells have efficiency levels up to 55% as compared to 35% of conventional power plants. The emissions are significantly lower (CO_2 and water vapour being the only emissions). Fuel cell systems are modular (*i.e.*, additional capacity can be added whenever required with relative ease) and can be set up wherever power is required.

7. **Co-Generation:** Co-generation is the concept of producing two forms of energy from one fuel. One of the forms of energy must always be heat and the other may be electricity or mechanical energy. In a conventional power plant, fuel is burnt in a boiler to generate high-pressure steam. This steam is used to drive a turbine, which in turn drives an alternator through a steam turbine to produce electric power. The exhaust steam is generally condensed to water which goes back to the boiler.

   As the low-pressure steam has a large quantum of heat which is lost in the process of condensing, the efficiency of conventional power plants is only around 35%. In a cogeneration plant, very high efficiency levels, in the range of 75% – 90%, can be reached. This is so, because the low-pressure exhaust steam coming out of the turbine is not condensed, but used for heating purposes in factories or houses.

   Since co-generation can meet both power and heat needs, it has other advantages as well in the form of significant cost savings for the plant and reduction in emissions of pollutants due to reduced fuel consumption.

   Even at conservative estimates, the potential of power generation from co-generation in India is more than 20,000 MW. Since India is the largest producer of sugar in the world, bagasse-based cogeneration is being promoted. The potential for cogeneration thus lies in facilities with joint requirement of heat and electricity, primarily sugar and rice mills, distilleries, petrochemical sector and industries such as fertilizers, steel, chemical, cement, pulp and paper, and aluminum.

Potential and Present Status of Renewable Sources of Energy in India

According to the 11th New and Renewable Energy five-year plan proposed by the government of India, from 2008-2012 the renewable energy market in India will reach an estimated US $19 billion. Investments of US $15 billion will be required in order to add the approximately 15,000 megawatts (MW) of renewable energy to the present installed capacity. The government of India has planned a subsidy support system of approximately US $1 billion in government funds. This amounts to adding renewable energy capacity at 1 Watt per US $1, with potential subsidy support of US $0.07/Watt.

The Indian government has also set specific targets for renewable energy: by 2012 it expects renewable energy to contribute 10% of total power generation capacity and have a 4–5% share in the electricity mix. This implies that growth in renewable energy will occur at a much faster pace than traditional power generation, with renewables making up 20% of the 70,000 MW of total additional energy planned from 2008-2012.

From 2002 to 2007, there was 3,075 MW of renewable grid-tied power planned, but the actual capacity addition exceeded 6,000 MW by 2006. A large share of this was the result of exceptional growth of wind energy in India. Wind energy is expected to add more than 10,000 MW of additional capacity by 2012, followed by small hydro (1,400 MW), co-generation (1,200 MW) and biomass (500 MW).

In the past five years, the government of India had withdrawn subsidies for grid-tied solar power because its costs were not considered viable. Plans for a 140 MW Integrated Solar Combined Cycle plant at Mathania, Rajasthan were canceled after an unfavorable technical and commercial review. But recently this sector has experienced a turn-around, with new technologies and manufacturing capacity in India (See Scaling Up Solar in India). With the success of grid-tied solar electricity in other countries, and new state-level initiatives for feed-in tariffs, the outlook for this segment is now positive. The government has set aside US $50 million to subsidize solar power when its costs match those of small hydro.

Financial assistance is available in various forms, such as direct installation subsidy, feed-in tariffs, tax rebates, and low interest loans. Increasingly Indians are seeing a shift away from subsidizing installed capacity and towards subsidizing power generation with feed-in tariffs and tax rebates. For example, for wind energy there is an income tax exemption for 10 years, 80% accelerated depreciation, sales tax exemption, and excise duty exemption.

Preferential tariffs are being worked out at the state level. While the new framework for supporting renewable energy generation is evolving, most of the direct subsidy has thus far been designed for installed capacity.

Projects in the states in northeastern part of India and the hill states of Jammu & Kashmir, Himachal Pradesh and Uttaranchal currently receive the best support, with subsidies there 20% to 50% higher than in other states of India. Some of the highest direct subsidy rates have been announced for installation of grid-tied wind and small hydro, and off-grid solar photovoltaic (PV) projects. The present subsidy for 1 MW grid-tied wind is $625, 000 and for 1 MW grid-tied small hydro it's $375, 000. See the chart below for more information.

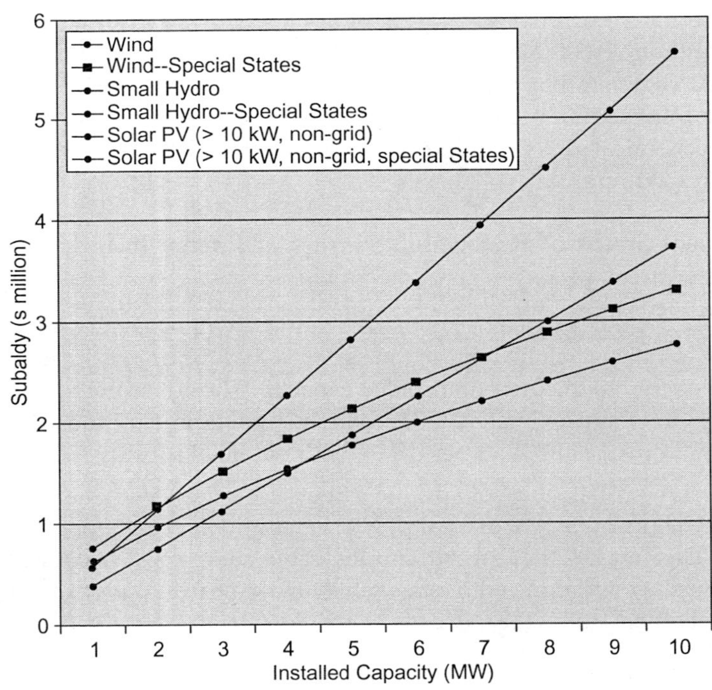

The Ministry of New and Renewable Energy has identified renewable energy R & D as an important factor for developing this sector. R & D subsidy is 100% of a project's cost in government R & D institutions, and 50% in the private sector. The R & D subsidy for the private sector may be enhanced for initial stages of technologies that have longer time-horizons.

Subsidy support is for firms with majority Indian stake. Firms with foreign majority stake can take advantage of subsidies if they have local Indian partnersv or if the R & D they provide is at lower costs and in shorter time than R & D provided by firms in India. Foreign firms can also received R & D subsidy support in exceptional cases when the larger national interest is evident and is served by that firm. Many Indian firms have foreign R & D collaborations or financial stakes in foreign R & D firms.

In the past a significant proportion of allocated funds have gone underutilized, possibly due to lack of availability of suitable projects and procedural delays. One area of concern is the lack of bankable projects in bagasse-based cogeneration and urban waste-to-energy space. In these cases, 50% of the direct subsidy amount will be given upfront, with the other half being distributed after the completion of the project.

With maturing technologies and integrated business models in renewable energy business, the suitability of projects is likely to improve, resulting in higher utilization of available government funds and faster market growth. This has been the case of wind energy in India. Integrated installation, operations and repair contracts are common in the wind energy business. This has provided a viable and investor-friendly business model.

Thin-film solar production is expected to lead to the development of PV-based electricity generation. Technologies in the wind, bio-mass, waste-to-energy and small hydro markets are at the venture capitalist stage, waiting for funding to begin commercialization and scaling up of thier projects.

Policies at state and local levels, like feed-in tariffs, tax incentives, technical training, financing mechanisms and revenue recovery can go a long way in building a successful renewable energy market in India. Improvements in energy efficiency can reduce recurring cost of energy using technology that is often priced even higher than the cost of renewable energy. This will make investments in this sector attractive.

Building renewable energy infrastructure also requires mapping the natural resource base. In a continuing effort to build the information infrastructure for renewable energy, GIS (geographic information systems) mapping of renewable sources will continue with government funding. In addition, classification and rating of potential location sites is considered a strategically important area. To realize the estimated total potential of 85,000 MW of renewable energy in India, keeping renewable energy business in sync with international standards is important. Therefore, equipment-related subsidies and incentives are often made contingent upon meeting international standards.

The Ministry of New and Renewable Energy in India has also provided funds for town and city level renewable energy planning. This local information infrastructure, comprising of local awareness, urban design, laws and smooth functioning processes, is an important step towards integrating renewable energy into economic life, and improving power delivery in the last kilometer. In addition to improving investment opportunities, this may also bring about greener, better designed and less polluted cities. With the expected growth in this sector, the availability of technically trained personnel may become a hurdle for further growth in the coming years. Thus, an opportunity exists for educational and training institutions to introduce new courses, curricula, and training for students to work in this emerging job market in urban and rural areas.

In recent years, India has emerged as one of the leading destinations for investors from developed countries. This attraction is partially due to the lower cost of manpower and good quality production. The expansion of investments has brought benefits of employment, development, and growth in the quality of life, but only to the major cities. This sector only represents a small portion of the total population. The remaining population still lives in very poor conditions.

India is now the eleventh largest economy in the world, fourth in terms of purchasing power. It is poised to make tremendous economic strides over the next ten years, with significant development already in the planning stages. This report gives an overview of the renewable energies market in India. We look at the current status of renewable markets in India, the energy needs of the country, forecasts of consumption and production, and we assess whether India can power its growth and its society with renewable resources.

Sources of Renewable Energy Available in India

We know where the non renewable energies – coal, oil and gas – are located and how these fuels are transported, combusted, and the power transmitted throughout the country over the power grid. Now, let's look at the renewable energies – hydro, solar, wind and biomass – and see where they are found.

(*a*) **Solar Energy:** Because of its location between the Tropic of Cancer and the Equator, India has an average annual temperature that ranges from 25°C – 27.5°C. This means that India has huge solar potential. The sunniest parts are situated in the south/east coast, from Calcutta to Madras.

Solar energy has serveral applications: photovoltaic (PV) cells are placed on the roof top of houses or commercial buildings, and collectors such as mirrors or parabolic dishes that can move and track the sun throughout the day are also used. This mechanism is being used for concentrated lighting in buildings.

Photovoltaic (PV) cells have a low efficiency factor, yet power generation systems using photovoltaic materials have the advantage of having no moving parts. PV cells find applications in individual home rooftop systems, community street lights, community water pumping, and areas where the terrain makes it difficult to access the power grid. The efficiency of solar photovoltaic cell with single crystal silicon is about 13% – 17%. High efficiency cell with concentrators are being manufactured which can operate with low sunlight intensities.

India has an expanding solar energy sector: 9 solar cell manufactures, 22 PV module manufactures, and 50 PV systems manufacturers. Therefore, technology resources exist in country and a growing market would lead to job growth in country.

(*b*) **Wind Energy:** India is surpassed only by Germany as one of the world's fastest growing markets for wind energy. By the mid 1990s, the subcontinent was installing more wind generating capacity than North America, Denmark, Britain, and the Netherlands.

The ten machines near Okha in the province of Gujarat were some of the first wind turbines installed in India. These 15-meter Vestas wind turbines overlook the Arabian Sea. Now, in 2006, there is an installed capacity of 4,430 MW; however, ten times that potential, or 46,092 MW, exists.

Advantages of Wind Power:

- It is one of the most environment friendly, clean and safe energy resources.
- It has the lowest gestation period as compared to conventional energy.
- Equipment erection and commissioning involve only a few months.
- There is no fuel consumption, hence low operating costs.
- Maintenance costs are low.
- The capital cost is comparable with conventional power plants. For a wind farm, the capital cost ranges between 4.5 crores to 5.5 crores, depending on the site and the wind electric generator (WEG) selected for installation.

Comparison Between Fossil Fuels and Wind

	Wind	*Fossil Fuel*
Availability	Usable as it exists	Have to be procured and made usable through laborious and environmentally damaging processes
Limitation on availability	Inexhaustible resource	Limited in reserves, expected to be completely exhausted in the coming 60 years
Transportation	Used where it is available or transported where needed	Has to be transported from its source site for further processing, exposing the environment to pollution from accidents
Environmental effect of use	Zero emission	Used in producing electricity, releasing green house gasses
Geo-political implications	Reduces our reliance on oil, safeguarding national security. Allows for self sufficiency. There is no adverse effect on global environment. The whole system is pollution free and environment friendly.	Over-reliance on oil as a resource has undermined India's energy security, *e.g.*, OPEC crises of 1973, Gulf War of 1991 and the Iraq War of 2003.

The Essential Requirements for a Wind farm

An area where a number of wind electric generators are installed is known as a wind farm. The essential requirements for establishment of a wind farm of optimal exploitation of the wind are the following:

- High wind resource at particular site.
- Adequate land availability.
- Suitable terrain and good soil condition.
- Maintenance access to site.
- Suitable power grid nearby.
- Techno-economic selection of specific turbines.
- Scientifically prepared layout.

Wind energy generation has limitations which will influence the extent and type of role it will ultimately play in overall generation of electricity in India.

Limitation of a Wind Farm

- Wind machines must be located where strong, dependable winds are available most of the time.
- Because winds do not blow strongly enough to produce power all the time. Energy from wind machines is considered "intermittent," that is, it comes and goes. Therefore, electricity from wind farms must have a back-up supply from another source.
- As wind power is "intermittent," utility companies can use it for only part of their total energy needs.
- Wind towers and turbine blades are subject to damage from high winds and lighting. Rotating parts, which are located high off the ground can be difficult and expensive to repair.
- Electricity produced by wind power sometimes fluctuates in voltage and power factor, which can cause difficulties in linking its power to utility system.
- The noise made by rotating wind machine blades can be annoying to nearby neighbors.
- Some environmental groups have complained about aesthetics and avian mortality from wind machines.

(*c*) **Hydropower:** The hydroelectric power refers to the energy produced from water (rainfall flowing into rivers, etc). Consequently, rainfall can be a good indicator to investors looking for a location to implement or build a new hydroelectric power plant in India.

Most hydropower plants are situated in regions of the major rainfall.

The dominant annual rainfall is located on the north/eastern part of India: Arunachal Pradesh, Assam, Nagaland, Manipur and Mizoram, and also on the west coast between Mumbai (Bombay) and Mahe.

India utilizes twelve primary hydroelectric power plants: Bihar (3), Punjab, Uttaranchal, Karnataka, Uttar Pradesh, Sikkim, Jammu & Kashmir, Gujarat, and Andhra Pradesh (2).

(*d*) **Biomass Energy:** Biomass includes solid biomass (organic, non-fossil material of biological origins), biogas (principally methane and carbon dioxide produced by anaerobic digestion of biomass and combusted to produce heat and/or power), liquid biofuels (bio-based liquid fuel from biomass transformation, mainly used in transportation applications), and municipal waste (wastes produced by the residential, commercial and public services sectors and incinerated in specific installations to produce heat and/or power).

The most successful forms of biomass are sugar cane bagasse in agriculture, pulp and paper residues in forestry and manure in livestock residues. It is argued that biomass can directly substitute fossil fuels, as more effective in decreasing atmospheric CO_2 than carbon sequestration in trees. The Kyoto Protocol encourages further use of biomass energy.

Biomass may be used in a number of ways to produce energy. The most common methods are:

- Combustion
- Gasification
- Fermentation
- Anaerobic digestion.

India is very rich in biomass. It has a potential of 19,500 MW (3,500MW from bagassebased cogeneration and 16,000 MW from surplus biomass). Currently, India has 537 MW commissioned and 536 MW under construction. The facts reinforce the idea of a commitment by India to develop these resources of power production.

Following is a list of some States with most potential for biomass production:

- Andhra Pradesh (200 MW)
- Bihar (200 MW)
- Gujarat (200 MW)
- Karnataka (300 MW)
- Maharashtra (1,000 MW)
- Punjab (150 MW)
- Tamil Nadu (350 MW)
- Uttar Pradesh (1,000 MW).

Forecasts: What are the General Forecasts for the Next Decades?

Around the world, a growing number of nations have recognized the economic, social, and environmental benefits of renewable energy and are enacting tax incentives and other policy measures favorable to renewable technologies. In Germany, Japan, Spain, and a handful of other countries, clear government commitments to renewable energy and strong, effective policies have overcome barriers and created demand for these technologies, leading to dramatic growth in renewable industries and driving down costs.

(*a*) **The position of India in the world potential renewable energy.** Thanks to its location and geography, India enjoys abundant potential to all of the renewable energies.

(*b*) **The electricity consumption and generation forecasts of India as part of the emerging economies.** Growth in net electricity consumption is expected to be most rapid among the emerging economies of the world, including India. According to the EIA, the annual average increase will be about 4.0 percent from 2002 to 2025.

Emerging economies are projected to more than double their net electricity consumption, from 4,645 billion kilowatt hours in 2002 to 11,554 billion in 2025. The projected growth in net electricity consumption for emerging market economies is driven in large party by gross domestic product (GDP) and population growth assumption.

Because of the links between reliable electricity supply, GDP growth, and living standards, many of the nations with emerging economies are attempting to increase access to reliable electricity supply.

(*c*) **Projected energy consumption of India for 2030.** Currently, 45 percent of households in India do not have access to electricity. New legislation has set a target of electrifying all households by 2010. As in the past, the ongoing challenge in providing electricity is the ability of the poor to pay. India announced plans in March, 2005, to continue subsidizing electricity consumption for rural and poor households that use less than 30 kilowatt hours per month.

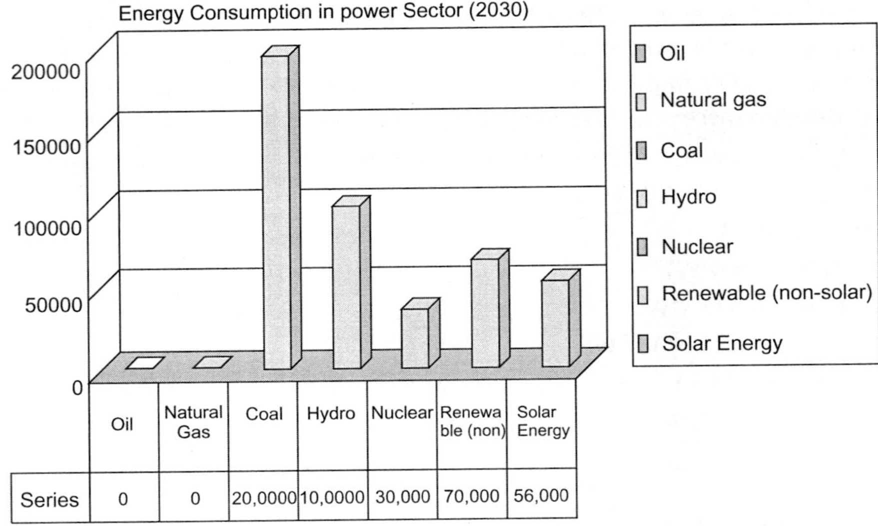

Estimates of Potential Capacities from Renewable Energy Sources (in MWs)

Source	*Approx. Potential*
Biomass energy	19,500
Solar energy	20,000
Wind energy	47,000
Small Hydropower	15,000
Ocean energy	50,000

Source: India Ministry of Non-Conventional Energy Sources.

The sum of these renewable resource potentials, 152,000 MW, is greater than the current total installed energy generating capacity of India.

3.2 GOVERNMENT REGULATIONS

What is the Current Commitment of the Government Regarding Renewable Energies?

India is one of the countries most involved in developing the use of renewable energies and is trying to make the opportunity for investors more attractive than costly.

(*a*) **Financing Sources and Incentives:** To promote renewable energy technologies in the country, the government has put in place some subsidies & fiscal incentives. The Indian Renewable Energy Development Agency has been set up under Ministry for Non-Conventional Energy Sources and is a specialized financing agency to promote and finance renewable energy projects. Following is a short list of new measures:

- Income tax breaks
- Accelerated depreciation
- Custom duty/duty free import concessions
- Capital/Interest subsidy
- Incentives for preparation of Detailed Project Reports (DPR) and feasibility reports

More details are as follows:

- 100 percent income tax exemption for any continuous block of power for 10 years in the first 15 years of operations
- providers of finance to such projects are exempt from tax on any income by way of dividends, interest or long-term capital gains from investment made in such projects on or after June 1, 1998 by way of shares or long-term finance
- accelerated 100-percent depreciation on specified renewable energy-based devices or projects
- accelerated depreciation of 80 percent in the first year of operations
- interest rate subsidies to promote commercialization of new technology
- lower customs and excise duties for specified equipment
- exemption or reduced rates of central and state taxes.

Ministry for Non-Conventional Energy Sources mix of fiscal and financial benefits:

- 2/3rd of the project cost subject to a maximum of Rs. 2.00 crore per 100 KW for procurement of modules, structures, power conditioning units, cabling etc., to the implementing agency. The balance cost on land, extension of grid lines, transformers, civil works, foundation and erection and commissioning, etc., is met by the implementing agency.
- Up to Rs. 1.0 lakh for the preparation of Detailed Project Report (DPR) for the grid interactive SPV power projects.
- 2.5 percent of its share of project cost, subject to a maximum of Rs. 5 lakhs for performance evaluation, monitoring report writing, etc., to the State Nodal Agency.
- Interest subsidy of up to 4 percent to Financial Institutions including IREDA, Nationalized Banks etc., for captive power projects of maximum capacity 200 KW by industry.

(*b*) **Environmental Legislation**

2001 Energy Conservation Act:

- Focus on energy efficiency
- Standards and labeling
- Designated consumers requirements
- Energy conservation building codes
- Energy conservation fund
- Bureau of Energy Efficiency.

2003 Electricity Act:

- Combined several existing pieces of legislation
- Intended to accelerate growth of power sector
- Targets additional 10 percent from renewable by 2012 (1000 MW/year capacity)
- Competitive market-based.
- Features include:
    - National Electricity policy
    - Delicensing of generation and captive generation
    - Public ownership of transmission companies
    - Open access in transmission
    - Freedom for distribution licenses
    - Establishment of State Electricity Regulatory Commissions
    - License-free generation and distribution in rural areas.

Provisions and activities impacting the power sector:

- Elimination of ceiling on foreign equity participation
- Streamlining the procedure for clearance of power projects
- Establishment of the Central Electricity Regulatory Commission
- Formulating an action plan to set up the National Grid.

State reforms impacting the power sector:

- unbundling the State Electricity Boards (SEB) into separate generation, transmission and distribution companies
- privatizing the generation, transmission and distribution companies
- setting up independent state electricity regulatory commissions
- making subsidy payments for subsidized categories of customers by state governments
- making tariff reforms by state governments
- enabling legislation and operational support extended to the SEB/utility
- improving operations of SEBs, particularly with regard to better management practices, reduction of transmission and distribution losses, better metering and reduction of power theft.

Summary and Conclusion: Could India meet all Energy needs with Renewable Energy?

India is a nation in transition. Considered an "emerging economy," increasing GDP is driving the demand for additional electrical energy, as well as transportation fuels. India is a nation of extremes. Poverty remains in areas with no energy services, while wealth grows in the new business hubs.

Coal fired generation currently provides two thirds of the generation capacity, and hydropower supplies the other third. Yet, India is blessed with vast resources of renewable energy in solar, wind, biomass and small hydro. In fact, the technical potential of these renewables exceeds the present installed generation capacity.

Unique in the world, India has the only Ministry that is dedicated to the development of renewable energies: the Ministry of New and Renewable Energy. This bodes well for the acceleration of renewable development throughout the nation -- both to meet the underserved needs of millions of rural residents and the growing demand of an energy hungry economy.

EXERCISE

1. What do you understand by renewable sources of Energy? Discuss various sources of renewable energy in details. [*Raj. Tech. Univ. Sem II, 2009*]
2. Write short notes on followings:
   (*i*) Solar Energy
   (*ii*) Biomass Energy
   (*iii*) Hydel Energy.
3. Write an essay on *Status of Renewable energy sources in India.*
4. Explain the 'Wind Energy' and write a note on the wind farm.
5. Compare Geothermal Energy and Hydrogen energy.

4

WATER POLLUTION

4.1 INTRODUCTION

When Neil Armstrong saw the Earth from the Moon, it appeared blue! This is because water covers more than two-thirds of the Earth's surface. But fresh water represents less than 0. 5% of the total water on Earth. The rest is either in the form of seawater or locked up in ice caps or the soil, which is why one often hears of water scarcity in many areas.

Water is continuously moving around the earth and constantly changing its form. It evaporates from land and water bodies and is also produced by all forms of life on Earth. This water vapour moves through the atmosphere, condenses to form clouds and precipitates as rain and snow. In time, the water returns to where it came from, and the process begins all over again. Although water is constantly moving, its total quantity on Earth's surface is constant.

Table 4.1 Chemical composition of underground water, river water and seawater : A comparison.

Parameter (ppm)	*Ground + Water (Well water)*	*Surface (River) water*	*Seawater*
Bicarbonate	339	119	140
Calcium	92	36	400
Chloride	9.6	13	19×10^3
Iron (III)	0.09	0.02	0.01
Magnesium	34	8.1	1.35×10^3
Nitrate	13	0.1	–
Potassium	1.4	1.2	380
Silica	10	1.2	6.0
Sodium	8.2	6.5	10.5×10^3
Sulphate	84	22	2.65×10^3
Total hardness (as $CaCO_3$)	369	123	–

Since water is referred as **"elixir of life"**, water pollution is by far the biggest environment problem, causing millions of death and so many illness perannum. Water pollution can be defined as the presence of foreign substances (organic, inorganic, biological or radiological) in water which tend to degrade its quality and constitute a hazard or impair its usefullness for drinking and domestic purpose or otherwise cause significant departures from the normal activities for various living communities.

National water commission stated that *water is polluted if it has not been of sufficient high quality and found suitable for the higher uses people wish to make of it. According to researchers water pollution changes the quality of surface and sub-soil waters to such a degree that its suitability either for human consumption or for the support of man's natural life processes decreases or ceases.*

4.2 FORMS OF WATER

Water is found in three different forms – liquid, solid or gas, depending on the temperature but it constantly changes from one form to another. Changes in temperature will determine which of these forms predominates in a particular area.

- **Liquid:** Water is usually encountered in the liquid state, because this is its natural state when temperatures are between 0°C and 100°C. 'Fresh' or drinking water is found as ground-water in underground aquifers, and on the surface in ponds, lakes, and rivers. Seas and oceans account for 97% of all water on Earth; but their waters contain dissolved salts and are therefore unfit to drink. In regions of young volcanic activity, hot water emerges from the earth in hot springs (examples are Garampani in Assam and Badrinath in Uttaranchal). How does this phenomenon occur? Surface water percolates downward through the rocks below the Earth's surface to high-temperature regions surrounding a magma reservoir, either active, or recently solidified but still hot. There the water is heated, becomes less dense, and rises back to the surface through fissures and cracks.
- **Solid:** Ice is the frozen form of water. It occurs when temperatures are below 0°C (32°F). For a given mass, ice occupies 9% more volume than water, which is why when water enters cracks in rocks and freezes it causes the rocks to crack and split. Being less dense than water, ice floats. This property of ice is vital to aquatic life in cold regions. As the temperature drops, ice forms a protective, insulating layer on the surfaces of streams, pools and other water bodies, allowing water to remain liquid in the layers beneath and life to survive. Glaciers, icebergs, and ice caps are all frozen water.
- **Gas:** Water is found in the atmosphere in its gaseous form, water vapour. Steam is nothing but vapourized water. In certain hot water springs called geysers, jets of steam and hot water rise one hundred feet or more from the ground. Geysers are found in Iceland, the North Island of New Zealand and in USA's Yellowstone National Park.

Worldwide, the consumption of water is doubling every 20 years – more than twice the rate of increase in population. A large amount of water is wasted in agriculture, industry, and urban areas. It has been estimated that with available technologies and better operational practices, agricultural water demand could be cut by about 50%, and that in urban areas by about 33% without affecting the quality or economics of life. But most governments do not have adequate laws or regulations to protect their water systems.

Due to the increase in population there has been a rise in the demand for food, space for housing, consumer products, etc., which has in turn resulted in increased industrialization, urbanization, and demands in agriculture thereby leading to both river and groundwater contamination.

Some Interesting Facts About Water:

The 'small and big' of water:

- The Pacific ocean is the biggest ocean covering approximately 32.6% of the Earth's surface
- The Arctic ocean is the smallest ocean

- 75% of the earth's surface is covered with water
- More than 97% of the earth's water is in its oceans
- About 2% of the available drinking water is frozen leaving only 1% for drinking
- The world's average rainfall is about 850 mm
- Water regulates the Earth's temperature. It also regulates the temperature of the human body, carries nutrients and oxygen to cells, cushions joints, protects organs and tissues, and removes waste
- 60% – 75% of the adult human body is water - 82% of blood is water; 70% of the brain and 90% of the lungs are made up of water
- Blood in animals and sap in plants is composed mainly of water
- To cook 1 cup of rice you need 2 cups of water but to wash the pan in which it has been cooked you need 4–5 litres of water
- A dripping tap can waste up to 6 litres of water in a day
- More than half the creatures on the Earth are found under water
  Life on earth probably originated in water
- In the summer our bodies require about 2 litres of water daily. Here is the water content of some foods (approximate) – 95% in tomato, 91% in spinach, 91% in milk, 85% in apples and 80% in potatoes
- 10% of the earth's surface is covered with ice.

There are more than one billion people particularly in North Africa and Western and South Asia, who lack access to a steady supply of clean water.

Access to water and sanitation, so crucial to human well-being and development, has now become a priority for the international community. To underscore the need for immediate action, the United Nations has designated 2003 as the International Year of Freshwater.

The importance given to water in ancient India is reflected in several hymns of the Vedas and epics and narratives from other valuable works such as the Arthasastra of Kautilya.

4.3 WATER CYCLE

The different water sources of the earth get their water supply from precipitation, while this precipitation in itself is the evaporation from these sources. Water is lost to the atmosphere as vapour from the earth, it condenses and then precipitates back in the form of rain, snow, hail, dew, frost or sleet. This is the hydrologic cycle that continues forever thereby maintaining a balance between the two.

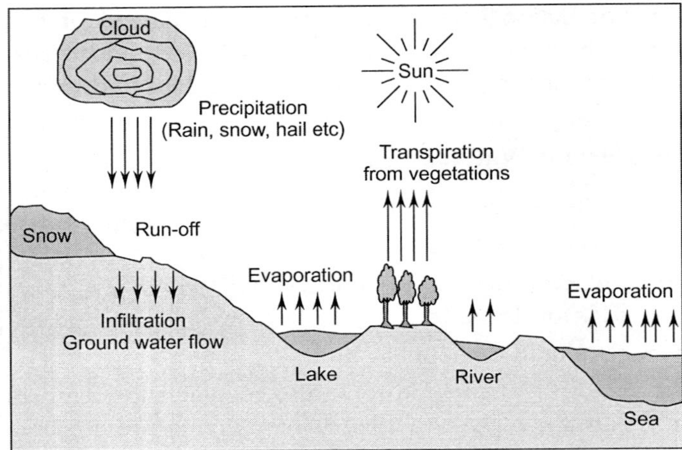

Fig. 4.1 Water Cycle

4.4 SOURCES OF WATER

Are you aware that only about 2.5 percent of the water on the Earth can be drunk? The balance 97.5 percent is salt water that we cannot use unless it is desalinated. Of the 2.5 percent that we can use, about three-fourths is frozen in the ice caps and glaciers, leaving just one quarter for household use. Freshwater is found in lakes, rivers, streams, underground, and glaciers. The world is heading towards a freshwater crisis due mainly due to its mismanagement. This crisis is already evident in many parts of the world, varying in scale and intensity depending on the time of the year, climate, and location.

Another reason for the depletion of freshwater resources, particularly groundwater resources, is a rapidly increasing human population.

Recognizing the importance of water resources to the planet's future, the United Nations General Assembly proclaimed the year 2003 the International Year of Freshwater. This will provide an opportunity to raise awareness, motivate people, and mobilize resources in order to manage water in a sustainable way.

(i) Rivers

Most ancient civilizations grew along the banks of rivers. Even today, millions of people all over the world live on the banks of rivers and depend on them for their survival.

All of us have seen a river—large or small, either flowing through our town, or somewhere else. Rivers are nothing more than surface water flowing down from a higher altitude to a lower altitude due to the pull of gravity. One river might have its source in a glacier, another in a spring or a lake. Rivers carry dissolved minerals, organic compounds, small grains of sand, gravel, and other material as they flow downstream. Rivers begin as small streams, which grow wider as smaller streams and rivers join them along their course across theland. Eventually they flow into seas or oceans. Most rivers With the exception of the Nile, flow towards the Equator. The flow in most rivers is not uniform, which means that sometimes there are floods and sometimes no water flows in them. Flood control projects attempt to reduce the variation in flow.

Unfortunately most of the world's major rivers are heavily polluted, but two of the world's largest river systems—the Amazon, that drains a vast area of South America and the Congo in sub-Saharan Africa—remain relatively healthy. This is because both have few industries and a small human population in their watersheds.

Some Facts about Rivers:

- The Nile, 6695 km long, and the Amazon 6437 km long, are the world's two longest rivers. Sometimes, measurements of their lengths can vary according to the criteria used for measurement.
- An Arab philosopher and physician Avicenna suggested, nearly 1000 years ago, that landscapes changed largely as a result of the action of running water. His views were largely ignored until the 16th century.
- The longest river in Asia is the Yangtze, which is 5472 km long.
- The world's highest waterfall are the Angel Falls in Venezuela, 979 m high, over 780 m of which is an uninterrupted drop.

Rivers in India

India has a large number of rivers that are lifelines for the millions living along their banks. These rivers can be categorized into four groups:

- Rivers that flow down from the Himalayas and are supplied by melting snow and glaciers. This is why these are perennial, that is, they never dry up during the year
- The Deccan Plateau rivers, which depend on rainfall for their water
- The coastal rivers, especially those on the west coast, which are short and do not retain water throughout the year
- The rivers in the inland drainage basin of west Rajasthan, which depend on the rains. These rivers normally drain towards silt lakes or flow into the sand.

The Major Rivers in India are Described here:

1. **Ganga:** This is considered the holiest of all the great rivers of India. It has its source at the Gangotri glacier, where it flows from the cave Goumukh, as the Bhagirathi, which then joins the River Alaknanda as it flows towards Devaprayag. The largest tributary of the river is the Ghaghara, which flows from the northern Nepal region and joins it before Patna in Bihar. Another major tributary is the Yamuna originating in the Yamunotri glacier, and flows through Delhi and Agra. Others are the Gomti, Gandak, Son, Kosi, Chambal, Sarda, etc. The Ganga is the lifeline for more than 500 million people living along its banks. The water of the river Ganges is considered so sacred that people keep it in their homes for use in prayers on important occasions such as at the time of death.

   Most cities along the river do not have sewage treatment plants and those that do have them can handle only part of the waste water. Millions of tonnes of untreated sewage are dumped daily into the river from the cities that lie along its banks. Bathing and washing also contribute to the pollution as most of the soap that is used is made from chemical substances. The river is also polluted by human and animal faeces. Industrial units that lie along the banks of the river discharge all the waste into the river and only a few of them have proper treatment facilities.
2. **Yamuna:** Rising from the Yamunotri glacier in the Tehri Garhwal District in the Himalayas, the Yamuna flows for about 1380 km almost parallel to the Ganga till they meet at Allahabad.

It flows through a number of important towns, Delhi, Mathura, Brindaban, and Agra to name some. Since ancient times, the Doab region, where the Ganga and the Yamuna flow, has been considered one of the most fertile areas in the subcontinent. Today however, this majestic river is polluted with domestic waste, silt, and industrial waste. The 22-km stretch between Wazirabad and the Okhla barrage in Delhi is only 2% of the catchment area, but it contributes about 80% of the river's total pollution load. The Hindon Canal also discharges waste from Uttar Pradesh in this stretch. Among the many casualties are birds and fish. There was a time when bird watchers had identified as many as 30 species of birds near the Yamuna, many of them exotic, such as the red-crested pochard and the godwit.

3. **Brahmaputra:** This mighty river rises in western Tibet in the Manasarovar region. It flows eastwards through the Himalayas, curves back across Arunachal Pradesh and Assam, then turns south to join the Padma and Ganga in Bangladesh and finally enters the Bay of Bengal. It is known by different names in different regions: as it flows through Tibet it is known as the Tsangpo. In the north-eastern states of Arunachal Pradesh and Assam, it is known as the Siang or Dihang. It is longer and more voluminous than the Ganga, and gushes down with enormous force for most of its course through the mountain regions and the forests of north-east India. A unique feature of this river is that it is navigable even at a height of 10,000 feet.

   When compared to the other major rivers in India, the Brahmaputra is less polluted but it has its own problems: petroleum refining units contribute most of the industrial pollution load into the basin along with other medium and small industries. The main problem facing the river basin is that of constant flooding. Floods have been occurring more often in recent years with deforestation, and other human activities being the major causes.

4. **Indus:** Known as the Sindhu in ancient times, the Indus was the cradle of India's great Indus Valley civilization. It has its source near Manasarovar close to southwestern Tibet, at an altitude of 16,000 feet and flows westward, through the Himalayas into Ladakh, and then through Sind and Punjab in Pakistan into the Arabian Sea.

   After flowing eleven miles beyond Leh, the Indus is joined by its first tributary, the Zanskar.

   When it enters the plains, its famous five tributaries—the Jhelum, Chenab, Ravi, Beas, and Sutlej—that give Punjab its name (land of the five rivers) join it.

   The name India is said to have its roots in Sindhu (Indus), the great river that constitutes the most imposing feature of that part of the subcontinent, home to some of the earliest civilizations. The river Sindhu has been invoked numerous times in the Vedic literature together with those of other gods and goddesses.

5. **Narmada:** This is the largest west–flowing river in India and originates from the Mekhala range in Shahdol district, Madhya Pradesh. It flows 1300 km west through the states of Madhya Pradesh and Gujarat before draining into the Gulf of Khambhat in the Arabian Sea. It is said to be one of the most beautiful rivers in India. In terms of its catchment area it is the seventh largest among the fourteen major river basins in the country. It is stated in the Matsya Purana that the mere sight of the river washes away all sins. With many short tributaries flowing into it from north and south, the Narmada basin forms a very important topographic feature of peninsular India.

   About 20% of the population along the river lives in the urban areas and the rest in the rural areas: Jabalpur, Hoshangabad, and Khargone in Madhya Pradesh, and Bharuch in Gujarat are some important cities located on the banks.

   The major cause of pollution is run-off from agricultural activities yet the pollution level is estimated to be lower than in other major rivers.

6. **Mahanadi:** This river is considered the lifeline of Orissa, through which it flows. It originates in south-eastern Madhya Pradesh near Raipur. In the upper drainage basin of the Mahanadi, which is centred on the Chhattisgarh Plain, periodic droughts contrast with the situation in the delta region where floods may damage the crops in what is known as the rice bowl of Orissa. The Hirakud Dam, constructed in the middle reaches of the Mahanadi, has helped in alleviating these adverse effects by creating a reservoir.
7. **Godavari:** This river originates in the Sahyadri range to the northeast of Mumbai in Maharashtra. It flows through that state and Andhra Pradesh before entering the Bay of Bengal. Its drainage basin is one of the largest in the country, second in size only to that of the Ganga; its delta on the east coast is also one of the country's main rice-growing areas. Despite the large catchment area the water available is only moderate because of the medium levels of annual rainfall.
8. **Kaveri:** The Kaveri is worshipped as a jeeva-nadi or perennial river. It has its origin in Talakaveri in the Kanara district of Karnataka and flows southwards through Tamil Nadu and Karnataka. The waters of the river have been a source of irrigation since antiquity; in the early 1990s, an estimated 95 percent of the Kaveri was diverted for agricultural use before it emptyied into the Bay of Bengal.

(ii) Groundwater

The importance of groundwater for the existence of human society cannot be overemphasized. Groundwater is the major source of drinking water in both urban and rural India. Besides, it is an important source of water for the agricultural and the industrial sector. Water utilization projections for 2000 put the groundwater usage at about 50%. Being an important and integral part of the hydrological cycle, its availability depends on the rainfall and recharge conditions. Till recently it had been considered a dependable source of uncontaminated water. The demand for water has increased over the years and this has led to water scarcity in many parts of the world. The situation is aggravated by the problem of water pollution or contamination. India is heading towards a freshwater crisis mainly due to improper management of water resources and environmental degradation, which has lead to a lack of access to safe water supply to millions of people. This freshwater crisis is already evident in many parts of India, varying in scale and intensity depending mainly on the time of the year.

> The National Water Policy (1987) states that water is a prime natural resource, basic human need, and precious national asset. It gives special attention to drinking water for both humans and animals over its other uses. The policy calls for controls on the exploitation of groundwater through regulation and an integrated and coordinated development of surface and groundwater. The central government has identified strategies for meeting drinking water needs and micro-watershed management and conducted pilot projects in different regions in the country. Even so, India is facing a freshwater crisis.

Groundwater crisis is not the result of natural factors; it has been caused by human actions. During the past two decades, the water level in several parts of the country has been falling rapidly due to an increase in extraction. The number of wells drilled for irrigation of both food and cash crops have rapidly and indiscriminately increased. India's rapidly rising population and changing lifestyles has also increased the domestic need for water. The water requirement for the industry also shows an overall increase. Intense competition among users—agriculture, industry, and domestic sectors—is driving the groundwater table lower. The quality of groundwater is getting severely affected because of the widespread pollution of surface water. Besides, discharge of untreated wastewater through bores and leachate from unscientific disposal of solid wastes also contaminates groundwater, thereby reducing the quality of freshwater resources. As far as the quality of groundwater is concerned, many states in the country have been identified as endemic to fluorosis due to abundance in naturally occurring fluoride-bearing minerals. These are Andhra Pradesh, Gujarat, Haryana, Orissa, Punjab, Rajasthan, TamilNadu, Uttar Pradesh, Karnataka, Madhya Pradesh, Maharastra, Bihar, and Delhi.

Nearly half million people in India suffer from ailments due to excess of fluoride in drinking water. In some districts of Assam and Orissa, groundwater has high iron content. About 31% of the total area of Rajasthan comes under saline groundwater. Groundwater is saline in almost all of the Bhakra Canal in Punjab and the lift canal system in south-western Haryana. Similarly high levels of arsenic in groundwater have been reported in the shallow aquifers in some districts of West Bengal. Certain places in Haryana, Gujarat, and Andhra Pradesh were also found to have dangerously high levels of mercury.

Causes of Groundwater Depletion and Contamination

Groundwater is an integral part of the environment, and hence cannot be looked upon in isolation. There has been a lack of adequate attention to water conservation, efficiency in water use, water re-use, groundwater recharge, and ecosystem sustainability. An uncontrolled use of the borewell technology has led to the extraction of groundwater at such a high rate that often recharge is not sufficient. The causes of low water availability in many regions are also directly linked to the reducing forest cover and soil degradation.

Pollution of groundwater resources has become a major problem today. The pollution of air, water, and land has an affect on the pollution and contamination of groundwater. The solid, liquid, and the gaseous waste that is generated, if not treated properly, results in pollution of the environment; this affects groundwater too due to the hydraulic connectivity in the hydrological cycle. For example, when the air is polluted, rainfall will settle many pollutants on the ground, which can then seep into and contaminate the groundwater resources. Water extraction without proper recharge and leaching of pollutants from pesticides and fertilizers into the aquifers has polluted groundwater supplies. In addition, leachates from agriculture, industrial waste, and the municipal solid waste have also polluted surface and groundwater. Some 45 million people the world over are affected by water pollution marked by excess fluoride, arsenic, iron, or the ingress of saltwater.

What can and should be done?

It is important to realize that groundwater is not a resource that could be utilized unmindfully simply because it is available in abundant quantities. Problems and issues such as water logging, salinity, agricultural toxins, and industrial effluents, all need to be properly looked into. Other than legislation and checks to conserve and improve the quality of groundwater, society itself plays a very important role. During the last decade there has been a rising awareness among the common people on the need for conservation and development of groundwater. Water use has to be integrated effectively with water regeneration, as was done in many traditional technologies.

Renovation of forest tanks in drought-prone regions will have a significant impact on wildlife and forest cover. Similarly, in some urban cities there is a need to regenerate groundwater aquifers because of the high degree of dependence on them for drinking water. Rainwater harvesting schemes have been taken up in many cities and even made compulsory in some of them. Temple tanks need to be renovated and urban wetlands protected. All these will contribute to a rise in the groundwater level and a reduction of salt water ingress. Community awareness and management of freshwater resources should be enhanced. The government should implement effective groundwater legislation and regulations through self-regulation by communities and local institutions. External support agencies should support freshwater resource management. Environmental restoration should be promoted along with household water security.

No single action whether community based, legislation, traditional water harvesting systems, or reliance on market forces will in itself alleviate the crisis in India. The effective answer to the freshwater crisis is to integrate conservation and development activities—from water extraction to water management—at the local level; making communities aware and involving them fully is therefore critical for success. All this will ultimately pave the way for combining conservation of the environment with the basic needs of people.

In India, the Water (Prevention and Control) Act was passed by the Parliament in 1974, and by 1990 all the states adopted the act. In 1986, the Environment Protection Act was passed by the Parliament. Under both these acts, the states and the central government developed environmental norms for air emissions and wastewater discharge for different types of sources.

(iii) Glaciers

Glaciers are large sheets of ice that flow down very high mountains and are often the source of snow-fed rivers. Glaciers, found mainly in places such as Antarctica and Greenland, cover almost 10% of the earth's landmass, varying in size. A glacier begins life as snowflakes. As more snow falls and gathers, the weight of the snow on top compresses the lower layers to form ice. The pile of snow and ice becomes thicker and heavier till the point when the layer of ice at the very bottom melts under the pressure. It re-freezes almost at once; but the process is repeated over and over, and as a result the entire mass begins to slide downhill slowly along the rock surface. A glacier has an enormous impact on the topography of the area, pushing aside boulders, cutting through rocks, and denting its path as it moves.

Most of the world's glaciers are found at the Poles, but they exist on all of the world's continents, even Africa. Australia doesn't have any glaciers; however, it is considered part of Oceania, which includes several Pacific island chains and the large islands of Papua New Guinea and New Zealand. Both of these islands have glaciers. Glaciers require very specific geographical and climatic conditions. Most are found in regions of high snowfall in winter and cool temperatures in summer. The amount of precipitation (whether in the form of snowfall, freezing rain, avalanches, or wind-drifted snow) is important to glacier survival. In areas such as Siberia and parts of Antarctica, the lack of adequate precipitation prevents glacier development.

Some Facts:

- At presently, about 10% of the world's land area is covered with glaciers.
- Glaciers store about 75% of the world's freshwater.
- Glacierized areas cover over 15,000,000 square kilometres.
- In the United States, glaciers cover over 75,000 square kilometres, with most of the glaciers located in Alaska.

- During the last Ice Age, glaciers covered 32% of the total land area.
- If all land ice melted, the sea level would rise approximately 70 metres worldwide.
- North America's longest glacier is the Bering Glacier in Alaska, measuring 204 kilometres long.
- The Malaspina Glacier in Alaska is the world's largest piedmont glacier, covering over 8,000 square kilometres and measuring over 193 kilometres across at its widest point.

In India, glaciers are found in the Himalayas. There are about 15,000 glaciers flowing through these mountains, covering about 17% of the mountain area and supporting numerous perennial rivers such as the Ganga, Indus, and Brahmaputra. Some well-known glaciers are described below.

1. **Dokriani:** This glacier is believed to be as old as the Himalayan Mountains. Dokriani 'Bamak' is a well-developed medium sized glacier of the Bhagirathi basin. The glacier is 5 km long and flows in a northwest direction terminating at an elevation of 3,800 m. It originates at an altitude of 13,000 feet in Uttaranchal's Garhwal district. It is one of the most studied glaciers in the world. A recent study says that it has been shrinking by a few metres every year.
2. **Gangotri:** This is located in Uttaranchal's Tehri Garhwal. One of the oldest glaciers in the Chaukhamba range, it is where the river Ganga originates. The Gangotri is not a single valley glacier, but a combination of several other glaciers that are fed to it and form a huge mass of ice. The glacier covers 28 km and terminates at Gaumukh (4,000 m).
3. **Pindari:** This is one of the most beautiful glaciers in the Kumaon hills and is known as the Jewel of Kumaon. It is located at a height of 13,000 feet above sea level between the Nanda Devi and Nandakot peaks and terminates at an altitude of 3,627 m. It is 5 km long, the snout is about 6 m high and 2.5 m wide and above the snout, the glacier extends for about 3 m. The Pinder River that emerges from the Pindari glacier drains the valley.
4. **Zemu:** The Zemu glacier is the largest and most famous glacier in the eastern Himalayas, with a length of about 26 km. It is located in northwestern Sikkim in a U-shaped valley at the base of the Kanchenjunga massif. The Teesta river has its source in this glacier. Many tributary glaciers feed the trunk glacier.
5. **Siachen:** This is the largest glacier in the world outside the Polar regions, stretching over a length of about 72 km. It lies in the extreme north-central part of Jammu and Kashmir near the border of India and Tibet, on the north-facing slopes of the Karakoram Range, and feeds the Shaksgam river that flows into Tibet. The glacier can be approached from Skardu in Ladakh. To the east of the Siachen lies a group of three glaciers known as the Rimo North, Central, and South. Between them, these glaciers have almost 700 square km of ice, which, at places, is 100 m deep. Altogether, the glaciers contain about 200 cubic kilometres of ice.

(iv) Lakes

A lake is a low-lying part of the Earth's surface in which rainwater, surface water run-off, outflow from a river, and water from other sources accumulates. There is a great variety of lakes on Earth: there are freshwater lakes and saltwater lakes, ranging in size from small fish-ponds to huge waterbodies such as Lake Superior in USA which is the world's largest freshwater lake. The Caspian Sea in Europe and the Sambhar lake in Rajasthan are examples of saltwater lakes. India has a large number of lakes spread all over the country, from Kashmir to Kerala and from Rajasthan to Assam. Among the better-known freshwater lakes in India are the Dal lake in Srinagar and the Nainital lake. Whether natural or manmade, all lakes are major sources of water. Lakes are homes to a large variety of aquatic life, with one notable exception: the Dead Sea, a saltwater lake in Israel does not have any form of life. There is too much salt in its water to sustain life.

Almost half of the world's lakes are degraded, depleted, and contaminated mainly by human activities. The main causes are inflow of domestic sewage, agricultural run-off, discharge of industrial effluents, over-fishing, introduction of exotic species and habitat degradation from population growth, expansion of cities.

As more water is withdrawn for human use and more of it is returned to lakes and rivers badly polluted there is less available to maintain vital freshwater ecosystems.

- The deepest lake is Lake Baikal in Siberia, Russia. The deepest point of the lake is known as the Olkhon Crevice; it has a depth of 5, 370 feet.
- The largest saline lake is the Caspian Sea, spread over parts of Iran. Russia, Turkmenistan, Azerbaijan, and Kazakhstan with a surface area of 1 43 560 square miles and an estimated volume of 21 500 cubic miles
- The most threatened lake in terms of size is the Aral Sea, which has shrunk due to extraction of water for irrigation.

Some Indian lakes:

1. **Keoladeo National Park:** This is an important habitat for waterfowl. It is the only wintering site in India for the central and western Asian population of Siberian cranes, a highly endangered species. The population of Siberian Cranes has decreased in the last two decades. In a programme now underway, reared Siberian crane chicks are being brought to Keoladeo from USA and Siberia. Loktak Lake. The measures undertaken for the conservation of Loktak Lake so far, include afforestation of indigenous species including fruit trees; control of silt; catchment area treatment; removal of floating lands locally called phumdis in some pockets of the lake; and generating awareness about the values and functions of the wetland. Several measures have been adopted to control the prolific growth of phumdis and water hyacinth. Weevils have been introduced for biological control of water hyacinth.
2. **Harike Lake:** This wetland sustains a large number of waterfowl. Demarcation of the wetland has been attempted, and fencing has been carried out of some ponds. Afforestation of catchment area has been carried out in some critical areas to control siltation. One of the major problems faced by this wetland is the prolific growth of water hyacinth.
3. **Wular Lake:** This wetland is the source of drinking water for Srinagar, and also acts as an absorption basin for floodwater. It is an important waterfowl habitat. However, Wular Lake is subject to heavy siltation due to loss of vegetal cover in the surrounding area.
4. **Sambhar Lake:** Located in the arid zone of Rajasthan, the Sambhar Lake is one of the largest inland saline lakes in India. This wetland is one of the most important wintering areas for flamingoes and pelicans. Salt extraction is one of the major activities in the wetland.
5. **The Bhoj Wetland in Bhopal, Madhya Pradesh:** It consists of two man-made lakes (upper and lower lakes). The upper lake was created in the 11th century by constructing an earthen dam across the Kolans river and the lower lake was constructed nearly two centuries ago.

   The wetland supports a wide variety of flora and fauna. Diverse flora provide ideal habitat for a large number of avifauna. Biotic interaction and natural selection have led to the development of a characteristic relationship between vegetation and the avifauna.
6. **Deepor Beel:** This is a permanent freshwater lake, in a former channel of the Brahmaputra river, south-west of Guwahati city. It is a large natural wetland having great biological and environmental importance and is also the only major storm water storage basin for Guwahati. The beel is endowed with rich floral and faunal diversity. In addition to a huge congregation of residential water birds, the Deepor ecosystem harbours a large number of migratory waterfowl each year. It is now threatened with large scale encroachment, brick making

factory and soil cutting within the beel ecosystem, and construction of railway line along the southern boundary of the beel.

7. **The Hussain Sagar Lake:** This is one of the largest man-made lakes in Asia, located in the heart of Hyderabad, contributing to its immense beauty. It is a sprawling artificial lake that holds water perennially. It was built during the reign of Ibrahim Qutub Shah in 1562, on a tributary of the river Musi.
8. **The Tso Morari Lake:** This is one of the largest in the Ladakh region and is almost like an inland sea. Situated at an elevation of about 4,900 m, it is about 22 km long, with a width varying from 5 to 7 km and a depth of more than 30 m at the deepest point.

The lake is probably a leftover from the Ice Age, formed by the melt waters of the ice masses left behind by the retreating glaciers. The waters from the surrounding areas drained into the lake. The huge amounts of water present in the beginning evaporated very fast in the desert-like atmosphere and what was fresh water became brackish and finally salty, unfit for human consumption.

One of the most spectacular lakes in Ladakh is the Pangong Tso, which lies across the Changla Pass from Leh. At an altitude of almost 4,500 metres, the Pangong Tso is only 8 km wide at its broadest point, but is an amazing 134km long. The Pangong is considered to be the longest lake in Ladakh. It is a saltwater lake formed in much the same way as the Tso Morari lake during the Ice Age.

(v) Urban Wetlands

Through the ages, urban wetlands have been the lifeline of most cities in India. They were preserved and looked after by the people as their main source of water supply for drinking and irrigation. These wetlands are found all over the country and are either natural or built by people. Over the years, they have gradually depleted, leading to a number of problems in urban areas such as flooding, water scarcity, and water logging.

Tanks were constructed in the catchment areas of cities that were not located near a river or a large lake. Rainwater or run-off would collect during the monsoons and be stored for the rest of the year in these tanks and lakes. The city would get its water supply from these. In southern India, the towns were built around a temple, which always had a tank at the centre. Wells were located in the tank, which served as a link with the aquifers. There were innumerable tanks dug by the rulers, most of which have been destroyed. In fact, in some areas tank building by the ruler was considered a noble deed.

In Karnataka, the Hoysalas built tanks all over the state. Some of these still exist. Festivals were organized around tanks; cleaning and de-silting the tanks were a part of the rituals. But over the years, human activity steadily destroyed these wetlands. In Bangalore, the city bus terminus has been built on the Dharmambudhi tank. A part of the Sampangi tank has given way to the Kanteerva stadium and the remaining to a housing colony. Another lake was filled up to house the city market.

- Pollution levels in the Hussain Sagar Lake in Hyderabad has been going up over the years. There are more than a thousand polluting industries located around the lake, a large number of them pharmaceutical industries. Untreated effluents from the industries and domestic sewage have deteriorated the water quality in the lake. The lake gets little chance to rejuvenate itself as a run-off happens only once a year during the monsoons, whereas pollutants flow in throughout the year.
- The Cholas were well known for the construction of tanks in the state of Tamil Nadu. But today, despite heavy rains, Chennai reels under severe water crisis along with flooding. A number of housing colonies have been built on tanks and over the years the numbers are increasing. This is the case in most large cities in the country.

- In Calcutta wetlands, both artificial and natural, have been a part of the city system. The main drainage for the city has been the Hooghly River; but the incline of the city is eastwards and this has led to the creation of the East Calcutta Wetlands. The Salt Lake is one of the largest wetlands in the city and serves as a flood cushion for the city. Calcutta has no sewage treatment plant. The sewage is treated in the sewage-fed fisheries; the process is cost-effective and lasting. A number of the fisheries purify this water through a series of sedimentation tanks and use the water for fish cultivation and agriculture.
- Indore's water needs were met by a large numbers of wells and tanks located all around the city. But over the last few years, the tanks have gradually dried up, including the largest of them, the Yashwant Sagar Lake. Vegetable and fruit cultivation along the banks of the lake has caused pollution due to run-offs from the fields. This has caused eutrophication and contamination.
- There are no rivers in the vicinity of Bhopal and the city depends entirely on lakes to meet the water needs of the city. Over the years, these lakes have become polluted due to the dumping of sewage into them. The government recently announced the Bhuj Wetland Project, a scheme to prevent the flow of liquid and solid waste into the lakes and also to de-silt them. More than 2000 ha of land around the upper lakes, which feed water to the lower ones, is to come under plantation.

Once upon a time, these urban wetlands maintained a steady supply of water for the city, recharged the groundwater, cooled the city, and prevented flood as they were a natural drainage system. Over the years, these very tanks and wetlands have been neglected, encroached upon either to accommodate more houses or to dump waste. This has caused large-scale water crisis and monsoon flooding and water logging, leading to misery and disease. These wetlands that have been protected for centuries are now being ravaged and destroyed in the name of development.

People in the urban areas must learn to live in harmony with nature in their own habitat. The wetlands, the green belts, and the flora and fauna of these areas have to be preserved for urban areas to survive and remain healthy.

Table 4.2 Physical properties of water

Property	*Unit*	*Value*
Boiling point	°C	100.00
Melting point	°C	0.00
Critical Temp.	°C	374.20
Molar heat of vaporization	KJ	40.67
Molar Entropy of vaporization	cal/deg.	26.10
Viscosity (at 20°C)	Centipoise	1.005
Surface Tension (at 20°C)	dynes/cm	73.00
Dielectric constant	–	80.54
Dipole moment	debye	1.84
Specific heat	cal/g/°C	1.00
Heat of evaporation	cal/g	540.00

4.5 WATER POLLUTION

When toxic substances enter lakes, streams, rivers, oceans, and other water bodies, they get dissolved or lie suspended in water or get deposited on the bed. This results in the pollution of water whereby the quality of the water deteriorates, affecting aquatic ecosystems. Pollutants can also seep down and affect the groundwater deposits.

Water pollution has many sources. The most polluting of them are the city sewage and industrial waste discharged into the rivers. The facilities to treat wastewater are not adequate in any city in India. Presently, only about 10% of the wastewater generated is treated; the rest is discharged as it is into our water bodies. Due to this, pollutants enter groundwater, rivers, and other water bodies. Such water, which ultimately ends up in our households, is often highly contaminated and carries disease-causing microbes. Agricultural run-off, or the water from the fields that drains into rivers, is another major water pollutant as it contains fertilizers and pesticides.

1. **Domestic Sewage** refers to wastewater that is discarded from households. Also referred to as sanitary sewage, such water contains a wide variety of dissolved and suspended impurities. It amounts to a very small fraction of the sewage by weight. But it is large by volume and contains impurities such as organic materials and plant nutrients that tend to rot. The main organic materials are food and vegetable waste, plant nutrient come from chemical soaps, washing powders, etc. Domestic sewage is also very likely to contain disease-causing microbes. Thus, disposal of domestic wastewater is a significant technical problem. Sewage generated from the urban areas in India has multiplied manifold since 1947.

   Today, many people dump their garbage into streams, lakes, rivers, and seas, thus making water bodies the final resting place of cans, bottles, plastics, and other household products. The various substances that we use for keeping our houses clean add to water pollution as they contain harmful chemicals. In the past, people mostly used soaps made from animal and vegetable fat for all types of washing. But most of today's cleaning products are synthetic detergents and come from the petrochemical industry. Most detergents and washing powders contain phosphates, which are used to soften the water among other things. These and other chemicals contained in washing powders affect the health of all forms of life in the water.

Biochemical Oxygen Demand, or BOD

The amount of organic material that can rot in the sewage is measured by the biochemical oxygen demand. BOD is the amount of oxygen required by micro-organisms to decompose the organic substances in sewage. Therefore, the more organic material there is in the sewage, the higher the BOD. It is among the most important parameters for the design and operation of sewage treatment plants. BOD levels of industrial sewage may be many times that of domestic sewage. Dissolved oxygen is an important factor that determines the quality of water in lakes and rivers. The higher the concentration of dissolved oxygen, the better the water quality. When sewage enters a lake or stream, micro-organisms begin to decompose the organic materials. Oxygen is consumed as micro-organisms use it in their metabolism. This can quickly deplete the available oxygen in the water. When the dissolved oxygen levels drop too low, many aquatic species perish. In fact, if the oxygen level drops to zero, the water will become septic. When organic compounds decompose without oxygen, it gives rise to the undesirable odours usually associated with septic or putrid conditions.

2. **Agricultural Run-off:** The use of land for agriculture and the practices followed in cultivation greatly affect the quality of groundwater. Intensive cultivation of crops causes chemicals from fertilizers (*e.g.*, nitrate) and pesticides to seep into the groundwater, a process commonly known as leaching. Routine applications of fertilizers and pesticides for agriculture and indiscriminate disposal of industrial and domestic wastes are increasingly being recognized as significant sources of water pollution. The high nitrate content in groundwater is mainly from irrigation run-off from agricultural fields where chemical fertilizers have been used indiscriminately.

3. **Industrial Effluents:** Wastewater from manufacturing or chemical processes in industries contributes to water pollution. Industrial wastewater usually contains specific and readily identifiable chemical compounds. During the last fifty years, the number of industries in India has grown rapidly. But water pollution is concentrated within a few subsectors, mainly in the form of toxic wastes and organic pollutants. Out of this a large portion can be traced to the processing of industrial chemicals and to the food products industry. In fact, a number of large and medium-sized industries in the region covered by the Ganga Action Plan do not have adequate effluent treatment facilities. Most of these defaulting industries are sugar mills, distilleries, leather processing industries, and thermal power stations. Most major industries have treatment facilities for industrial effluents. But this is not the case with small-scale industries, which cannot afford enormous investments in pollution control equipment as their profit margin is very slender.

Eutrophication

When freshwater is artificially supplemented with nutrients, it results in an abnormal increase in the growth of water plants. This is known as eutrophication. The discharge of waste from industries, agriculture, and urban communities into water bodies generally stretches the biological capacities of aquatic systems. Chemical run-off from fields also adds nutrients to water. Excess nutrients cause the water body to become choked with organic substances and organisms. When organic matter exceeds the capacity of the micro-organisms in water that break down and recycle the organic matter, it encourages rapid growth, or blooms, of algae. When they die, the remains of the algae add to the organic wastes already in the water; eventually, the water becomes deficient in oxygen. Anaerobic organisms (those that do not require oxygen to live) then attack the organic wastes, releasing gases such as methane and hydrogen sulphide, which are harmful to the oxygen-requiring (aerobic) forms of life. The result is a foul-smelling, waste-filled body of water. This has already occurred in such places as Lake Erie and the Baltic Sea, and is a growing problem in freshwater lakes all over India. Eutrophication can produce problems such as bad tastes and odours as well as green scum algae. Also the growth of rooted plants increases, which decreases the amount of oxygen in the deepest waters of the lake. It also leads to the death of all forms of life in the waler bodies.

4. **Effects of Water Pollution:** The effects of water pollution are not only devastating to people but also to animals, fish, and birds. Polluted water is unsuitable for drinking, recreation, agriculture, and industry. It diminishes the aesthetic quality of lakes and rivers. More seriously, contaminated water destroys aquatic life and reduces its reproductive ability. Eventually, it is a hazard to human health. Nobody can escape the effects of water pollution.

   The individual and the community can help minimize water pollution. By simple housekeeping and management practices the amount of waste generated can be minimized.

5. **Oil Spills:** Among the most dangerous of all water pollutants is fuel oil. Oil spills from tankers at sea or leaks from underground storage tanks on land are very difficult to control

as oil tends to spread very fast, affecting a large area in a very short time. They are a major menace to the environment as they cause severe damage to surrounding ecosystems. Oil spills at sea decrease the oxygen level in the water and cause grave harm to the creatures living in the sea. Since crude oil is lighter than water, it floats on the surface and poses the threat of swift-spreading fire. Oil spills are rather common as oil tankers can meet with accidents either in deep sea or offshore, while waiting to be loaded or unloaded. During the Gulf War, several oil reservoirs were destroyed at sea and the effect on marine life was devastating. Seabirds and fish were washed ashore, their bodies covered with oil. On land crude is transported through pipelines or tankers which can get damaged and spew out crude oil over the land, thereby contaminating it.

6. **Health Impacts of Water Pollution:** It is a well-known fact that clean water is absolutely essential for healthy living. Adequate supply of fresh and clean drinking water is a basic need for all human beings on the earth, yet it has been observed that millions of people worldwide are deprived of this.

   Freshwater resources all over the world are threatened not only by over exploitation and poor management but also by ecological degradation. The main source of freshwater pollution can be attributed to discharge of untreated waste, dumping of industrial effluent, and run-off from agricultural fields. Industrial growth, urbanization and the increasing use of synthetic organic substances have serious and adverse impacts on freshwater bodies. It is a generally accepted fact that the developed countries suffer from problems of chemical discharge into the water sources mainly groundwater, while developing countries face problems of agricultural run-off in water sources. Polluted water like chemicals in drinking water causes problem to health and leads to water-borne diseases which can be prevented by taking measures can be taken even at the household level.

7. **Groundwater and its Contamination:** Many areas of groundwater and surface water are now contaminated with heavy metals, POPs (Persistent Organic Pollutants), and nutrients that have an adverse affect on health. Water-borne diseases and water-caused health problems are mostly due to inadequate and incompetent management of water resources. Safe water for all can only be assured when access, sustainability, and equity can be guaranteed. Access can be defined as the number of people who are guaranteed safe drinking water and sufficient quantities of it. There has to be an effort to sustain it, and there has to be a fair and equal distribution of water to all segments of the society.

   Urban areas generally have a higher coverage of safe water than the rural areas. Even within an area there is variation: areas that can pay for the services have access to safe water whereas areas that cannot pay for the services have to make do with water from hand pumps and other sources. In the urban areas water gets contaminated in many different ways, some of the most common reasons being leaky water pipe joints in areas where the water pipe and sewage line pass close together. Sometimes the water gets polluted at source due to various reasons and mainly due to inflow of sewage into the source.

   Ground water can be contaminated through various sources and some of these are mentioned below.

   Pesticides: Run-off from farms, backyards, and golf courses contain pesticides such as DDT that in turn contaminate the water. Leechate from landfill sites is another major contaminating source. Its effects on the ecosystems and health are endocrine and reproductive damage in wildlife. Groundwater is susceptible to contamination, as pesticides are mobile in the soil. It is a matter of concern as these chemicals are persistent in the soil and water.

Table 4.3 Pesticides—their structures and uses

Commercial Name	*Structure*	*Uses*
I. Chlorinated hydrocarbons		
1. D.D.T.	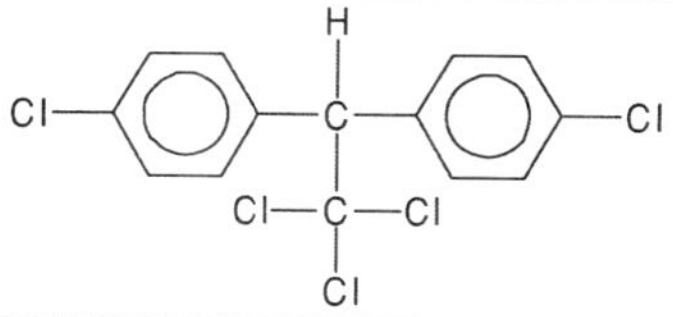	Useful against cotton soybean and peanuts pests mosquito control.
2. Aldrin	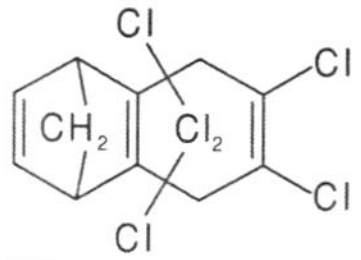	Soil insecticide for control of nuts and cotton pests.
3. Dieldrin	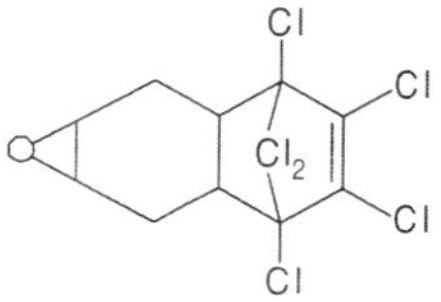	
4. Lindane	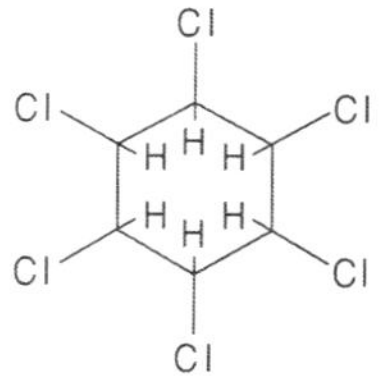 1, 2, 3, 4, 5, 6-hexa chloro cyclohexane	Control of cotton insects.
5. Chlordane	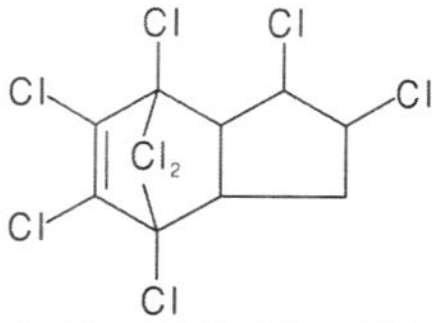	Effective against termites
6. Methoxychlor	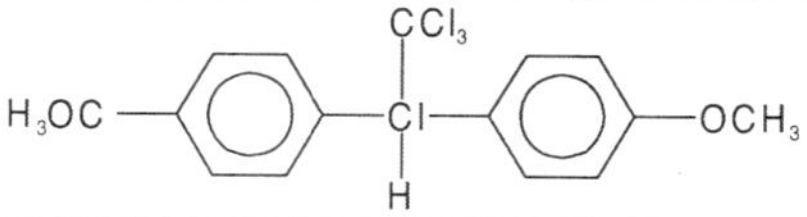	Popular DDT substitute
II. Carbamates carbaryl		
7. Baygon	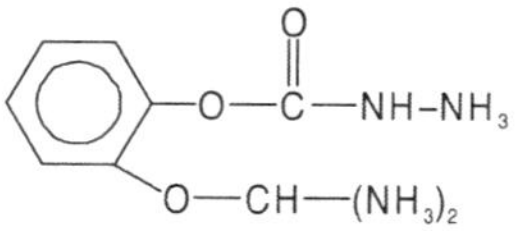 2-isopropyl-N-methyl carbamate	Control of flies mosquitoes, ants and cockroaches
8. Serin	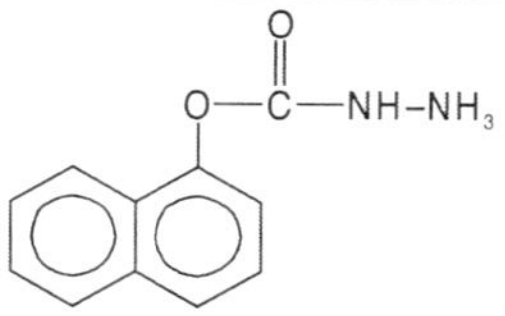 1-naphthyl-N-methyl-carbamate	Control fruits and garden insects.

Commercial Name	*Structure*	*Uses*
III. Organophosphates		
9. Parathion	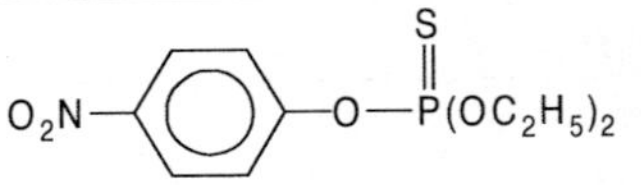 $O_2N-C_6H_4-O-P(=S)(OC_2H_5)_2$ o, o-diethyl-o-p-nitrophenyl phosphorothionate	Mosquito control and control fruit and vegetable pests.
10. Malathion	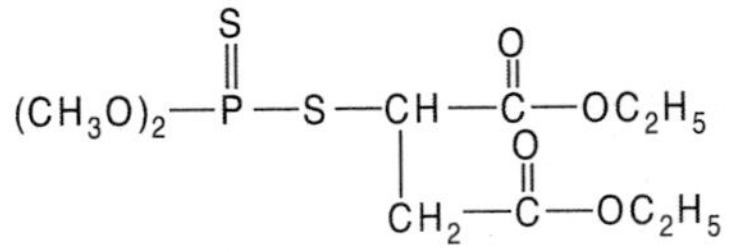 $(CH_3O)_2-P(=S)-S-CH(-C(=O)-OC_2H_5)-CH_2-C(=O)-OC_2H_5$ O, O-dimethyl-5-(1,2-dicarbo ethoxyethyl) phosphorodithioate	
IV. Chlorophenoxyacids		
11. 2,4-D	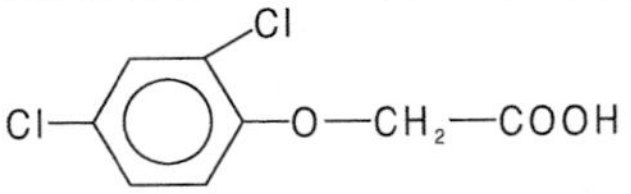 $Cl-C_6H_3(Cl)-O-CH_2-COOH$ 2,4-dichlorophenoxy acetic acid	Control of board leave weed. It is a herbicide.

Sewage: Untreated or inadequately treated municipal sewage is a major source of groundwater and surface water pollution in the developing countries. The organic material that is discharged with municipal waste into the watercourses uses substantial oxygen for biological degradation thereby upsetting the ecological balance of rivers and lakes. Sewage also carries microbial pathogens that are the cause of the spread of disease.

Nutrients: Domestic wastewater, agricultural run-off, and industrial effluents contain phosphorus and nitrogen, fertilizer run-off, manure from livestock operations, which increase the level of nutrients in water bodies and can cause eutrophication in the lakes and rivers and continue on to the coastal areas. The nitrates come mainly from the fertilizer that is added to the fields. Excessive use of fertilizers cause nitrate contamination of groundwater, with the result that nitrate levels in drinking water is far above the safety levels recommended. Good agricultural practices can help in reducing the amount of nitrates in the soil and thereby lower its content in the water.

Synthetic organics: Many of the 100 000 synthetic compounds in use today are found in the aquatic environment and accumulate in the food chain. POPs or Persistent Organic Pollutants, represent the most harmful element for the ecosystem and for human health, for example, industrial chemicals and agricultural pesticides. These chemicals can accumulate in fish and cause serious damage to human health. Where pesticides are used on a large-scale, groundwater gets contaminated and this leads to the chemical contamination of drinking water.

Acidification: Acidification of surface water, mainly lakes and reservoirs, is one of the major environmental impacts of transport over long distance of air pollutants such as sulphur dioxide from power plants, other heavy industry such as steel plants, and motor vehicles. This problem is more severe in the US and in parts of Europe.

8. **Chemicals in Drinking Water:** Chemicals in water can be both naturally occurring or introduced by human interference and can have serious health effects.

   (*i*) ***Fluoride:*** Fluoride in the water is essential for protection against dental caries and weakening of the bones, but higher levels can have an adverse effect on health. In India, high fluoride content is found naturally in the waters in Rajasthan.

(*ii*) ***Arsenic:*** Arsenic occurs naturally or is possibly aggrevated by over powering aquifers and by phosphorus from fertilizers. High concentrations of arsenic in water can have an adverse effect on health. A few years back, high concentrations of this element was found in drinking water in six districts in West Bengal. A majority of people in the area was found suffering from arsenic skin lesions. It was felt that arsenic contamination in the groundwater was due to natural causes. The government is trying to provide an alternative drinking water source and a method through which the arsenic content from water can be removed.

(*iii*) ***Lead:*** Pipes, fittings, solder, and the service connections of some household plumbing systems contain lead that contaminates the drinking water source.

(*iv*) ***Recreational use of water:*** Untreated sewage, industrial effluents, and agricultural waste are often discharged into the water bodies such as the lakes, coastal areas and rivers endangering their use for recreational purposes such as swimming and canoeing.

(*v*) ***Petrochemicals:*** Petrochemicals contaminate the groundwater from underground petroleum storage tanks.

(*vi*) ***Other heavy metals:*** These contaminants come from mining waste and tailings, landfills, or hazardous waste dumps.

(*vii*) ***Chlorinated solvents:*** Metal and plastic effluents, fabric cleaning, electronic and aircraft manufacturing are often discharged and contaminate groundwater.

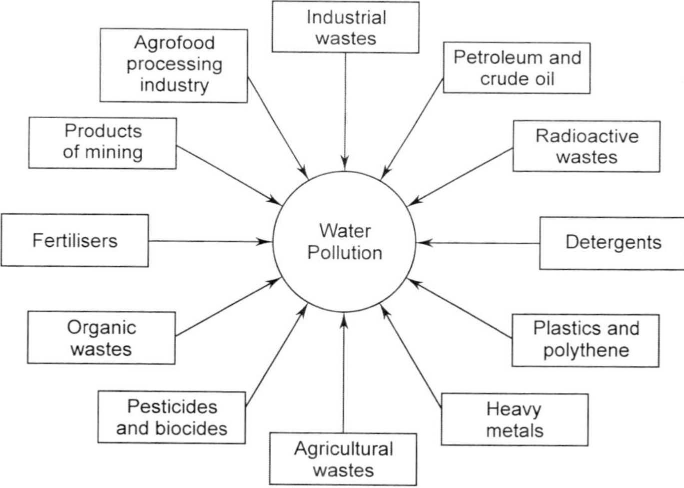

Fig. 4.2 Important Water Pollutants

Disease: Water-borne diseases are infectious diseases spread primarily through contaminated water. Though these diseases are spread either directly or through flies or filth, water is the chief medium for spread of these diseases and hence they are termed as water-borne diseases.

Cause	*Water-borne diseases*
Bacterial infections	Typhoid Cholera Paratyphoid fever Bacillary dysentery
Viral infections	Infectious Hepatitis (jaundice) Poliomyelitis
Protozoal infections	Amoebic dysentery

Most intestinal (enteric) diseases are infectious and are transmitted through faecal waste. Pathogens — which include virus, bacteria, protozoa, and parasitic worms — are disease-producing agents found in the faeces of infected persons. These diseases are more prevalent in areas with poor sanitary conditions. These pathogens travel through water sources and interfuses directly through persons handling food and water. Since these diseases are highly infectious, extreme care and hygiene should be maintained by people looking after an infected patient. Hepatitis, cholera, dysentery, and typhoid are the more common water-borne diseases that affect large populations in the tropical regions.

A large number of chemicals that either exist naturally in the land or are added due to human activity dissolve in the water, thereby contaminating it and leading to various diseases.

Pesticides: The organophosphates and the carbonates present in pesticides affect and damage the nervous system and can cause cancer. Some of the pesticides contain carcinogens that exceed recommended levels. They contain chlorides that cause reproductive and endocrinal damage.

Lead: Lead is hazardous to health as it accumulates in the body and affects the central nervous system. Children and pregnant women are most at risk.

Fluoride: Excess fluorides can cause yellowing of the teeth and damage to the spinal cord and other crippling diseases.

Nitrates: Drinking water that gets contaminated with nitrates can prove fatal especially to infants that drink formula milk as it restricts the amount of oxygen that reaches the brain causing the 'blue baby' syndrome. It is also linked to digestive tract cancers. It causes algae to bloom resulting in eutrophication in surface water.

Petrochemicals: Benzene and other petrochemicals can cause cancer even at low exposure levels.

Chlorinated solvents: These are linked to reproduction disorders and to some cancers.

Arsenic: Arsenic poisoning through water can cause liver and nervous system damage, vascular diseases and also skin cancer.

Other heavy metals: Heavy metals cause damage to the nervous system and the kidney, and other metabolic disruptions.

Salts: It makes the freshwater unusable for drinking and irrigation purposes.

Exposure to polluted water can cause diarrhoea, skin irritation, respiratory problems, and other diseases, depending on the pollutant that is in the water body. Stagnant water and other untreated water provide a habitat for the mosquito and a host of other parasites and insects that cause a large number of diseases especially in the tropical regions. Among these, malaria is undoubtedly the most widely distributed and causes most damage to human health.

Preventive measures: Water-borne epidemics and health hazards in the aquatic environment are mainly due to improper management of water resources. Proper management of water resources has become the need of the hour as this would ultimately lead to a cleaner and healthier environment. In order to prevent the spread of water-borne infectious diseases, people should take adequate precautions. The city water supply should be properly checked and necessary steps taken to disinfect it. Water pipes should be regularly checked for leaks and cracks. At home, the water should be boiled, filtered, or other methods and necessary steps taken to ensure that it is free from infection.

Minamata: Environmental Contamination with Methyl Mercury. In Minamata, Japan, inorganic mercury was used in the industrial production of acetaldehyde. It was discharged into the nearby bay as wastewater and was ingested by organisms in the bottom sediments. Fish and other creatures in the sea were soon contaminated and eventually residents of this area who consumed the fish suffered from MeHg (methyl mercury) intoxication, later known as the Minamata disease. The disease was first detected in 1956 but the mercury emissions continued until 1968. But even after the emission of mercury stopped, the bottom sediment of the polluted water contained high levels of this mercury.

Various measures were taken to deal with this disease. Environmental pollution control, which included cessation of the mercury process; industrial effluent control, environmental restoration of the bay; and restrictions on the intake of fish from the bay. This apart research and investigative activities were promoted assiduously, and compensation and help was offered by the Japanese Government to all those affected by the disease.

The Minamata disease proved a turning point, towards progress in environment protection measures. This experience clearly showed that health and environment considerations must be integrated into the process of economic and industrial development from an early stage.

9. **Corals:** Corals are small, sedentary marine animals that occur in dense colonies in warm shallow waters of oceans. Reef-building corals are scattered throughout the tropical and subtropical western Atlantic and Indo-Pacific oceans, generally between 30 degrees North and 30 degrees South latitudes.

   Coral reefs are formed by the skeleton remains of many generations of stony corals. Massive reef structures are built over thousands of years by tiny coral polyps aided by minute algae (zooxanthellae) that live in their tissues, calcifying algae, and other organisms that secrete calcium carbonate and adhesives. Reef-building corals are generally found at depths of less than 46 m, where there is sunlight and clear water through which the sunlight penetrates better. Reef-building corals, along with the algae, require warm ocean temperatures (20–28° C) and are therefore found along the eastern shores of major land masses where the water is warmer. These reefs are amongst the earth's oldest living communities of plants and animals. They vary in shape, size and colour.

Coral Ecosystem

Coral reefs are sometimes referred to as 'tropical rainforests of the deep' since they are one of the most diverse, productive, and beautiful marine ecosystems in the world. The extraordinary diversity of reefs makes them biologically important and, like rainforests, they have provided valuable scientific insights into the nature of underwater ecology. It is a diverse collection of species that interact with each other and the physical environment. The sun is the initial source of energy for this ecosystem. They are considered to be one of the most sensitive to any change. When they are environmentally stressed they lose much of the algae that gives them the colour along with other pigments. When this happens the corals appear white in colour and are referred to as bleached.

Excessive growth and accumulation of phytoplankton and seaweed would be detrimental to coral vitality and diversity, and low nutrient conditions are needed to prevent this. Diverse and abundant populations of grazing fish and invertebrates also keep the growth down.

Table 4.4 Treatment methods for removal of chemical contamination

S. No.	*Characteristic Factor*	*Treatments used*	*Limitations*
1.	pH	pH can be corrected by (*i*) Addition of acids or alkalies (usually H_2SO_4), HCl, CO_2, $Ca(OH)_2$, Na_2CO_3, KOH or NaOH. (*ii*) Neutralizing filter containing calcite ($CaCO_3$) and/or corosex (MgO) will combat low pH problems, if within the range of 5–6.	T.D.S. Value increases. Downflow filters require frequent back washing to prevent" cementing of the bed".
2.	Chloride (Cl^-)	Chloride can be removed by (*i*) Filtering the water through a bed of molecular carbon or adding small percentage of activated carbon in water, (*ii*) Ion exchange methods. (*iii*) Distillation, (*iv*) Reverse osmosis, and (*v*) Electrodialysis.	Chemical used for regeneration are hazardous. Safe practice. Requirement of heat energy is a problem. This method is expensive and require well trained operators. -do-
3.	Nitrate (NO_3)	Some methods for nitrate removal are given below: (*i*) Chemical Reduction (*ii*) Bacterial de-nitrification (*iii*) Ion-exchange-some strong base and weak base ion exchange resins are nitrate selective and can reduce nitrate concentration.	Very rigorous control of treatment conditions has made the process unfeasible. Safe practice.
4.	Fluoride (F)	Commonly used-de-fluoridation methods are given below: (*i*) Precipitation methods—They include the use of lime and alum, aluminium sulphate, poly aluminum chloride, magnesite, semicalcined dolomite or calcium chloride etc. Nalgonda technique developed by NEERI (1987) has been used extensively in India at domestic and community level. It involves the addition of lime and alum	It cannot be used where total dissolved solids are in excess of 1,500 mg/1. or for waters having total hardness of more than 250 mg/1.

	followed by rapid mixing, flocculation, sedimentation and filtration	
	(*ii*) Adsorption and ion-exchange methods—Defluoron-2, activated alumina and bone char are most commonly used.	Steep decline in sorption capacity of defuoron-2 and activated alumina with water alkalinity. Operation, maintenance of the treatment process and regeneration of defluoron-2 is difficult. Sulphate and arsenic can interfere with the treatment processes of activated alumina and bone char. Chemicals used for regeneration of activated alumina and bone char are hazardous.
	(*iii*) Adsorption methods—They include clay, activated car-bon and charcoal as adsorbents,	Charcoal has little capacity for fluoride removel.
	(*iv*) Electro-coagulation, reverse osmosis and electrodealysis.	All the three methods are expensive. They require well trained operators and pretreatment is usually necessary in these methods.
Total Dissolved Solids (T.D.S)	Desalination—Water can be desalined by processes given below (*i*) Thermal process—	
	(*a*) Distillation	Requirement of heat energy poses a big problem to distill water on a large scale. This process also has problem of scale formation.
	(*b*) Solar humidification	Solar still is expensive to build and requires a large land area. The process needs additional energy for pumping water to and from the facility.
	(*c*) Freezing	The mechanical difficulties of handling mixture of ice and water prevents the construction of large scale commercial plant. In hot climate heat leakage into the facility is a problem.
	(*ii*) Membrane process—Reverse osmosis, electrodialysis and ultrafiltration	Reverse osmosis and electrodialysis plants require high initial, operative and maintenance cost and high skill for regular maintenance. Another problem related to these is brine disposal.
	(*iii*) Ion-exchange process	Ion exchange method is economical for water with salt content above 1000 mg/1.

6.	Total Hardness (T.H)	Softening of water—water can be softened by the following processes	
		(*i*) Boiling	It reduces only calcium carbonate hardness.
		(*ii*) Lime treatment	
		(*iii*) Lime-soda process.	Re-carbonation is necessary otherwise deposition takes places in distribution pipes. Disposal of large amount of sludge poses a problem. Hardness can not be reduced to less than about 30 mg/l.
		(*iv*) Ion-exchange process	
		(*a*) Zeolite or base-exchange process	Mineral acids if present in water, destroy the zeolite bed. Water containing large quantities of Mn^{2+}, Fe^{2+} and turbidity cannot be treated efficiently by this method.
		(*b*) Demineralization process by ion-exchange resins.	The equipment used in this process is costly and more expensive chemicals are needed. If water contains turbidity then the output of the process is reduced.

Creatures Found on Coral Reefs

The coral reef ecosystem is a diverse collection of species that interact with each other and the physical environment. The numerous species residing and depending on coral reefs represent a bank containing the genetic diversity necessary for adaptation to changes in the environment. Sponges have been an important part of the coral reef ecosystem. Sea anemones provide shelter to the fish and other creatures in the reefs. Fishes play a vital role in the reefs food web, acting as both predators and prey. Bryozoans are microscopic invertebrates that form branching colonies over coral skeletons and reef debris, cementing the reef structure.

The reef is also home to a variety of worms, shrimps, crabs, lobsters, starfish, sea cucumbers, and sea urchins. Octopuses, squids, clams, scallops, marine snails, and also some species of sharks, skates, and rays live on or near the reef. Some sea turtles frequent reef areas. Green, loggerhead, and hawksbill sea turtles live in the warm waters of the Great Barrier Reef.

The Predators

The crown of thorns, a starfish, is a well- known predator of coral. Large numbers of these starfish can devastate reefs, leaving behind only the calcium carbonate skeletons. Parrotfish use chisel-like teeth to nibble on hard corals and eat the algae within the coral. Eels are one of the reefs largest predators and feed on the small fishes, octopuses, shrimps, and crabs.

Types of Reefs

There are three types of reefs: the fringing reef, the barrier reef, and the atoll.

- Fringing reefs border shorelines of continents and islands in tropical seas.
- Barrier reefs occur farther offshore. The Great Barrier Reef off northern Australia in the Indo-Pacific is the largest barrier reef in the world. This reef stretches more than 2000 km.
- Atolls are reefs that surround a central lagoon. The result is several low coral islands around a lagoon. Atolls commonly occur in the Indo-Pacific region.

Importance of Corals and Coral Reefs:

- Corals remove and recycle carbon dioxide. Excessive amounts of this gas contribute to global warming.
- Reefs shelter land from harsh ocean storms and floods by breaking the force of the waves, thereby allowing mangroves and seagrass to flourish.
- Reefs provide resources for fisheries.
- They attract millions of tourists every year.
- The coral reef is an intricate ecosystem and contains a diverse collection of organisms. Without the reef, these organisms would die.
- Coral skeletons are being used as bone substitutes in reconstructive bone surgery.
- The coral reef provides a living laboratory for both students and scientists.

Major Threats to Coral Reefs:

- Deforestation, construction, and other activities have led to silt or sand covering the corals, smothering them, and preventing light from passing through.

- Mangroves and seagrass that normally act as filters for sediment are being rapidly destroyed.
- Prawn harvesters have destroyed large areas of corals to create artificial prawn farms.
- Fishing with explosives have reduced nearby coral to lifeless rubble.
- Overfishing makes this problem even worse because the fish that would normally eat the algae have been captured and killed.
- Commercial fishing fleets often use cyanide and other poisons to stun and capture valuable reef fish. This poisons not only the fish but the coral polyps and other creatures in the area as well. In the Philippines, less than 10% of the coral reefs remain healthy due to extensive fishing.
- Trash dumped into the water can also kill coral reef life.
- Fertilizers and sewage dumped into coastal waters encourage rapid algae growth which chokes coral polyps, cutting off their supply of light and oxygen. This appears to be the case in some parts of the Great Barrier Reef.
- Careless boating, diving, fishing, and other recreational uses of coral reef areas can cause damage to coral reefs.
- Coral reefs are also threatened by global warming. There has been an unprecedented increase in the number of coral bleaching events during the past 2 decades (which have had some of the warmest years in history). When ocean temperatures get too high, coral lose the symbiotic algae inside them, causing them to turn white, or 'bleach', and eventually die.

While coral reefs are sensitive to environmental changes, they appear to be able to recover effectively from physical disturbance or temporary pollution events provided the water quality is generally high. For example, the corals in Kaneohe Bay, Hawaii, for the most part recovered from severe overgrowth of algae after sewage inputs were diverted away from the Bay.

Conservation Measures

The world has woken up to the magnitude of the problem and has taken steps to halt this degradation of one of the richest ecosystems.

The Convention on International Trade in Endangered Species of Wild Fauna and Flora, or CITES, has classified many corals as threatened species.

The establishment of marine sanctuaries or preserves may help ensure the availability of this ecosystem in the years to come. The Great Barrier Reef Marine Park was established in 1975 and is the largest project undertaken to preserve the coral reef in the world. Initiatives for the preservation of coral reefs in India have been undertaken on a large scale in the Wandur Marine National Park in the Andaman Nicobar islands. People on their part can play a role in preventing this depletion. Corals should not be collected, either alive or dead. All waste should be treated before it is released into the sea and no waste is to be dumped directly into the water. Once this awareness is built in the people the well being of these reefs can be ensured

Table 4.5 Analytical methods for chemical parameters in natural water supplies

Chemical Parameters	*Effect on water*	*Methods of Analysis*
Acidity	Adverse effect on aquatic life	Titration
Alkalinity	Algal productivity	Titration
Ammonia	Productivity, pollutant	Spectrophotometry; AAS (Atomic Abs. Spectrophotometry)
Beryllium	Toxic pollutant	AAS. Fluorimetry
Boron	Toxic pollutant	AAS
Calcium	Hardness (Hard water)	AAS; Titration
Chloride	Saline water contamination	Titration; Potentiometry
Chlorine	Water treatment	Spectrophotometry
Chromium	Toxic pollutant (Cr VI)	Spectrophotometry; Potentiometry
Copper	Plant growth	AAS
Cyanide	Toxic pollutant	Spectrophotometry; Potentiometry
Fluoride	Toxic at high-levels; Water treatment	Spectrophotometry; Potentiometry
Hardness	Water quality	Titration
Iron	Water quality; Staining	Spectrophotometry AAS
Lead	Toxic pollutant	AAS; Voltmeter
Magnesium	Hardness	Titration; AAS
Mercury	Toxic pollutant	Flameless atomic absorption
Nitrate	Productivity	Spectrophotometry;
Nitrite	Toxicity	Potentiometry
Nitrogen (Organic-bound)	Organic pollution Indicator	Spectrophotometry; titration
Oils and Grease	Industrial pollution	Gravimetry
Oils	Water quality pollution indicator	Microbiological titration
Oxygen demand (Chemical) COD	Water quality pollution indicator (Organic lead)	Titration procedure by oxidation
Pesticides	Water pollution	Gas chromatography
pH	Water quality	Potentiometry
Phenols	Toxic pollutant	Distillation followed by Spectrophotometry
Phosphate	Productivity, eutrophication	Spectrophotometry
Selenium	Toxic pollutant	Spectrophotometry
Selenium	Toxic pollutant	Spectrophotometry; neutron activation
Silica	Water quality	Spectrophotometry gravimetry
Silver	Water pollution	AAS

Strontium	Water quality	AAS; Flame photometry
Sulphate	Water quality	Gravimetry; Turbidimetry
Sulphide	Water pollution	Titration, Spectrophotometry; Potentiometry
Surfactants	Water pollution	Spectrophotometry
Tannin and Ligin	Water pollution	Spectrophotometry
Vanadium	Water pollution	Spectrophotometry; AAS
Zinc	Water pollution	AAS

4.6 WATER CONSERVATION

Our ancient religious texts and epics give a good insight into the water storage and conservation systems that prevailed in those days.

Over the years rising populations, growing industrialization, and expanding agriculture have pushed up the demand for water. Efforts have been made to collect water by building dams and reservoirs and digging wells; some countries have also tried to recycle and desalinate (remove salts) water. Water conservation has become the need of the day. The idea of groundwater recharging by harvesting rainwater is gaining importance in many cities.

In the forests, water seeps gently into the ground as vegetation breaks the fall. This groundwater in turn feeds wells, lakes, and rivers. Protecting forests means protecting water 'catchments'. In ancient India, people believed that forests were the 'mothers' of rivers and worshipped the sources of these water bodies.

Some Ancient Indian Methods of Water Conservation

The Indus Valley Civilization, that flourished along the banks of the river Indus and other parts of western and northern India about 5, 000 years ago, had one of the most sophisticated urban water supply and sewage systems in the world. The fact that the people were well acquainted with hygiene can be seen from the covered drains running beneath the streets of the ruins at both Mohenjodaro and Harappa. Another very good example is the well-planned city of Dholavira, on Khadir Bet, a low plateau in the Rann in Gujarat. One of the oldest water harvesting systems is found about 130 km from Pune along Naneghat in the Western Ghats. A large number of tanks were cut in the rocks to provide drinking water to tradesmen who used to travel along this ancient trade route. Each fort in the area had its own water harvesting and storage system in the form of rock-cut cisterns, ponds, tanks and wells that are still in use today. A large number of forts like Raigad had tanks that supplied water.

- In ancient times, houses in parts of western Rajasthan were built so that each had a rooftop water harvesting system. Rainwater from these rooftops was directed into underground tanks. This system can be seen even today in all the forts, palaces and houses of the region.
- Underground baked earthen pipes and tunnels to maintain the flow of water and to transport it to distant places, are still functional at Burhanpur in Madhya Pradesh, Golkunda and Bijapur in Karnataka, and Aurangabad in Maharashtra.

Rainwater Harvesting

In urban areas, the construction of houses, footpaths and roads has left little exposed earth for water to soak in. In parts of the rural areas of India, floodwater quickly flows to the rivers, which then dry up soon after the rains stop. If this water can be held back, it can seep into the ground and recharge the groundwater supply.

This has become a very popular method of conserving water especially in the urban areas. Rainwater harvesting essentially means collecting rainwater on the roofs of building and storing it underground for later use. Not only does this recharging arrest groundwater depletion, it also raises the declining water table and can help augment water supply. Rainwater harvesting and artificial recharging are becoming very important issues. It is essential to stop the decline in groundwater levels, arrest sea-water ingress, *i.e.*, prevent sea-water from moving landward, and conserve surface water run-off during the rainy season.

Town planners and civic authority in many cities in India are introducing bylaws making rainwater harvesting compulsory in all new structures. No water or sewage connection would be given if a new building did not have provisions for rainwater harvesting. Such rules should also be implemented in all the other cities to ensure a rise in the groundwater level.

Realizing the importance of recharging groundwater, the CGWB (Central Ground Water Board) is taking steps to encourage it through rainwater harvesting in the capital and elsewhere. A number of government buildings have been asked to go in for water harvesting in Delhi and other cities of India.

All you need for a water harvesting system is rain, and a place to collect it! Typically, rain is collected on rooftops and other surfaces, and the water is carried down to where it can be used immediately or stored. You can direct water run-off from this surface to plants, trees or lawns or even to the aquifer.

Some of the benefits of rainwater harvesting are as follows:

- Increases water availability
- Checks the declining water table
- Is environmentally friendly
- Improves the quality of groundwater through the dilution of fluoride, nitrate, and salinity
- Prevents soil erosion and flooding especially in urban areas

Rainwater Harvesting: a success story

Once Cherrapunji was famous because it received the largest volume of rainfall in the world it still does but ironically, experiences acute water shortages. This is mainly the result of extensive deforestation and because proper methods of conserving rainwater are not used. There has been extensive soil erosion and often, despite the heavy rainfall and its location in the green hills of Meghalaya, one can see stretches of hillside devoid of trees and greenery. People have to walk long distances to collect water. In the area surrounding the River Ruparel in Rajasthan, the story is different — this is an example of proper water conservation. The site does not receive even half the rainfall received by Cherrapunji, but proper management and conservation have meant that more water is available than in Cherrapunji. The water level in the river began declining due to extensive deforestation and agricultural activities along the banks and, by the 1980s, a drought-like situation began to spread. Under the guidance of some NGOs (non-government organizations), the women living in the area were encouraged to take the initiative in building johads (round ponds) and dams to hold back rainwater. Gradually, water began coming back as proper methods of conserving and harvesting rainwater were followed. The revival of the river has transformed the ecology of the place and the lives of the people living along its banks. Their relationship with their natural environment has been strengthened. It has proved that humankind is not the master of the environment, but a part of it. If human beings put in an effort, the damage caused by us can be undone.

Agriculture

Conservation of water in the agricultural sector is essential since water is necessary for the growth of plants and crops. A depleting water table and a rise in salinity due to overuse of chemical fertilizers and pesticides has made matters serious. Various methods of water harvesting and recharging have been and are being applied all over the world to tackle the problem. In areas where rainfall is low and water is scarce, the local people have used simple techniques that are suited to their region and reduce the demand for water.

- In India's arid and semi-arid areas, the 'tank' system is traditionally the backbone of agricultural production. Tanks are constructed either by bunding or by excavating the ground and collecting rainwater.
- Rajasthan, located in the Great Indian Desert, receives hardly any rainfall, but people have adapted to the harsh conditions by collecting whatever rain falls. Large bunds to create reservoirs known as khadin, dams called johads, tanks, and other methods were applied to check water flow and accumulate run-off. At the end of the monsoon season, water from these structures was used to cultivate crops. Similar systems were developed in other parts of the country. These are known by various local names ¾ jal talais in Uttar Pradesh, the haveli system in Madhya Pradesh, ahar in Bihar, and so on.

Reducing Water Demand

Simple techniques can be used to reduce the demand for water. The underlying principle is that only part of the rainfall or irrigation water is taken up by plants, the rest percolates into the deep groundwater, or is lost by evaporation from the surface. Therefore, by improving the efficiency of water use, and by reducing its loss due to evaporation, we can reduce water demand. There are numerous methods to reduce such losses and to improve soil moisture. Some of them are listed below.

- Mulching, *i.e.*, the application of organic or inorganic material such as plant debris, compost, etc., slows down the surface run-off, improves the soil moisture, reduces evaporation losses and improves soil fertility.
- Soil covered by crops, slows down run-off and minimizes evaporation losses. Hence, fields should not be left bare for long periods of time.
- Ploughing helps to move the soil around. As a consequence it retains more water thereby reducing evaporation.
- Shelter belts of trees and bushes along the edge of agricultural fields slow down the wind speed and reduce evaporation and erosion.

- Planting of trees, grass, and bushes breaks the force of rain and helps rainwater penetrate the soil.
- Fog and dew contain substantial amounts of water that can be used directly by adapted plant species. Artificial surfaces such as netting-surfaced traps or polyethylene sheets can be exposed to fog and dew. The resulting water can be used for crops.
- Contour farming is adopted in hilly areas and in lowland areas for paddy fields. Farmers recognize the efficiency of contour-based systems for conserving soil and water.
- Salt-resistant varieties of crops have also been developed recently. Because these grow in saline areas, overall agricultural productivity is increased without making additional demands on freshwater sources. Thus, this is a good water conservation strategy.
- Transfer of water from surplus areas to deficit areas by inter-linking water systems through canals, etc.
- Desalination technologies such as distillation, electro-dialysis and reverse osmosis are available.
- Use of efficient watering systems such as drip irrigation and sprinklers will reduce the water consumption by plants.

Water Conservation

The most important step in the direction of finding solutions to issues of water and environmental conservation is to change people's attitudes and habits 3/4 this includes each one of us. Conserve water because it is the right thing to do. We can follow some of the simple things that have been listed below and contribute to water conservation.

- Try to do one thing each day that will result in saving water. Don't worry if the savings are minimal 3/4 evely drop counts! You can make a difference.
- Remember to use only the amount you actually need.
- Form a group of water—conscious people and encourage your friends and neighbours to be part of this group. Promote water conservation in community newsletters and on bulletin boards. Encourage your friends, neighbours and co-workers to also contribute.
- Encourage your family to keep looking for new ways to conserve water in and around your home.
- Make sure that your home is leak-free. Many homes have leaking pipes that go unnoticed.
- Do not leave the tap running while you are brushing your teeth or soaping your face.
- See that there are no leaks in the toilet tank. You can check this by adding colour to the tank. If there is a leak, colour will appear in the toilet bowl within 30 minutes. (Flush as soon as the test is done, since food colouring may stain the tank.)
- Avoid flushing the toilet unnecessarily. Put a brick or any other device that occupies space to cut down on the amount of water needed for each flush.
- When washing the car, use water from a bucket and not a hosepipe.
- Do not throw away water that has been used for washing vegetables, rice or dals3/4use it to water plants or to clean the floors, etc.
- You can store water in a variety of ways. A simple method is to place a drum on a raised platform directly under the rainwater collection source. You can also collect water in a bucket during the rainy season.

What we can do to conserve water?

1. Use only as much water as you require.
2. Close the taps well after use.
3. While brushing your teeth do not leave the tap running, open it only when you require it.
4. See that there are no leaking taps. Get a plumber to come in and seal all leaks.
5. Use a washing machine that does not consume too much water.
6. Do not leave the taps running while washing dishes and clothes.
7. Install small shower heads to reduce the flow of the water.
8. Water in which the vegetables and fruits have been washed can be used to water the flowers and ornamental potted plants.
9. At the end of the day if you have water left in your water bottle do not throw it away, pour it over some plants.

4.7 WATER QUALITY PARAMETERS

Here we provide general information on water quality parameters:

1. Temperature
2. Specific Conductance
3. pH
4. Dissolved Oxygen
5. Total Organic Carbon
6. Hardness
7. Alkalinity
8. Nitrate and Nitrite
9. Ammonia
10. Phosphate
11. Fecal Coliform Bacteria
12. Nitrites
13. Turbidity
14. Total Suspended Solids
15. Total Dissolved Solids

1. Temperature

Temperature of water is a very important factor for aquatic life. It controls the rate of metabolic and reproductive activities, and determines which fish species can survive. Temperature also affects the concentration of dissolved oxygen and can influence the activity of bacteria and toxic chemicals in water.

Factors Affecting Temperature:

(*i*) ***Riparian Vegetation:*** Riparian vegetation, or trees and plants growing along the banks of a river or creek, provide shade, preventing the sun from heating up the water. If the sun shines directly on water, the water can warm up very quickly, and to very high temperatures.

(*ii*) ***Flow Rate:*** During dry seasons, there is less water in a river or creek, and it flows more slowly. This allows the water to warm up more quickly, and to warmer temperatures.

(*iii*) ***Paved Surfaces:*** As business and housing developments are built, previously open lands are covered with buildings and pavement. This covered area is called "impermeable surface." Less rainwater is soaked into the ground, and more of it runs over land into streams during storms. This run-off also moves faster into the stream than would natural run-off because it travels through straight concrete or plastic storm drain pipes. This increased volume and velocity of run-off scours the stream channel and widens it. During dry weather between storms, the channels have a very shallow flow. These wide, shallow streams heat up much more quickly than do more natural narrow, deep ones.

Additionally, black surfaces, like many streets and parking lots, absorb heat, and rainwater moving over these surfaces during storm events becomes warmer.

(*iv*) ***Industrial Discharge:*** Some industries use water as a coolant during processing. This water is sometimes discharged to a stream or lake. When this water is discharged to a creek, it is much warmer than the water in the creek, and the temperature of the creek becomes higher. This phenomenon is called thermal pollution. This can cause the period of ice cover on the water to be shortened, and can increase the metabolic rate of plants and animals, producing an increase in oxygen demand.

(*v*) ***Sewage Outflow:*** Water discharged from treatment plants is warm, because of the wastewater entering the plant (coming from our homes and businesses) is warm. This is another form of thermal pollution.

Temperature preferences among aquatic species vary widely, but all species tolerate slow, seasonal changes better than rapid changes.

Other Information about Temperature

Respiration of organisms is temperature-related; respiration rates can increase by 10% or more per 1 °C temperature rise. Therefore, increased temperature not only reduces oxygen availability, but also increases oxygen demand, which can add to physiological stress of organisms (Giller and Malmqvist, 1998).

2. Specific Conductance (SC)

Specific Conductance (SC) is a measure of how well water can conduct an electrical current. Conductivity increases with increasing amount and mobility of ions. These ions, which come from the breakdown of compounds, conduct electricity because they are negatively or positively charged when dissolved in water. Therefore, SC is an indirect measure of the presence of dissolved solids such as chloride, nitrate, sulfate, phosphate, sodium, magnesium, calcium, and iron, and can be used as an indicator of water pollution.

Measurement of SC

Specific conductance measures how well water can conduct an electrical current for a unit length and unit cross-section at a certain temperature. More specifically, it is defined as the reciprocal (opposite) of the resistance in ohms measured between opposite faces of a centimetre cube of an aqueous solution at a specified temperature" (Hem, 1985). That is, conductance = 1 / resistance

Specific conductance is measured using a sensor which measures resistance. Resistance means how well something can *resist* an electrical current, and is reported in ohms. The unit of conductance was originally ohm spelled backwards - "mho." More recently, however, the name "siemen" has been used to match the term used by the International System of Units. So both "mho" and "siemen" are sometimes seen in water quality reports. One siemen is equal to one mho. Because SC in natural waters is usually much less than 1 siemen/cm, SC is usually reported in microsiemens (1/1, 000, 000 siemen) per centimetre, or μS/cm. SC is affected by temperature, so for consistency SC values are converted to what they would be at room temperature (25 °C).

Factors Affecting SC:

(*i*) ***Geology and Soil in the Watershed:*** Some rock and soil release ions very easily when water flows over them; for example, if acidic water flows over rocks containing calcite ($CaCO_3$), such as calcareous shales, calcium (Ca^{2+}) and carbonate (CO_3^{2-}) ions will dissolve into the water. Therefore, SC will increase. However, some rocks, such as quartz (SiO_2), are very resistant, and don't dissolve easily when water flows over them (and even if SiO_2 does dissolve, it is not conductive). SC of waters draining areas where the geology only consists of quartz or other resistant rocks will be low (unless other factors are involved).

(*ii*) ***Acid Mine Drainage:*** Drainage from operating and abandoned mine sites can contribute iron, sulfate, copper, cadmium, arsenic, and/or other constituents if minerals containing these constituents are present and are exposed to air and water. This increase in ions will increase the SC.

(*iii*) ***Agricultural Run-off:*** Run-off from farms can contain fertilizers, which contain phosphate and nitrate. In addition, irrigation return flow can leach salts from the soil.

(*iv*) ***Road Run-off:*** Run-off from roads can contain leaked automobile fluids and salts from chemicals used in road de-icing, such as sodium chloride (NaCl) and magnesium chloride ($MgCl_2$).

3. pH

pH represents the effective concentration (activity) of hydrogen ions (H^+) in water. This concentration could be expressed in the same kind of units as other dissolved species, but H^+ concentrations are much smaller than other species in most waters. The activity of hydrogen ions can be expressed most conveniently in logarithmic units. pH is defined as the negative logarithm of the activity of H^+ ions:

$$pH = -\log [H^+]$$

where $[H^+]$ is the concentration of H^+ ions in moles per litre (a mole is a unit of measurement, equal to 6.022×10^{23} atoms). Because H^+ ions associate with water molecules to form hydronium (H_3O^+) ions, pH is often expressed in terms of the concentration of hydronium ions. In pure water at 22 °C (72 °F), H_3O^+ and hydroxyl (OH^-) ions exist in equal quantities; the concentration of each is 1.0×10^{-7} moles per litre (mol/L). Therefore, pH of pure water $= -\log (1.0 \times 10^{-7}) = -(-7.00) = 7.00$. Because pH is defined as

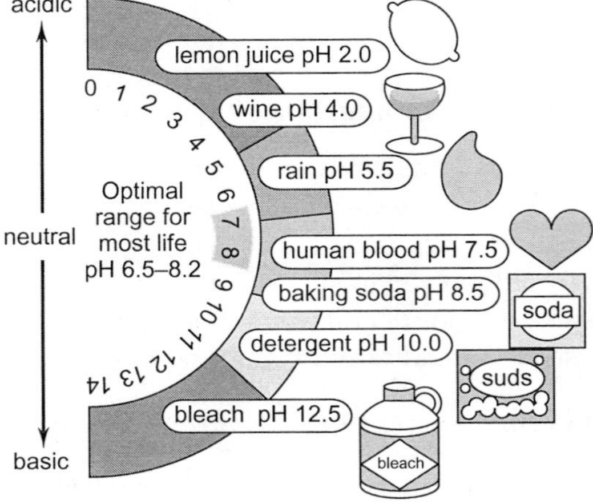

Fig. 4.3

$-\log [H^+]$, pH decreases as $[H^+]$ increases (which will happen if acid is added to the water). Since pH is a log scale based on 10, the pH changes by 1 for every power of 10 change in $[H^+]$. A solution of pH 3 has an H^+ concentration 10 times that of a solution of pH 4. The pH scale ranges from 0 to 14. However, pH values less than 0 and greater than 14 have been observed in very rare concentrated solutions.

Measurement of pH

The pH of water can be measured with a pH metre, which is an electronic device with a probe. The probe contains an acidic aqueous solution enclosed by a glass membrane that allows migration of H^+ ions. The electrical potential of the glass electrode depends on the difference in $[H^+]$ between the reference solution and the solution into which the electrode is dipped. pH can also be measured with pH paper or by adding a reagent (indicator solution) to the water sample and recording the color change.

Factors Affecting pH

(*i*) ***The Concentration of Carbon Dioxide in the Water:*** Carbon dioxide (CO_2) enters a water body from a variety of sources, including the atmosphere, run-off from land, release from bacteria in the water, and respiration by aquatic organisms. This dissolved CO_2 forms a weak acid. Natural, unpolluted rainwater can be as acidic as pH 5.6, because it absorbs CO_2 as it falls through the air. Because plants take in CO_2 during the day and release it during the night, pH levels in water can change from daytime to night. For an example of how pH typically varies over a daily cycle, select here.

(*ii*) ***Geology and Soils of the Watershed:*** Acidic and alkaline compounds can be released into water from different types of rock and soil. When calcite ($CaCO_3$) is present, carbonates (HCO_3, CO_3^{-2}) can be released, increasing the alkalinity of the water, which raises the pH. When sulfide minerals, such as pyrite, or "fool's gold," (FeS_2) are present, water and oxygen interact with the minerals to form sulfuric acid (H_2SO_4). This can significantly drop the pH of the water. Drainage water from forests and marshes is often slightly acidic, due to the presence of organic acids produced by decaying vegetation.

(*iii*) ***Drainage from Mine Sites:*** Mining for gold, silver, and other metals often involves the removal of sulfide minerals buried in the ground. When water flows over or through sulfidic waste rock or tailings exposed at a mine site, this water can become acidic from the formation of sulfuric acid. In the absence of buffering material, such as calcareous rocks, streams that receive drainage from mine sites can have low pH levels.

(*iv*) ***Air Pollution:*** Air pollution from car exhaust and power plant emissions increases the concentrations of nitrogen oxides (NO_2, NO_3) and sulfur dioxide (SO_2) in the air. These pollutants can travel far from their place of origin, and react in the atmosphere to form nitric acid (HNO_3) and sulfuric acid (H_2SO_4). These acids can affect the pH of streams by combining with moisture in the air and falling to the earth as acid rain or snow.

Very high (greater than 9.5) or very low (less than 4.5) pH values are unsuitable for most aquatic organisms. Young fish and immature stages of aquatic insects are extremely sensitive to pH levels below 5 and may die at these low pH values. High pH levels (9–14) can harm fish by denaturing cellular membranes.

Changes in pH can also affect aquatic life indirectly by altering other aspects of water chemistry. Low pH levels accelerate the release of metals from rocks or sediments in the stream. These metals can affect a fish's metabolism and the fish's ability to take water in through the gills, and can kill fish fry.

> **Other Information about pH**
>
> The term "pH" was originally derived from the French term "pouvoir hydrogene," in English, this means "hydrogen power. " The term pH is always written with a lower case *p* and an upper case H.

4. Dissolved Oxygen (DO)

Dissolved Oxygen (DO) is found in microscopic bubbles of oxygen that are mixed in the water and occur between water molecules. DO is a very important indicator of a water body's ability to support aquatic life. Fish "breathe" by absorbing dissolved oxygen through their gills. Oxygen enters the water by absorption directly from the atmosphere or by aquatic plant and algae photosynthesis. Oxygen is removed from the water by respiration and decomposition of organic matter.

Measurement of DO

Dissolved oxygen can be measured with an electrode and metre or with field test kits. The electronic metre does not measure oxygen directly; rather, it uses electrodes to measure the partial pressure of oxygen in the water, which is converted to oxygen mass weight concentration. The field test kits (such as a drop bottle, a microburet, or a digital titrator) involve adding a solution of known strength to a treated sample of water from the stream. The amount of solution required to change the color of the sample reflects the concentration of DO in the sample. The amount of oxygen dissolved in water is expressed as a concentration, in milligrams per litre (mg/1) of water.

Dissolved oxygen levels are also often reported in percent saturation. Temperature affects DO concentrations, and calculating the percent saturation will factor out the effect of temperature. The "saturation level" is the maximum concentration of dissolved oxygen that would be present in water at a specific temperature, in the absence of other factors. Scientists have determined the saturation DO level for various temperatures. Saturation levels also vary with elevation. Percent saturation is calculated by dividing the measured dissolved oxygen concentration by the saturation level and multiplying by 100.

This equation is shown as:

$$\% \text{ Saturation} = (\text{DO} / \text{Saturation Level}) \times 100.$$

Factors Affecting DO

(*i*) ***Volume and Velocity of Water Flowing in the Water Body:*** In fast-moving streams, rushing water is aerated by bubbles as it churns over rocks and falls down hundreds of tiny waterfalls. These streams, if unpolluted, are usually saturated with oxygen. In slow, stagnant waters, oxygen only enters the top layer of water, and deeper water is often low in DO concentration due to decomposition of organic matter by bacteria that live on or near the bottom of the reservoir.

Dams slow water down, and therefore can affect the DO concentration of water downstream. If water is released from the top of the reservoir, it can be warmer because the dam has slowed the water, giving it more time to warm up and lose oxygen. If dams release water from the bottom of a reservoir, this water will be cooler, but may be low in DO due to decomposition of organic matter by bacteria.

(*ii*) ***Climate/Season:*** The colder the water, the more oxygen can be dissolved in the water. Therefore, DO concentrations at one location are usually higher in the winter than in the summer. During dry seasons, water levels decrease and the flow rate of a river slows down. As the water moves slower, it mixes less with the air, and the DO concentration decreases. During rainy seasons, oxygen concentrations tend to be higher because the rain interacts with oxygen in the air as it falls.

More sunlight and warmer temperatures also bring increased activity levels in plant and animal life; depending on what organisms are present, this may increase or decrease the DO concentration.

(*iii*) ***The Type and Number of Organisms in the Water Body:*** During photosynthesis, plants release oxygen into the water. During respiration, plants remove oxygen from the water. Bacteria and fungi use oxygen as they decompose dead organic matter in the stream. The type of organisms present (plant, bacteria, fungi) affect the DO concentration in a water body. If many plants are present, the water can be supersaturated with DO during the day, as photosynthesis occurs. Concentrations of oxygen can decrease significantly during the night, due to respiration. DO concentrations are usually highest in the late afternoon, because photosynthesis has been occurring all day. For an example of how DO can vary from day to night, select here.

(*iv*) ***Altitude:*** Oxygen is more easily dissolved into water at low altitudes than at high altitudes, because of higher atmospheric pressure.

(*v*) ***Dissolved or Suspended Solids:*** Oxygen is more easily dissolved into water with low levels of dissolved or suspended solids. Waters with high amounts of salt, such as the ocean (which contains about 35 grams of salt for each 1000 grams of water) have low concentrations of DO. Freshwater lakes, streams, and tap water generally contain much less salt, so DO concentrations are higher. As the amount of salt in any body of water increases, the amount of dissolved oxygen decreases. An increase in salt concentration due to evaporation of water from an ecosystem tends to reduce the dissolved oxygen available to the ecosystem's inhabitants.

Run-off from roads and other paved surfaces can bring salts and sediments into stream water, increasing the dissolved and suspended solids in the water.

(*vi*) ***Amount of Nutrients in the Water:*** Nutrients are food for algae, and water with high amounts of nutrients can produce algae in large quantities. When these algae die, bacteria decompose them, and use up oxygen. This process is called eutrophication. DO concentrations can drop too low for fish to breathe, leading to fish kills. However, nutrients can also lead to increased plant growth. This can lead to high DO concentrations during the day as photosynthesis occurs, and low DO concentrations during the night when photosynthesis stops and plants and animals use the oxygen during respiration. For an example of how DO can vary from day to night, select here.

Nitrate and phosphate are nutrients. Nitrate is found in sewage discharge, fertilizer run-off, and leakage from septic systems. Phosphate is found in fertilizer and some detergents.

(vii) ***Organic Wastes:*** Organic wastes are the remains of any living or once-living organism. Organic wastes that can enter a body of water include leaves, grass clippings, dead plants or animals, animal droppings, and sewage. Organic waste is decomposed by bacteria; these bacteria remove dissolved oxygen from the water when they breathe. If more food (organic waste) is available for the bacteria, more bacteria will grow and use oxygen, and the DO concentration will drop.

Directly downstream from where sewage effluent is discharged to a river, DO content often decreases, because of the increase in growth rate of bacteria that consume the organic matter contained in the effluent. The degree and extent of the DO "sag" depends on the Biological Oxygen Demand (BOD) of the effluent (how much oxygen the effluent can consume) (Giller and Malmqvist, 1998).

(viii) ***Riparian Vegetation:*** Shading tends to lower average summer temperature and reduce the daily duration of higher temperature. Removing trees reduces shade on the creek, allowing the sun to warm the water. This can affect DO concentrations in different ways. As mentioned above, in general, as water temperature increases, DO drops. Also, the bare soil exposed from removing the tree can erode, increasing the amount of dissolved and suspended solids in the water. This also leads to a decrease in DO concentrations. However, direct sunlight, along with increased nutrients can increase the growth rate of aquatic plants. These plants release oxygen to the water during the day, but then remove oxygen from the water at night. This can cause DO concentrations to become very high during the day, then very low during the night. For an example of how DO can vary from day to night, select here.

(ix) ***Groundwater Inflow:*** The amount of groundwater entering a river or stream can influence oxygen levels. Groundwater usually has low concentrations of DO, but it is also often colder than stream water. Therefore, groundwater may at first lower the DO concentration, but as groundwater cools the stream or river, the ability of the water to hold oxygen improves.

Very high DO concentrations can also be harmful to aquatic life. Fish in waters containing excessive dissolved gases may suffer a condition in which bubbles of oxygen block the flow of blood through blood vessels, causing death. Abrupt changes in dissolved oxygen induce stress and subsequently make fish more susceptible to disease.

The ideal dissolved oxygen concentration for many fish is between 7 and 9 mg/1; the optimal DO for adult brown trout is 9–12 mg/1. Most fish cannot survive at concentrations below 3 mg/1 of dissolved oxygen.

5. Total Organic Carbon (TOC)

Organic matter plays a major role in aquatic systems. It affects biogeochemical processes, nutrient cycling, biological availability, chemical transport and interactions. It also has direct implications in the planning of wastewater treatment and drinking water treatment. Organic matter content is typically measured as total organic carbon and dissolved organic carbon, which are essential components of the carbon cycle. Organic matter in water consists of thousands of components, including macroscopic particles, colloids, dissolved macromolecules, and specific compounds.

Measurement of TOC

The TOC in Boulder Creek samples is analyzed at the U. S. Geological Survey on a Sievers Model 800 Carbon Analyzer. TOC concentration is not directly measured; the Analyzer measures total carbon (TC) and total inorganic carbon (TIC) and subtracts TIC from TC to obtain TOC. An oxidizer and an acid are added to the sample. The acid reacts with bicarbonate and carbonate ions present in the sample to release carbon dioxide (CO_2). The CO_2 released from bicarbonate and carbonate ions represents the TIC in the sample. The sample is then subjected to ultra-violet (UV) radiation, which reacts with the oxidant and breaks down all remaining carbon bonds in the sample to release CO_2. The CO_2 released from both the acid reaction and the UV radiation represents all the carbon (TC) released from the sample. TOC is then obtained by subtracting TIC from TC.

The TOC of a water body is affected by several factors, including:

Vegetation

Climate

Treated Sewage

6. Hardness

Hardness is measure of polyvalent cations (ions with a charge greater than +1) in water. Hardness generally represents the concentration of calcium (Ca^{2+}) and magnesium (Mg^{2+}) ions, because these are the most common polyvalent cations. Other ions, such as iron (Fe^{2+}) and manganese (Mn^{2+}), may also contribute to the hardness of water, but are generally present in much lower concentrations. Waters with high hardness values are referred to as "hard," while those with low hardness values are "soft". Affects the amount of soap that is needed to produce foam or lather. Hard water requires more soap, because the calcium and magnesium ions form complexes with soap, preventing the soap from sudsing. Hard water can also leave a film on hair, fabrics, and glassware. Hardness of the water is very important in industrial uses, because it forms scale in heat exchange equipment, boilers, and pipe lines. Some hardness is needed in plumbing systems to prevent corrosion of pipes.

Hardness mitigates metals toxicity, because Ca^{2+} and Mg^{2+} help keep fish from absorbing metals such as lead, arsenic, and cadmium into their bloodstream through their gills. The greater the hardness, the harder it is for toxic metals to be absorbed through the gills.

Measurement of Hardness

Hardness is generally measured by titration. A buffer and a color indicator are added to a volume of water. An acid (the titrant) is then added to the water, and it reacts with the Ca^{2+} and Mg^{2+} in the water. The volume of acid required to change the color of the sample reflects the Ca^{2+} and Mg^{2+} concentration of the sample. The more acid needed, the more Ca^{2+} and Mg^{2+} in the sample. Hardness is generally expressed in units of milligrams per liter (mg/1) or parts per million (ppm) of $CaCO_3$ (calcium carbonate). It can also be expressed in "grains per gallon" (gpg); one gpg equals approximately 17 mg/L.

Hardness can also be calculated from measurements of calcium and magnesium using the following formula:

Hardness, mg equivalent/L $CaCO_3$ = ([Ca, mg/l]*2. 497) + ([Mg, mg/l]*4. 116).

Factors Affecting Hardness:

(*i*) ***Geology:*** Soft waters are mainly derived from the drainage of igneous rocks, because these rocks don't weather very easily and so don't release many cations. Hard water is often derived from the drainage of calcareous (calcite-rich) sediments, because calcite ($CaCO_3$) dissolves, releasing the calcium. Calcium, magnesium, and other polyvalent cations such as iron and manganese may be added to a natural water system as it passes through soil and rock containing large amounts of these elements in mineral deposits.

(*ii*) ***Mining:*** Drainage from operating and abandoned mine sites can contribute calcium, magnesium, iron, manganese, and other ions if minerals containing these constituents are present and are exposed to air and water. This can increase the hardness of a stream.

(*iii*) ***Industrial Discharge:*** Some industrial processes may also produce significant amounts of calcium and manganese that are later discharged into streams.

(*iv*) ***Sewage Outflow:*** The effluent from Wastewater Treatment Plants (WWTPs) can add hardness to a stream. The wastewater from our houses contains calcium, magnesium, and other cations from the cleaning agents, food residue, and human waste that we put down our drains. Most of these cations are removed from the water at the WWTP before being discharged to the stream, but treatment can't eliminate everything.

Other Information about Hardness

When hardness equals alkalinity, the only cations present in significant concentrations in the water are calcium and magnesium. When hardness is greater than alkalinity, the waters may contain considerable amounts of other cations.

Calcium is an important component of plant cell walls and the shells and bones of many aquatic organisms. Low calcium levels can cause osmotic problems and affect shell or cuticle secretion in invertebrates (such as crayfish and snails). Magnesium is an essential nutrient for plants and a component of chlorophyll.

7. Alkalinity

Alkalinity is a measure of the buffering capacity of water, or the capacity of bases to neutralize acids. Measuring alkalinity is important in determining a stream's ability to neutralize acidic pollution from rainfall or wastewater. Alkalinity does not refer to pH, but instead refers to the ability of water to resist change in pH. The presence of buffering materials help neutralize acids as they are added to the water. These buffering materials are primarily the bases bicarbonate (HCO_3^-), and carbonate (CO_3^{2-}), and occasionally hydroxide (OH^-), borates, silicates, phosphates, ammonium, sulfides, and organic ligands.

Waters with low alkalinity are very susceptible to changes in pH. Waters with high alkalinity are able to resist major shifts in pH. As increasing amounts of acid are added to a water body, the pH of the water decreases, and the buffering capacity of the water is consumed. If natural buffering materials are present, pH will drop slowly to around 6; then a rapid pH drop occurs as the bicarbonate buffering capacity (CO_3^{2-} and HCO_3^-) is used up. At pH 5.5, only very weak buffering ability remains, and the pH drops further with additional acid. A solution having a pH below 4.5 contains no alkalinity, because there are no CO_3^{2-} or HCO_3^- ions left.

Alkalinity not only helps regulate the pH of a water body, but also the metal content. Bicarbonate and carbonate ions in water can remove toxic metals (such as lead, arsenic, and cadmium) by precipitating the metals out of solution.

Measurement of Alkalinity

Alkalinity is measured by titration. An acid of known strength (the titrant) is added to a volume of a treated sample of water. The volume of acid required to bring the sample to a specific pH level reflects the alkalinity of the sample. The pH end point is indicated by a color change. Alkalinity is expressed in units of milligrams per litre (mg/1) of $CaCO_3$ (calcium carbonate).

Factors Affecting Alkalinity:

(*i*) ***Geology and Soils:*** Carbonates are added to a water system if the water passes through soil and rock that contain carbonate minerals, such as calcite ($CaCO_3$). Where limestone and sedimentary rocks and carbonate-rich soils are predominant, (such as the eastern part of the Boulder Creek watershed) waters will often have high alkalinity. Where igneous rocks (such as granite) and carbonate-poor soils are predominant (such as the western part of the Boulder Creek watershed) waters will have low alkalinity.

(*ii*) ***Changes in pH:*** Because alkalinity and pH are so closely related, changes in pH can also affect alkalinity, especially in a poorly buffered stream. See the section on pH for more information on factors affecting pH.

(*iii*) ***Sewage Outflow:*** The effluent from Wastewater Treatment Plants (WWTPs) can add alkalinity to a stream. The wastewater from our houses contains carbonate and bicarbonate from the cleaning agents and food residue that we put down our drains.

Other Information about Alkalinity

Above pH 8.3, alkalinity is mostly in the form of carbonate (CO_3^{2-}); below 8.3, alkalinity is present mostly as bicarbonate (HCO_3^-).

Nitrogen is required by all organisms for the basic processes of life to make proteins, to grow, and to reproduce. Nitrogen is very common and found in many forms in the environment. Inorganic forms include nitrate (NO_3), nitrite (NO_2), ammonia (NH_3), and nitrogen gas (N_2). Organic nitrogen is found in the cells of all living things and is a component of proteins, peptides, and amino acids. Nitrogen is most abundant in Earth's environment as N_2 gas, which makes up about 78 percent of the air we breathe.

8. Ammonia—the Nitrogen Cycle

Nitrogen is recycled continually by plants and animals. This recycling of nitrogen through the environment is called the "nitrogen cycle."

Most organisms (including humans) can't use nitrogen in the gaseous form N_2 for their nutrition, so they are dependent on other organisms to convert nitrogen gas to nitrate, ammonia, or amino acids. "Fixation" is the conversion of gaseous nitrogen to ammonia or nitrate. The most common kind of fixation is "biological fixation" which is carried out by a variety of organisms, including blue-green algae, the soil bacteria *Azobacter*, and the association of legume plants and the bacteria *Rhizobium*. Additionally, nitrogen can be fixed by some inorganic processes. For example,

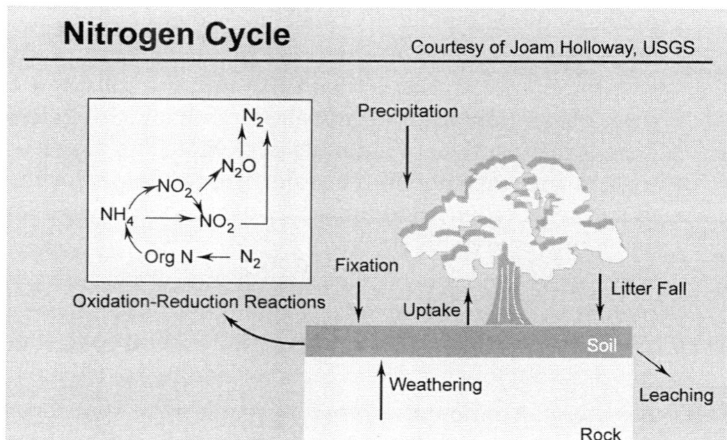

Fig. 4.4 Nitrogen Cycle

"high-energy fixation" occurs in the atmosphere as a result of lightning, cosmic radiation, and meteorite trails. Atmospheric nitrogen and oxygen combine to form nitrous oxides (NOx), which fall to the earth as nitrate.

When plants and animals die, proteins (which contain organic nitrogen) are broken down by bacteria to form ammonia (NH_3). This process is called "ammonification." Ammonia is then broken down by other bacteria (*Nitrosomonas*) to form nitrite (NO_2), which is then broken down by another type of bacteria (*Nitrobacter*) to form nitrate (NO_3). This conversion of ammonia to nitrate and nitrite is called "nitrification." Nitrates can then be used by plants in order to grow. Completing the nitrogen cycle, nitrates are reduced to gaseous nitrogen by the process of "denitrification." This process is performed by organisms such as fungi and the bacteria Pseudomonas. These organisms breakdown nitrates to obtain oxygen.

Common Forms of Nitrogen in Water

(*i*) ***Nitrate and Nitrite:*** **Nitrate** (NO_3) is highly soluble (dissolves easily) in water and is stable over a wide range of environmental conditions. It is easily transported in streams and ground-water. Nitrates feed plankton (microscopic plants and animals that live in water), aquatic plants, and algae, which are then eaten by fish. **Nitrite** (NO_2) is relatively short-lived in water because it is quickly converted to nitrate by bacteria.

Excessive concentrations of nitrate and/or nitrite can be harmful to humans and wildlife. Nitrate is of most concern for humans. Nitrate is broken down in our intestines to become nitrite. Nitrite reacts with hemoglobin in human blood to produce methemoglobin, which limits the ability of red blood cells to cany oxygen. This condition is called methemoglobin-emia or "blue baby" syndrome (because the nose and tips of ears can appear blue from lack of oxygen). It is especially serious for infants, because they lack the enzyme necessary to correct this condition. Wells contaminated by sewage or agricultural run-off are a major concern in some areas, because of the possibility of water high in nitrite/nitrates and the subsequent increased risk of blue baby disease. High nitrate and nitrite levels can also cause methemoglobinemia in livestock and other animals.

High concentrations of nitrate and/or nitrite can produce "brown blood disease" in fish. Nitrite enters the bloodstream through the gills and turns the blood a chocolate-brown color. As in humans, nitrite reacts with hemoglobin to form methemoglobin. Brown blood cannot carry sufficient amounts of oxygen, and affected fish can suffocate despite adequate oxygen concentration in the water. This accounts for the gasping behaviour often observed in fish with brown blood disease, even when oxygen levels are relatively high (Mississippi State University, 1998).

If excessive amounts of phosphorus and nitrates are added to the water, algae and aquatic plants can be produced in large quantities. When these algae die, bacteria decompose them, and use up oxygen. This process is called eutrophication. Dissolved oxygen concentrations can drop too low for fish to breathe, leading to fish kills.

(*ii*) ***Ammonia:*** **Ammonia,** another inorganic form of nitrogen, is the least stable form of nitrogen in water. Ammonia is easily transformed to nitrate in waters that contain oxygen and can be transformed to nitrogen gas in waters that are low in oxygen. Ammonia is found in water in two forms — the ammonium ion (NH_4^+/), and dissolved, unionized (no electrical charge) ammonia gas (NH_3). Total ammonia is the sum of ammonium and unionized ammonia. The dominant form depends on the pH and temperature of the water. The reaction between the two forms is shown by this equation:

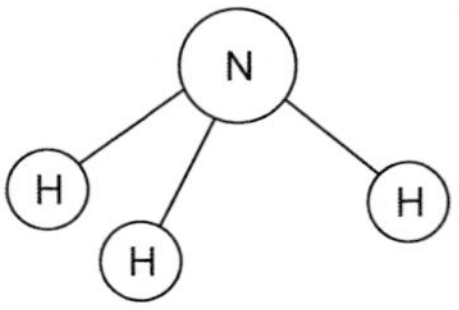

N
H H H

$$NH_3 + H_2O \rightarrow NH_4^+ + OH^-$$

The form of ammonia changes easily when pH changes. As pH increases, H^+ concentration decreases, and OH^- concentrations increase. This makes the equation above move left, increasing the amount of aqueous NH_3. When the pH is below 8.75, NH_4^+ predominates. At pH 9.24, about half of aqueous NH_3 is transformed to NH_4^+. Above pH 9.75, NH_3 predominates (Hem, 1985). Unionized ammonia (NH_3) is much more toxic to aquatic organisms than the ammonium ion (NH_4^+).

Toxic concentrations of ammonia in humans may cause loss of equilibrium, convulsions, coma, and death. Ammonia concentrations can affect hatching and growth rates of fish; changes in tissues of gills, liver, and kidneys may occur during structural development.

Measurement of Nitrogen Forms

There are many ways of measuring nitrogen forms. Total nitrogen can be determined by adding chemicals to convert all of the nitrogen forms in a sample to nitrate, and then measuring nitrate concentration. Nitrate and nitrite can be measured together or separately. Nitrate and nitrite are most often measured using a colorimetric method, which means the color of treated sample reflects the concentration of the parameter. A chemical is added to the water sample, and the. darker the color of the sample, the more nitrate and/or nitrite present. This test can be done visually, comparing the treated sample to a set of reference colors. However, it is more accurate to use an electronic colorimeter, which uses a light source and a photodetector to find the concentration based on how much light is absorbed by the sample. If nitrate and nitrite are reported separately, concentrations are given as nitrite as nitrogen ($NO_2 - N$) and nitrate as nitrogen ($NO_3 - N$). If they are reported together, concentrations are given as nitrite plus nitrate as nitrogen ($NO_2 + NO_3 - N$). Nitrate and nitrite can be measured in the field using a portable colorimeter, such as a Hach© kit.

Total ammonia (ammonium ion (NH_4^+) plus unionized ammonia gas (NH_3)) is often measured in a laboratory by titration. Ammonia and organic nitrogen compounds are separated by distillation, then an acid (the titrant) is added to a volume of the ammonia portion. The volume of acid required to change the color of the sample reflects the ammonia concentration of the sample. The more acid needed, the more ammonia in the sample. Ammonia is the least stable form of nitrogen, so it can be difficult to measure accurately. The proportion of unionized ammonia can be calculated, using formulas that contain factors for pH and temperature.

Factors Affecting Nitrate+Nitrite Concentrations:

(*i*) ***Wastewater and Septic System Effluent:*** Human waste is significant contributor of nitrogen to water. Ammonia, nitrite, and nitrate are decomposition products from urea and protein, which are in human waste. Ammonia is an ingredient in many household cleaning products and is sometimes used to remove carbonate from hard water. Therefore, these nitrogen species go down the drains in our houses and businesses, and can enter streams from Wastewater Treatment Plant (WWTP) effluent, illegal sanitary sewer connections, and poorly functioning septic systems.

Nutrients in sewage effluent have been among the primary targets of pollution-control legislation, beginning with the Clean Water Act in 1972. Organic forms of nitrogen have largely been controlled by upgrading treatment plants, and advanced treatment processes have been used to decrease ammonia discharge. However, these processes result in an increase in nitrate discharge, so the total nitrogen discharge does not change. Therefore, concerns about fish toxicity have decreased, but the potential for eutrophication has not changed (Mueller and Helsel, 1999).

(*ii*) ***Fertilizer Runoff:*** Fertilizer is a major influence on nitrogen concentrations in the environment. Commercial nitrogen fertilizers are applied either as ammonia or nitrate, but ammonia is rapidly converted to nitrate in the soil. Animal manure is also used as a nitrogen fertilizer in some areas. Organic nitrogen and urea in the manure are converted to ammonia and, ultimately, to nitrate in the soil. Nitrate that is not used by plants washes from farmlands and residential and commercial lawns into storm drains and nearby streams, or seeps into groundwater.

(*iii*) ***Animal Waste:*** A significant amount of nitrogen is released in the wastes produced by animals. This can be a serious problem in waters near cattle feedlots, hog farms, dairies, and barnyards. Ducks and geese contribute a heavy load of nitrogen if they are present in large numbers. Excretions of aquatic organisms are very rich in ammonia, a decay product of animal proteins, but the amount of nitrogen they add to waters is usually small. Through the process of nitrification, ammonia is oxidized to nitrite and then to nitrate in water.

(*iv*) ***Fossil Fuels:*** The burning of fossil fuels such as gasoline and coal in cars, trucks, and power plants produces many by products. Coal and petroleum generally contain about 1 percent nitrogen (Hem, 1985). Part of the nitrogen is converted to the gas nitric oxide (NO) during the burning of the fuel. Nitric oxide is converted by sunlight and photochemical processes in air to nitrogen oxide gases (NO and NO_2, which are commonly referred together as NO_x), which are a major component of smog. Nitrogen oxide gases are a major contributor to acid rain.

(*v*) ***Industrial Discharge:*** Many industries use nitrogen during processing. Nitrite is sometimes used as a corrosion inhibitor in industrial process water. Ammonia is used in the production of nitric acid, urea and other nitrogen compounds, and in the production of ice and in refrigerating plants. Ammonia is also used in cleaning supplies and to remove carbonate from hard water. Water from industries is usually discharged to a Wastewater treatment plant (WWTP), and may end up in a downstream water body if not completely removed in the WWTP.

9. Phosphorus

Phosphorus is a nutrient required by all organisms for the basic processes of life. Phosphorus is a natural element found in rocks, soils and organic material. Phosphorus clings tightly to soil particles and is used by plants, so its concentrations in clean waters is generally very low. However, phosphorus is used extensively in fertilizer and other chemicals, so it can be found in higher concentrations in areas of human activity. Many seemingly harmless activities added together can cause phosphorus overloads.

Phosphorus exists in water in either a particulate phase or a dissolved phase. Particulate matter includes living and dead plankton, precipitates of phosphorus, phosphorus adsorbed to particulates, and amorphous phosphorus. The dissolved phase includes inorganic phosphorus and organic phosphorus. Phosphorus in natural waters is usually found in the form of phosphates (PO_4^{-3}). Phosphates can be in inorganic form (including orthophosphates and polyphosphates), or organic form (organically bound phosphates).

(i) **Organic phosphate** is phosphate that is bound to plant or animal tissue. Organic phosphates are formed primarily by biological processes. They are contributed to sewage by body waste and food residues, and also may be formed from orthophosphates in biological treatment processes or by receiving water biota. Organic phosphates may occur as a result of the breakdown of organic pesticides which contain phosphates. They may exist in solution, as loose fragments, or in the bodies of aquatic organisms.

(ii) **Inorganic phosphate** is phosphate that is not associated with organic material. Types of inorganic phosphate include orthophosphate and polyphosphates. **Orthophosphate** is sometimes referred to as "reactive phosphorus." Orthophosphate is the most stable kind of phosphate, and is the form used by plants. Orthophosphate is produced by natural processes and is found in sewage. **Polyphosphates** (also known as metaphosphates or condensed phosphates) are strong complexing agents for some metal ions. Polyphosphates are used for treating boiler waters and in detergents. In water, polyphosphates are unstable and will eventually convert to orthophosphate. Phosphates are not toxic to people or animals unless they are present in very high levels. Digestive problems could occur from extremely high levels of phosphate. In freshwater lakes and rivers, phosphorus is often found to be the growth-limiting nutrient, because it occurs in the least amount relative to the needs of plants. If excessive amounts of phosphorus and nitrogen are added to the water, algae and aquatic plants can be produced in large quantities. When these algae die, bacteria decompose them, and use up oxygen. This process is called eutrophication. Dissolved oxygen concentrations can drop too low for fish to breathe, leading to fish kills. The loss of oxygen in the bottom waters can free phosphorus previously trapped in the sediments, further increasing the available phosphorus.

Measurement of Phosphorus

There are several forms of phosphorus which can be measured. Total phosphorus (TP) is a measure of all the forms of phosphorus, dissolved or particulate, that are found in a sample. Soluble reactive phosphorus (SRP) is a measure of orthophosphate, the filterable (soluble, inorganic) fraction of phosphorus, the form directly taken up by plant cells. Both phosphorus and orthophosphate are often measured using a colorimetric method, which means the color of treated sample reflects the concentration of the parameter. If total phosphorus is being measured, all forms of phosphorus are converted to dissolved orthophosphate with acid, persulfate, and heat. A chemical is then added to the water sample. The darker the color of the sample becomes, the more phosphorus present. This test can be done visually, comparing the treated sample to a set of reference colors. However, it is more accurate to use an electronic colorimeter, which uses a light source and a photodetector to find the concentration based on how much light is absorbed by the sample.

Factors Affecting Phosphorus Concentrations:

(i) ***Wastewater and Septic System Effluent:*** Domestic and industrial sewage are very important sources of phosphorus to surface water. Organic phosphates are formed primarily by biological processes. They are contributed to sewage by body waste and food residues. Phosphorus is essential in metabolism so is always present in animal waste. Orthophosphates and polyphosphates can be contributed by detergents, as discussed below.

(ii) ***Detergents:*** Orthophosphates and certain polyphosphates are major constituents of many commercial cleaning preparations. In the 1950s and 1960s, sodium phosphate was used often as a "builder" in households detergent to increase cleaning power. The extensive use of detergents led to major eutrophication problems, and in the 1960s efforts were made by governments, detergent manufacturers, and consumers to reduce the use of phosphates in detergents. As a result, phosphorus concentrations in many streams and lakes decreased. This was due to limits on the phosphate content of detergent, and also additional treatment used in waste water treatment plants to remove phosphorus. Many states have a ban on phosphates in detergents.

(*iii*) ***Fertilizers:*** Fertilizers generally contain phosphorus in the form of orthophosphate. Phosphate is not very mobile in soil; it tends to remain attached to solid particles rather than dissolving in water. However, if too much fertilizer is applied, the phosphates are carried into surface waters with storm runoff and also with melting snow. Soil erosion of fertilized fields and lawns can also carry a considerable amount of particulate phosphate to streams.

(*iv*) ***Animal Waste:*** Phosphorus is essential in metabolism, so is present in animal waste. Therefore, phosphate runoff can be an issue in waters near cattle feedlots, hog farms, dairies, and barnyards.

(*v*) ***Development/Paved Surfaces:*** Development can cause soil erosion, which will release phosphorus. If swamps and wetlands are drained for development, phosphorus that was buried can be exposed. During the building phase, and after everything has stabilized, phosphorus concentrations in stormwater can increase because natural filters such as trees, shrubs, and puddles have been eliminated.

(*vi*) ***Industrial Discharge:*** Polyphosphates are often added to water to prevent iron oxides or calcium carbonates from forming. If this water is released to streams or lakes, polyphosphates can enter the water body, and will convert to orthophosphate.

(*vii*) ***Phosphate Mining:*** Phosphate mining, concentrating, and processing are sources of phosphate to rivers in some areas. The most common phosphorus-containing mineral is apatite ($Ca_5F(P0_4)_3$). There are no significant sources of phosphate minerals in the Boulder Creek Watershed, so this is not a problem in our area.

(*viii*) ***Drinking Water Treatment:*** Small amounts of orthophosphates or certain polyphosphates are added to some water supplies during treatment.

(*ix*) ***Forest Fires:*** Forest fires can cause soil erosion, which will release phosphorus bound to soil particles.

(*x*) ***Synthetic Materials:*** Organophosphates are commonly used as construction materials, flame retardant and plasticizers. Reduced forms of phosphorus are present in certain synthetic organic chemicals, including some that are used in insecticides.

Eutrophication

Eutrophication is a process that results from accumulation of nutrients in lakes or other water bodies. Eutrophication is a natural process, but can be greatly accelerated by human activities that increase the rate at which nutrients enter the water.

Algae growth is limited by the available supply of phosphorus or nitrogen, so if excessive amounts of these nutrients are added to the water, algae and aquatic plants can grow in large quantities. When these algae die, they are decomposed by bacteria, which use dissolved oxygen. This process is called "eutrophication. " Dissolved oxygen concentrations can drop too low for fish to breathe, leading to fish kills. Excessive amounts of algae grow into scum on the water surface, decreasing recreational value and clogging water-intake pipes. Rapid decomposition of dense algae scums with associated organisms can give rise to foul odors.

In freshwater lakes and rivers, phosphorus is often the growth limiting nutrient, because it occurs in the least amount relative to the needs of plants. In estuaries and coastal waters, nitrogen is generally the growth limiting nutrient.

"Eutrophic" waters are characterized by high nutrient concentrations, resulting in high productivity of plant growth. Such waters are often shallow, with algal blooms and periods of oxygen deficiency. Slightly or moderately eutrophic water can support a complex web of plant and animal life. However, such waters are generally undesirable for drinking water and other needs. Waters with extreme nutrient concentrations are called "hypereutrophic." "Oligotrophic" waters are characterized by extremely low nutrient concentrations, resulting in moderate plant productivity. Oligotrophic lakes are those low in nutrient materials and consequently poor areas for the development of extensive aquatic plants and animals. Such lakes are often deep, with sandy bottoms and very limited plant growth, but with high dissolved-oxygen levels.

Some scientists have categorized trophic status according to phosphorus concentration. Lakes with phosphorus concentrations below 0.010 mg/L are classified as oligotrophic, phosphorus concentrations between 0.010 and 0.020 mg/L are indicative of mesotrophic lakes, and eutrophic lakes have phosphorus concentrations exceeding 0.020 mg/L (Muller and Helsel, 1999).

10. Total and Fecal Coliform Bacteria

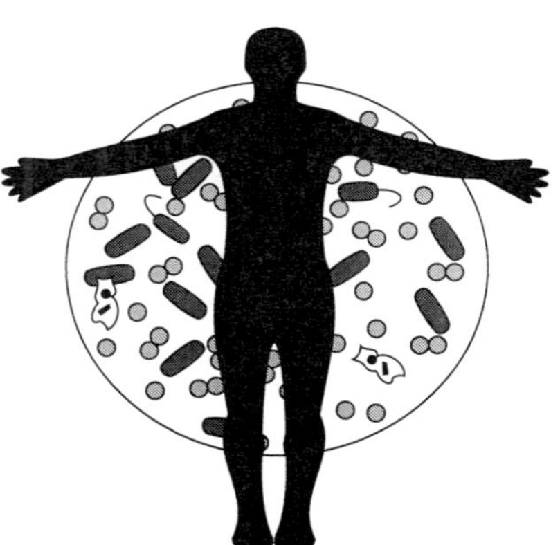

The coliform bacteria group consists of several genera of bacteria belonging to the family *enterobacteriaceae*. These mostly harmless bacteria live in soil, water, and the digestive system of animals. Fecal coliform bacteria, which belong to this group, are present in large numbers in the feces and intestinal tracts of humans and other warm-blooded animals, and can enter water bodies from human and animal waste. If a large number of fecal coliform bacteria (over 200 colonies/100 millilitres (ml) of water sample) are found in water, it is possible that pathogenic (disease or illness causing) organisms are also present in the water. Fecal coliform by themselves are usually* not pathogenic; they are indicator organisms, which means they may indicate the presence of other pathogenic bacteria. Pathogens are typically present in such small amounts it is impractical monitor them directly. Swimming in waters with high levels of fecal coliform bacteria increases the chance of developing illness (fever, nausea or stomach cramps) from pathogens entering the body through the mouth, nose, ears, or cuts in the skin. Diseases and illnesses that can be contracted in water with high fecal coliform counts include typhoid fever, hepatitis, gastroenteritis, dysentery and ear infections. Fecal coliform, like other bacteria, can usually be killed by boiling water or by treating it with chlorine. Washing thoroughly with soap after contact with contaminated water can also help prevent infections.

Fecal coliform, like other bacteria, can usually be killed by boiling water or by treating it with chlorine. Washing thoroughly with soap after contact with contaminated water can also help prevent infections.

Measurement of Fecal Coliform

Bacteria are single-celled organisms that can only be seen with the aid of a very powerful microscope. However, coliform bacteria form colonies as they multiply, which may grow large enough to be seen. By growing and counting colonies of coliform bacteria from a sample of water, it is possible to determine approximately how many bacteria were originally present. There are several ways coliform bacteria are grown and measured. Methods commonly used include the most probable number (MPN) method and the membrane filter (MF) method. In the MPN method, a "presumptive test" is performed first. A series of fermentation tubes that contain lauryl tryptose broth are inoculated with the water sample and incubated for 24 hours at 35 °C. Fermentation tubes are arranged in 3 or more rows, with 5 or 10 tubes per row, with varying dilutions of the samples in the tubes. The fermentation tube contains an inverted tube to trap gases that are produced by the coliform bacteria. After 24 hours, the fermentation tube is examined for gas production. If there is no gas production, the samples are incubated for another 24 hours and reexamined. If gas production is observed by the end of 48 hours, the presumptive test is positive; coliform bacteria are present in the sample. A "confirmed test" is then performed to determine if fecal coliform bacteria are present. For the confirmed test, some of the content of the fermentation tube is transferred with a sterile loop to a fermentation tube containing another broth. The sample is incubated in a water bath at 44.5 °C for 24 hours. Gas production in the fermentation tube after 24 hours is considered a positive reaction, indicating fecal coliform. Based on which dilutions showed positive for coliform and/or fecal coliform, a table of most probable numbers is used to estimate the coliform content of the sample. The results are reported as most probable number (MPN) of coliform per 100 ml (American Public Health Association, 1998).

The MF method is more rapid than the MPN method, but the results are not as reliable for samples that contain many non-coliform bacteria, high turbidity, and/or toxic substances such as metals or phenols. The water sample is filtered through a sterile membrane filter. The filter is transferred to a sterile petri dish and placed on a nutrient pad saturated with broth. The plates are inverted, placed in watertight plastic bags, and incubated in a water bath at 44.5 degrees C for 24 hours. Colonies produced by fecal coliform bacteria are blue, and are counted using a microscope or magnifying lens. The fecal coliform density is recorded as the number of organisms per 100 ml.

Sometimes the unit of colony producing units per 100 millilitres of water (CPU/100 ml) is used; this is equal to the number of organisms per 100 ml.

Factors Affecting Fecal Coliform:

(*i*) ***Wastewater and Septic System Effluent:*** Fecal coliform is present in human waste, so the bacteria goes down the drains in our houses and businesses, and can enter streams from illegal or leaky sanitary sewer connections, poorly functioning septic systems, and poorly functioning Wastewater Treatment Plant (WWTP) effluent.

(*ii*) ***Animal Waste:*** A significant amount of fecal coliform is released in the wastes produced by animals. This can be a serious problem in waters near cattle feedlots, hog farms, dairies, and barnyards that have poor animal keeping practices and waste is not properly contained. In urban areas, fecal coliform can be contributed to surface water by dog, cat, raccoon, and human waste when it is carried into storm drains, creeks, and lakes during storms.

(*iii*) ***Sediment Load:*** High amounts of sediment are often related to high concentrations of pathogenic bacteria. The bacteria can attach to sediment particles, escaping invertebrate predators (Murdoch and Cheo, 1996). Fast-running water can carry more sediment, so higher levels of bacteria can occur during high run-off events. Bacteria are much more abundant on soils than in water.

(*iv*) ***Temperature:*** Bacteria grow faster at higher temperatures. The growth rate slows drastically at very low temperatures.

(*v*) ***Nutrients:*** High levels of nutrients can increase the growth rate of bacteria.

Other Information About Fecal Coliform:

The fecal coliform group includes all of the rod-shaped bacteria that are non-sporeforming, *Gram-Negative*, lactose-fermenting in 24 hours at 44.5 °C, and which can grow with or without oxygen.

Fecal coliform is a type of fecal bacteria. Another type of fecal bacteria is *Fecal Streptococcus*. Fecal Streptococcus is a group of bacteria normally present in large numbers in the intestinal tracts of warm-blooded animals other than humans.

11. Turbidity

Turbidity is a measure of the cloudiness of water- the cloudier the water, the greater the turbidity. Turbidity in water is caused by suspended matter such as clay, silt, and organic matter and by plankton and other microscopic organisms that interfere with the passage of light through the water (American Public Health Association, 1998). Turbidity is closely related to total suspended solids (TSS), but also includes plankton and other organisms.

Turbidity itself is not a major health concern, but high turbidity can interfere with disinfection and provide a medium for microbial growth. It also may indicate the presence of microbes.

Measurement of Turbidity

Turbidity is a measure of how much of the light travelling through water is scattered by suspended particles. The scattering of light increases with increasing suspended solid and plankton content. Turbidity in slow moving, deep waters can be measured using a device called a Secchi disk. A Secchi disk is a black and white, 20 cm diameter disk. The disk is lowered into the water until it just disappears from sight. The depth at which the disk disappears is called the Secchi depth, and is recorded in metres.

A Secchi disk does not work in shallow, fast-moving streams. In these waters, a turbidimeter (sometimes called a nephelometer) is used. A turbidimeter measures the scattering of light, and provides a relative measure of turbidity in Nephelometric Turbidity Units (NTUs). A less expensive method of measuring turbidity is to evaluate the fuzziness of a mark at the bottom of a clear tube when a water sample is poured in the tube. Units are reported in Jackson Turbidity Units (JTUs). This method can only be used in highly turbid waters.

Factors Affecting Turbidity

Because one of the primary factors affecting turbidity is total suspended solids, the factors affecting TSS will also affect turbidity. In addition, organic matter contributes to turbidity.

(*i*) ***High Flow Rates:*** The flow rate of a water body is a primary factor influencing turbidity concentrations. Fast running water can carry more particles and larger-sized sediment. Heavy rains can pick up sand, silt, clay, and organic particles from the land and carry it to surface water. A change in flow rate also can affect turbidity; if the speed or direction of the water current increases, particulate matter from bottom sediments may be resuspended (WATERSHEDSS).

(*ii*) ***Soil Erosion:*** Soil erosion is caused by disturbance of a land surface. Soil erosion can be caused by ***Building and Road Construction, Forest Fires, Logging,*** and ***Mining***. The eroded soil particles can be carried by stormwater to surface water. This will increase the turbidity of the water body.

(*iii*) ***Urban Run-off:*** During storm events, soil particles and debris from streets and industrial, commerical, and residential areas can be washed into streams. Because of the large amount of pavement in urban areas, natural settling areas have been removed, and sediment is carried through storm drains to creeks and rivers.

(*iv*) ***Wastewater and Septic System Effluent:*** The effluent from Wastewater Treatment Plants (WWTPs) can add suspended solids and organic material to a stream. The wastewater from our houses contains food residue, human waste, and other solid material that we put down our drains. Most of the solids and organic mateial are removed from the water at the WWTP before being discharged to the stream, but treatment can't eliminate everything.

(*v*) ***Decaying Plants and Animals:*** As plants and animals present in a water body die and decay, suspended organic particles are released and can contribute to turbidity.

(*vi*) ***Bottom-Feeding Fish:*** Bottom-feeding fish (such as carp) can stir up sediments as they remove vegetation. These sediments can contribute to turbidity.

(*vii*) ***Algal Blooms:*** Algal blooms can contribute to turbidity. Algal production is enhanced when nutrients are released from bottom sediments during seasonal turnovers and changes in water current. (WATERSHEDSS).

(*viii*) ***Flooding:*** As flood waters recede, they will bring along inorganic and organic particles from the land surface, and contribute this to the stream.

12. Total Suspended Solids (TSS)

The term "total solids" refers to matter suspended or dissolved in water or wastewater, and is related to both specific conductance and turbidity. Total solids (also referred to as **total residue**) is the term used for material left in a container after evaporation and drying of a water sample. **Total Solids** includes both **total suspended solids,** the portion of total solids retained by a filter and **total dissolved solids,** the portion that passes through a filter (American Public Health Association, 1998).

Total solids can be measured by evaporating a water sample in a weighed dish, and then drying the residue in an oven at 103 to 105 °C. The increase in weight of the dish represents the total solids. Instead of total solids, laboratories often measure total suspended solids and/or total dissolved solids. Total Suspended Solids (TSS) are solids in water that can be trapped by a filter. TSS can include a wide variety of material, such as silt, decaying plant and animal matter, industrial wastes, and sewage. High concentrations of suspended solids can cause many problems for stream health and aquatic life.

High TSS can block light from reaching submerged vegetation. As the amount of light passing through the water is reduced, photosynthesis slows down. Reduced rates of photosynthesis causes less dissolved oxygen to be released into the water by plants. If light is completely blocked from bottom dwelling plants, the plants will stop producing oxygen and will die. As the plants are decomposed, bacteria will use up even more oxygen from the water. Low dissolved oxygen can lead to fish kills. High TSS can also cause an increase in surface water temperature, because the suspended particles absorb heat from sunlight. This can cause dissolved oxygen levels to fall even further (because warmer waters can hold less DO), and can harm aquatic life in many other ways, as discussed in the **temperature** section. (Mitchell and Stapp, 1992; KanCRN website)

The decrease in water clarity caused by TSS can affect the ability of fish to see and catch food. Suspended sediment can also clog fish gills, reduce growth rates, decrease resistance to disease, and prevent egg and larval development. When suspended solids settle to the bottom of a water body, they can smother the eggs of fish and aquatic insects, as well as suffocate newly hatched insect larvae. Settling sediments can fill in spaces between rocks which could have been used by aquatic organisms for homes. (Mitchell and Stapp, 1992; GREEN Hands-On Centre website) High TSS in a water body can often mean higher concentrations of bacteria, nutrients, pesticides, and metals in the water. These pollutants may attach to sediment particles on the land and be carried into water bodies with storm water. In the water, the pollutants may be released from the sediment or travel farther downstream (Federal Interagency Stream Restoration Working Group, 1998).

High TSS can cause problems for industrial use, because the solids may clog or scour pipes and machinery.

Measurement of Total Suspended Solids

To measure TSS, the water sample is filtered through a pre-weighed filter. The residue retained on the filter is dried in an oven at 103 to 105 °C until the weight of the filter no longer changes. The increase in weight of the filter represents the total suspended solids. TSS can also be measured by analyzing for total solids and subtracting total dissolved solids.

Factors Affecting Total Suspended Solids:

(*i*) ***High Flow Rates:*** The flow rate of the water body is a primary factor in TSS concentrations. Fast running water can carry more particles and larger-sized sediment. Heavy rains can pick up sand, silt, clay, and organic particles (such as leaves, soil, tire particles) from the land and carry it to surface water. A change in flow rate can also affect TSS; if the speed or direction of the water current increases, particulate matter from bottom sediments may be resuspended. (WATERSHEDSS website)

(*ii*) ***Soil Erosion:*** Soil erosion is caused by disturbance of a land surface. Soil erosion can be caused by *Building and Road Construction, Forest Fires, Logging,* and **Mining**. The eroded soil particles can be carried by stormwater to surface water. This will increase the TSS of the water body.

(*iii*) ***Urban Run-off:*** During storm events, soil particles and debris from streets and industrial, commerical, and residential areas can be washed into streams. Because of the large amount of pavement in urban areas, infiltration is decreased, velocity increases, and natural settling areas have been removed. Sediment is carried through storm drains directly to creeks and rivers.

(*iv*) ***Wastewater and Septic System Effluent:*** The effluent from Wastewater Treatment Plants (WWTPs) can add suspended solids to a stream. The wastewater from our houses contains food residue, human waste, and other solid material that we put down our drains. Most of the solids are removed from the water at the WWTP before being discharged to the stream, but treatment can't eliminate everything.

(*v*) ***Decaying Plants and Animals:*** As plants and animals decay, suspended organic particles are released and can contribute to the TSS concentration.

(*vi*) ***Bottom-Feeding Fish:*** Bottom-feeding fish (such as carp) can stir up sediments as they remove vegetation. These sediments can contribute to TSS.

13. Total Dissolved Solids (TDS)

Total Dissolved Solids (TDS) are solids in water that can pass through a filter (usually with a pore size of 0. 45 micrometers). TDS is a measure of the amount of material dissolved in water. This material can include carbonate, bicarbonate, chloride, sulfate, phosphate, nitrate, calcium, magnesium, sodium, organic ions, and other ions. A certain level of these ions in water is necessary for aquatic life. Changes in TDS concentrations can be harmful because the density of the water determines the flow of water into and out of an organism's cells. However, if TDS concentrations are too high or too low, the growth of many aquatic life can be limited, and death may occur.

Similar to TSS, high concentrations of TDS may also reduce water clarity, contribute to a decrease in photosynthesis, combine with toxic compounds and heavy metals, and lead to an increase in water temperature.

TDS is used to estimate the quality of drinking water, because it represents the amount of ions in the water. Water with high TDS often has a bad taste and/or high water hardness, and could result in a laxative effect.

Measurement of Total Dissolved Solids

To measure TDS, the water sample is filtered, and then the filtrate (the water that passes through the filter) is evaporated in a pre-weighed dish and dried in an oven at 180° C, until the weight of the dish no longer changes. The increase in weight of the dish represents the total dissolved solids, and is reported in milligrams per litre (mg/1).

The TDS concentration of a water sample can be estimated from specific conductance if a linear correlation between the two parameters is first established. Depending on the chemistry of the water, TDS (in mg/1) can be estimated by multiplying specific conductance (in micromhos/cm) by a factor between 0.55 and 0.75.

TDS can also be determined by measuring individual ions and adding them up.

Factors Affecting Total Dissolved Solids:

(*i*) ***Geology and Soil in the Watershed:*** Some rock and soil release ions very easily when water flows over them; for example, if acidic water flows over rocks containing calcite ($CaCO_3$), such as calcareous shales, calcium (Ca^{2+}) and carbonate (CO_3^{2-}) ions will dissolve into the water. Therefore, TDS will increase. However, some rocks, such as quartz-rich granite, are very resistant to dissolution, and don't dissolve easily when water flows over them. TDS of waters draining areas where the geology only consists of granite or other resistant rocks will be low (unless other factors are involved).

(*ii*) ***Urban Run–off:*** During storm events, pollutants such as salts from streets, fertilizers from lawns, and other material can be washed into streams and rivers. Because of the large amount of pavement in urban areas, natural settling areas have been removed, and dissolved solids are carried through storm drains to creeks and rivers.

(*iii*) ***Fertilizer Run–off:*** Fertilizer can dissolve in stormwater and be carried to surface water during storms, and contribute to TDS.

(*iv*) ***Wastewater and Septic System Effluent:*** The effluent from Wastewater Treatment Plants (WWTPs) adds dissolved solids to a stream. The wastewater from our houses contains both suspended and dissolved solids that we put down our drain. Most of the suspended solids are removed from the water at the WWTP before being discharged to the stream, but WWTPs only remove some of the TDS. Important components of the TDS load from WWTPs include phosphorus, nitrogen, and organic matter.

(*v*) ***Soil Erosion:*** Soil erosion is caused by disturbance of a land surface. Soil erosion can be caused by ***Building and Road.***

(*vi*) ***Construction, Forest Fires, Logging, and Mining:*** The eroded soil particles may contain soluble components that can dissolve and be carried by stormwater to surface water. This will increase the TDS of the water body.

(*vii*) ***Decaying Plants and Animals:*** As plants and animals decay, dissolved organic particles are released and can contribute to the TDS concentration.

4.8 WATER TREATMENT METHODS

Today, people are becoming more aware of the importance to drinking water, with that awareness comes the concern about the quality of water they are drinking, water that looks drinkable can contain harmful substances that may cause illness if ingested. Communities get their water from rivers and lakes (surface water), from springs and wells (groundwater) or from both surface and ground, these sources of freshwater makes up less than 1% of the total water resources on earth. About 97% of the water on earth is found in the oceans. Most of the rest is frozen in the polar ice caps and in glaciers.

Water purification is a process of removing harmful substances and odor from a raw water source. Water that looks drinkable can contain harmful substances that may cause illness if ingested. Surface water may be polluted by sewage, industrial wastes, or dead plants and animals. Groundwater may be polluted by leaks from sewage or industrial wastes. Both surface water and groundwater may also contain other solids and may have objectionable tastes and odors.

Methods of Water Treatment:

1. *Intake Filtering* — In this method raw or untreated water usually enters a water-treatment plant through a pipeline or tunnel. A series of screens prevents debris from entering the water-treatment system.
2. *Chemical Treatment* — In this method the raw water is pumped into a discharge basin and then given several types of chemical treatment. Disinfectants like chlorine, chlorine dioxide, iodine, or ozone are added in controlled amounts to kill bacteria and other organisms. When chlorine is used, activated carbon or ammonia is usually also added to remove the taste of chlorine from the water. **Fluorides** are sometimes added to the water to help prevent tooth decay. Water containing calcium or magnesium (hard water) is treated to remove these minerals, this process is called **water softening**. To prevent the corrosion of pipes, lime or caustic soda is sometimes added to acidic water.
3. *Sedimentation* — This method is often used as a first stage in modern wastewater treatment plant. This process consists of allowing matter suspended in the water to settle. Coagulant like aluminum sulfate is added to the water to speed up the sedimentation process by causing small particles to cling together, forming clumps called **floe**. Water that has been treated with coagulant is gently stirred in a mixing basin before it passes into settling basin. The stirring promotes the formation of the floe. In the settling basin, the water flows slowly and the froc gradually settles forming sludge (thick mud) at the bottom of the basin.

4. *Filtration* — After separating most floe, the water is filtered as the final step to remove remaining suspended particles, water passes on to filter beds. They are usually made up of a layer of gravel 18 to 24 inches (45 to 60 cm) thick covered with sand or crushed anthracite in a layer 24 to 30 inches (60 to 75 cm) thick. As the water drains, the filter trap remaining particles suspended in the water. Drains carry the filtered water to a reservoir. To clean the filter beds, water is passed quickly upward through the filter from below (opposite of the normal direction) with filtered water to remove embedded particles. The wash water is then removed through special drains.
5. *Aeration* — This is done either before or after filtration, the water may be aerated (exposed to the air) to remove unpleasant odors and tastes caused by decomposing organic matters, industrial wastes and some gases. It is also effective in removing **radon** (gaseous element formed by the radioactive decay of radium) from water. In the process, oxygen in the air combines with foul-smelling substances to change them into odorless compounds.

Things to Consider when Selecting a Purification Solution

The list of water purification solutions below is designed for people who would like to have protection at home or work from actual or potential contaminants in the water they drink. In order to determine which of the solutions is best for your needs you might want to consider the following:

- How much purified water you would like to use per day for drinking and cooking (a family of four will probably use 4 to 8 gallons per day).
- Which contaminants are actually in your water (and which ones might occasionally show up). The only way to know this for sure is to request a report from your water supplier or have your water tested.
- Which contaminants you are interested in removing (the information on the other pages of this site should help). The plan I adopted was to go with a solution that would provide the best protection from most of the potential contaminants (for the lowest cost and least maintenance) - even though most of the contaminants are not currently present in my tap water.
- The cost of the solution you decide upon, both the **initial cost** (which may be high for some filtration systems, ion exchange units, distillers, etc.,) and the **ongoing costs** (which might be high for distillers, bottled water, etc.). Look at the total amount you will spend over the next 10 to 40 years based on the amount of pure water you would like to be able to use for drinking, cooking, washing foods, etc.).
- The value of the product. Is the product certified to perform as advertised (or in the case of bottled water, is it certified to be free of contaminants). Does the product provide you with safe, good tasting water at a reasonable cost. Is there minimal maintenance required.

Water Treatment Methods for Domestic Use

Most of the methods of water treatment methods discussed below are Point Of Use (POU) devices. *POU methods treat water at the point where is is used* frequently at the kitchen sink. Only the water that is actually used for drinking, cooking, beverage preparation, etc., is treated. This has the advantage of economy - only a few hundred gallons of water need to be treated per year instead of many thousands if all of the water entering the home were to be treated.

Most people who use water supplied by a municipal water company only need to worry about POU treatment, because it is your water company's responsibility to provide biologically and chemically safe water that has most objectionable taste and odor causing substances removed. As you have read elsewhere on my site and in the local news, this is unfortunately not always the case. Most people using public water do not need to employ Point of Entry treatment devices or the more expensive POU devices like distillation and reverse osmosis. It is important, though, to obtain and read the annual Water Quality Report that the water company is required to make available to you. The contaminants most people using public water would be liable to experience at harmful or unacceptable levels are:

(*i*) Residual disinfectants (chlorine and/or chloramine, for example) added to keep water safe during distribution.

(*ii*) Disinfection byproducts, like the trihalomethanes.

(*iii*) Lead (as discussed elsewhere, many homes leach lead into the water from pipes and/or fixtures).

(*iv*) Brief, accidental contamination by microbes (E. coli, giardia, Cryptosporidia, etc.,) or other contaminants.

(*v*) If you live in an agricultural region, unacceptable levels of nitrates or organic compounds (even if they are below regulated levels).

Point Of Entry (POE), or whole house water treatment (where all water entering the home is treated) is indicated when the water has problems that affect all areas of the home. The most common example is a water softening ion exchange system that removes calcium and magnesium ions (and some other ions) from the water. Hard water, while quite healthy to drink, can cause scale buildup in pipes and on fixtures, interfere with the effectiveness of soap, and shorten the life of appliances, like dish washers and hot water heaters. Other POE water treatment systems are designed to remove iron and manganese, adjust pH levels, add chlorine or other disinfectant, etc. *People using water from a private well, spring, or surface source are most likely to require POE treatment.* Check the Water Treatment Table for some POE methods of treatment (particularly the notes toward the bottom). A high quality POU treatment system should also be considered if you have a well or use water from a spring or surface source, because each type of POE treatment generally removes only one contaminant type (the one or two POE treatment methods someone might use would only remove the most obvious, annoying, or dangerous contaminant(*s*)). Other contaminants that might be present like lead (from the home's plumbing and pipes) and chlorine residue (if the water is disinfected) would best be removed by a POU device.

Treatment Methods

1. Boiling

In an emergency, boiling is the best way to purify water that is unsafe because of the presence of protozoan parasites or bacteria.

If the water is cloudy, it should be filtered before boiling. Filters designed for use when camping, coffee filters, towels (paper or cotton), cheesecloth, or a cotton plug in a funnel are effective ways to filter cloudy water.

Place the water in a clean container and bring it to a full boil and continue boiling for at least 3 minutes (covering the container will help reduce evaporation). If you are more than 5,000 feet above sea level, you must increase the boiling time to at least 5 minutes (plus about a minute for every additional 1,000 feet). Boiled water should be kept covered while cooling.

The advantages of Boiling Water include:

(*i*) Pathogens that might be lurking in your water will be killed if the water is boiled long enough.

(*ii*) Boiling will also drive out some of the Volatile Organic Compounds (VOCs) that might also be in the water. This method works well to make water that is contaminated with living organisms safe to drink, but because of the inconvenience, *boiling is not routinely used to purify drinking water except in emergencies.*

The disadvantages of Boiling Water include:

- Boiling should not be used when toxic metals, chemicals (lead, mercury, asbestos, pesticides, solvents, etc.), or nitrates have contaminated the water.
- Boiling may concentrate any harmful contaminants that do not vaporize as the relatively pure water vapor boils off.
- Energy is needed to boil the water

2. Distillation

In many ways, distillation is the reverse of boiling. To remove impurities from water by distillation, the water is usually boiled in a chamber causing water to vaporize, and the pure (or mostly pure) steam leaves the non volatile contaminants behind. The steam moves to a different part of the unit and is cooled until it condenses back into liquid water. The resulting distillate drips into a storage container.

Salts, sediment, metals — anything that won't boil or evaporate — remain in the distiller and must be removed. Volatile organic compounds (VOCs) are a good example of a contaminant that will evaporate and condense with the water vapor. A vapor trap, carbon filter, or other device must be used along with a distiller to ensure the more complete removal of contaminants.

The advantages of Distillation include:

1. A good distillation unit produces very pure water. This is one of the few practical ways to remove nitrates, chloride, and other salts that carbon filtration cannot remove.
2. Distillation also removes pathogens in the water, mostly by killing and leaving them behind when the water vapor evaporates. If the water is boiled, or heated just short of boiling, pathogens would also be killed.
3. As long as the distiller is kept clean and is working properly the high quality of treated water will be very consistent regardless of the incoming water—no drop in quality over time.
4. No filter cartridges to replace, unless a carbon filter is used to remove volatile organic compounds.

The disadvantages of Distillation include:

- Distillation takes time to purify the water. It can take two to five hours to make a gallon of distilled water.
- Distillers uses electricity all the time the unit is operating.
- Distillers requires periodic cleaning of the boiler, condensation compartment, and storage tank.

- Countertop Distillation is one of the more expensive home water treatment methods, using $0. 25 to $0. 35 of electrical energy per gallon of distilled water produced - depending on local electricity costs. The cost of ownership is high because you not only have the initial cost of the distillation unit to consider, but you also must pay for the electrical energy for each gallon of water produced. If it cost you $0. 25 to distill each gallon, and you purified 10 gallons per week, you would pay $130 for your 520 gallons of distilled water each year.
- Most home distillation units require electricity, and will not function in an emergency situation when electrical power is not available.

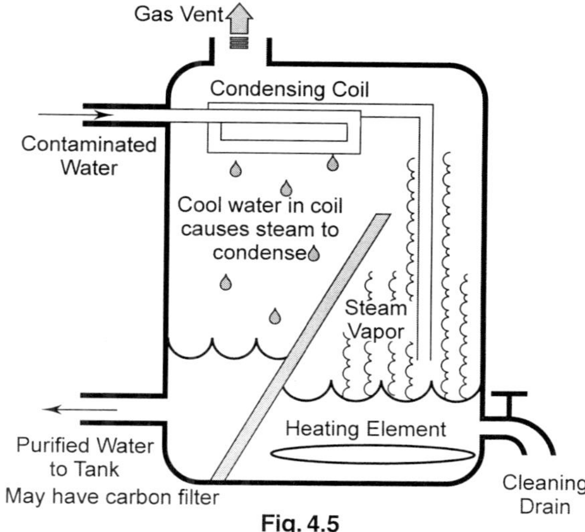

Fig. 4.5

3. Reverse Osmosis (RO)

Water pressure is used to force water molecules through a membrane that has extremely tiny pores, leaving the larger contaminants behind. Purified water is collected from the "clean" side of the membrane, and water containing the concentrated contaminants is flushed down the drain from the "contaminated" side. The average RO system is a unit consisting of a sediment/chlorine pre filter, the reverse-osmosis membrane, a water storage tank, and an activated-carbon post filter. They cost from about $150 to over $ 1,500 for point of use systems.

The advantages of Reverse Osmosis include:

1. Reverse osmosis significantly reduces salt, most other inorganic material present in the water, and some organic compounds. With a quality carbon filter to remove any organic materials that get through the filter, the purity of the treated water approaches that produced by distillation.
2. Microscopic parasites (including viruses) are usually removed by properly functioning RO units, *but any defect in the membrane would allow these organisms to flow undetected into the "filtered" water — they are not recommended for use on biologically unsafe water.*
3. Though slower than a carbon or sediment water filter, RO systems can typically purify more water per day than distillers and are less expensive to operate and maintain.
4. Reverse Osmosis systems also do not use electricity, although because they require relatively high water pressure to operate, they may not work well in some emergency situations.

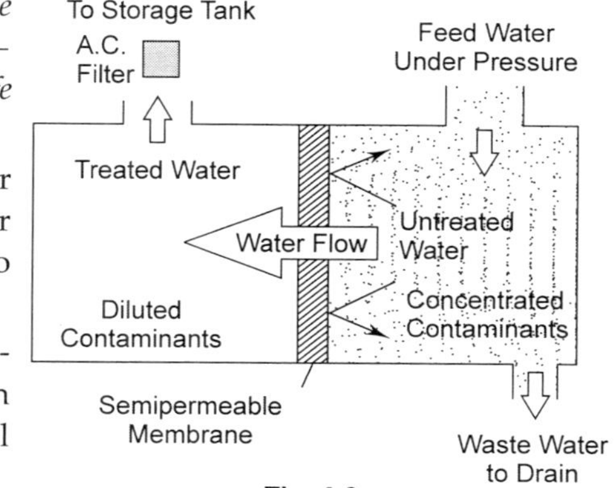

Fig. 4.6

The disadvantages of Reverse Osmosis include:

1. Point of Use RO units make only a few gallons of treated water a day for drinking or cooking.
2. RO systems waste water. Two to four gallons of "waste" water are flushed down the drain for each gallon of filtered water produced.
3. Some pesticides, solvents and other volatile organic chemicals (VOCs) are not completely removed by RO. A good activated carbon post filter is recommended to reduce these contaminants.
4. Many conditions affect the RO membrane's efficiency in reducing the amount of contaminant in the water. These include the contaminant concentration, chemical properties of the contaminants, the membrane type and condition, and operating conditions (like pH, water temperature, and water pressure).
5. Although RO filters do not use electricity, they depend on a relatively high water pressure to force the water molecules through the membrane. In an emergency situation where water pressure has been lost, these systems will not function. * However, if a high quality activated carbon filter is used for the post filter, it could probably be disconnected and used to siphon water through in an emergency to reduce many contaminants.
6. RO systems require maintenance. The pre and post filters and the reverse osmosis membranes must be changed according to the manufacturer's recommendation, and the storage tank must be cleaned periodically.
7. Damaged membranes are not easily detected, so it is hard to tell if the system is functioning normally and safely.

A reverse-osmosis system is a good treatment option for people who have unacceptably high levels of dissolved inorganic contaminants in their drinking water which cannot be removed effectively or economically by other methods. Water from shallow wells in agricultural areas that contains high nitrate levels is a good example of a situation where RO would make sense. Most people, however, who are using municipal water that is subject to EPA regulations usually have acceptably low levels of nitrates. Maximum nitrate levels should be able to be determined by requesting a water quality report from your local water provider.

4. Water Filters

The topic of water filters is complicated because there are so many models available (over 2,500 different models manufactured by more than 500 companies), and because there are so many types of filtration strategies and combinations of strategies used. The basic concept behind nearly all filters, however, is fairly simple. The contaminants are physically prevented from moving through the filter either by screening them out with very small pores and/or, in the case of carbon filters, by trapping them within the filter matrix by attracting them to the surface of carbon particles (the process of adsorption).

- There are two main types of filters (sediment and activated carbon), and sometimes they are combined into a single unit. A third type, which will be considered as a separate topic, is reverse osmosis.
- You will read about micron or sub micron filtration. This is a measure of how good the filter is at removing particles from the water — smaller is better. A micron is a unit of measure—one micron is about 1/100 the diameter of a human hair. A filter that removes particles down

to 5 microns will produce fairly clean-looking water, but most of the water parasites, bacteria, cryptosporidia, giardia, etc., will pass through the pores. A filter must trap particles one micron or smaller to be effective at removing cryptosporidia or giardia cysts. Viruses cannot be effectivel removed by any filtration method. In theory, reverse osmosis will remove viruses, but a small flaw in the membranes would allow viruses to pass undetected into the 'filtered' water.

- A benefit of all home filtration systems is that they are passive. That is, they require no electricity to filter the water, and normal home water pressure is used to force the water though the filter. The only routine maintenance required is periodic replacement of the filtration element. As long as the cost of the replacement filter elements is reasonable, owning a even a high-end water filter can be very inexpensive if you look at the long term costs and compare it with other solutions.

Sediment Filters — Solid Particles are Strained Out of the Water:

1. *Fiber Filters:* These filters contain cellulose, rayon or some other material spun into a mesh with small pores. If you take a piece of cloth and pour water containing sand through it you will get the picture. Suspended sediment (or turbidity) is removed as water pressure forces water through tightly wrapped fibers. Some small organic particles that cause disagreeable odors and taste may also be removed. These filters come in a variety of sizes and meshes from fine to coarse, with the lower micron rating being the finer. The finer the filter, the more particles are trapped and the more often the filter must be changed. Fiber filters are often used as pre-filters to reduce the suspended contaminants that could clog carbon or RO filters. Fiber filters will not remove contaminants that are dissolved in the water, like chlorine, lead, mercury, trihalomethanes or other organic compounds.
2. *Ceramic Filters:* Ceramic filters are much like fiber filters and use a process where water is forced through the pores of a ceramic filtration media. This provides mechanical filtration only. This type of filter can reduce asbestos fibers, cysts (if the pores are one micron or smaller), some bacteria (with pore sizes in the 0. 2 – 0. 8 micron range**) and other particulate matter.
    - Ceramic filters will not remove contaminants that are dissolved in the water, like chlorine, lead, mercury, trihalomethanes or other organic compounds, nor will they remove viruses. These filters may be used as a back-end to an activated carbon filter to provide a more thorough removal of contaminants.
3. *Activated Carbon Filters:* Activated carbon (AC) is particles of carbon that have been treated to increase their surface area and increase their ability to adsorb a wide range of contaminants - activated carbon is particularly good at adsorbing organic compounds. You will find two basic kinds of carbon filters **granular activated carbon** (GAC) and **solid block activated carbon** (SBAC). Contaminant reduction in AC filters takes place by two processes, **physical removal** of contaminant particles, blocking any that are too large to pass through the pores (obviously, filters with smaller pores are more effective), and a process called **adsorption** by which a variety of dissolved contaminants are attracted to and held (adsorbed) on the surface of the carbon particles. The characteristics of the carbon material (particle and pore size, surface area, surface chemistry, density, and hardness) influence the efficiency of adsorption.

   AC is a highly porous material; therefore, it has an extremely high surface area for contaminant adsorption. One reference mentions *The equivalent surface area of 1 pound of AC ranges from 60 to 150 acres (over 3 football fields). Another article states, Under a scanning electron microscope the activated carbon looks like a porous bath sponge. This high concentration of pores within a relatively small volume produces a material with a phenomenal surface area: one tea spoon of activated carbon would exhibit a surface area equivalent to that of a football field.*

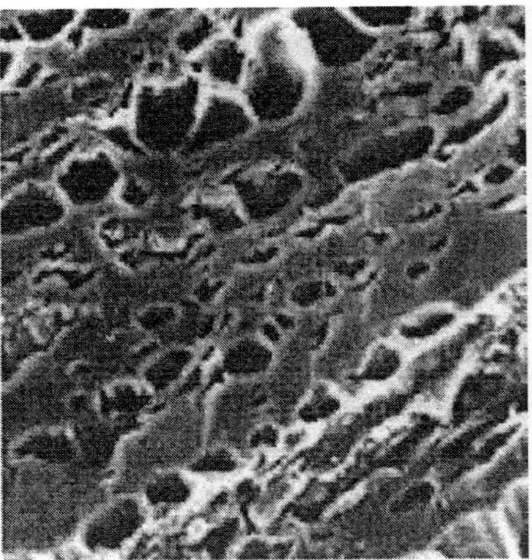

Fig. 4.7 Picture of Granular Activated Carbon

AC is made of tiny clusters of carbon atoms stacked upon one another. The carbon source is a variety of materials, such as peanut shells, coconut husks, or coal. The raw carbon source is slowly heated in the absence of air to produce a high carbon material. The carbon is activated by passing oxidizing gases through the material at extremely high temperatures. The activation process produces the pores that result in such high adsorptive properties. An article about Activated Carbon states that *Activated carbon is one of the best tools which can be used to reduce risks to human health and provide an aesthetically pleasing product at reasonable cost.* The article also describes how AC works and has some of the best scanning electron microscope pictures of actual AC granules I have seen.

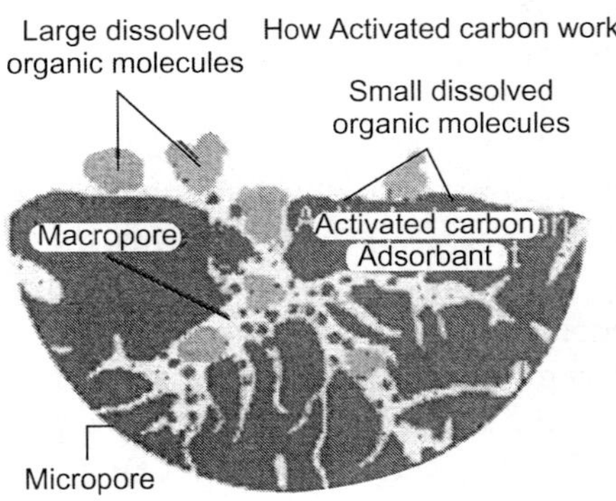

Fig. 4.8

The adsorption process depends on the following factors: 1) physical properties of the AC, such as pore size distribution and surface area; 2) the chemical nature of the carbon source, or the amount of oxygen and hydrogen associated with it; 3) chemical composition and concentration of the contaminant; 4) the temperature and pH of the water; and 5) the flow rate or time exposure of water to AC. Interesting article: What is Activated Carbon, Adsorption, History of AC.

- **The effectiveness of carbon filters to reduce contaminants** is affected by the factors affecting adsorption listed above and three additional characteristics of the filter, **contact time** between the water and the carbon material, the **amount of carbon** in the filter, and **pore size.**
- The length of **contact time** between the water and the carbon material, governed by the rate of water flow and the amount/volume of activated carbon, has a significant effect on adsorption of contaminants. More contact time results in greater adsorption.
- The **amount of carbon** present in a cartridge or filter affects the amount and type of contaminant removed. Less carbon is required to remove taste and odor-producing chemicals than to remove trihalomethanes.
- **Pore size** characteristics will be discussed in greater detail on the GAC and SB AC pages, but GAC filters contain loose granules of activated carbon while in SB AC filters, the activated carbon is in the form of very small particles bound into a solid, matrix with very small pores.
- Because of the filter characteristics discussed above, the **most effective Point of Use activated carbon filters are large SB AC filtration systems,** and the least effective are the small, pour-through pitcher filters.
- Activated carbon filter cartridges will, over time, become less effective at reducing contaminants as the pores clog with particles (slowing water flow) and the adsorptive surfaces in the pores become filled with contaminants (typically not affecting flow rate). **There is often no noticeable indication that a carbon filter is no longer removing contaminants,** so it is important to replace the cartridge according to the manufacturer's instructions. The overall water quality (turbidity or presence of other contaminants) also affects the capacity of activated carbon to adsorb a specific contaminant.
- It is important to note, particularly when using counter-top and faucet-mount carbon filtration systems, that **hot water should NEVER be run through a carbon filter**. I have seen warnings about possible damage to the filter from hot water. Perhaps more importantly, hot water will tend to release trapped contaminants into the water flow potentially making the water coming out of the filter more contaminated than the water going in.

(a) Granular Activated Carbon (GAC)

In this type of filter, water flows through a bed of loose activated carbon granules which trap some particulate matter and remove some chlorine, organic contaminants, and undesirable tastes and odors. The three main problems associated with GAC filters are: channeling, dumping, and an inherently large pore size. Most of the disadvantages discussed below are not the fault of the activated carbon filtration media, rather, the problem is the design of the filters and the use of loose granules of activated carbon.

The advantages of GAC filters include:

- Simple GAC filters are primarily used for aesthetic water treatment, since they can reduce chlorine and particulate matter as well as improve the taste and odor of the water.
- Loose granules of carbon do not restrict the water flow to the extent of Solid Block Activated Carbon (SBAC) filters. This enables them to be used in situations, like whole house filters, where maintaining a good water flow rate and pressure is important.
- Simple, economical maintenance. Typically an inexpensive filter cartridge needs to be changed every few months to a year, depending on water use and the manufacturer's recommendation.
- GAC filters do not require electricity, nor do they wastewater.

- Many dissolved minerals are not removed by activated carbon. In the case of calcium, magnesium, potassium, and other beneficial minerals, the taste of the water can be improved and some (usually small) nutrient value can be gained from the water.

  The bottom line is that GAC filters are effective and valuable water treatment devices, but their limitations always need to be considered. A uniform flow rate, not to exceed the manufacture's specifications, needs to be maintained for optimal performance, and the filter cartridge must be changed after treating the number of gallons the filter is rated for.

The disadvantages of GAC filters include:

- Water flowing through the filter is able to **"channel"** around the carbon granules and avoid filtration. Water seeks the path of least resistance. When it flows through a bed of loose carbon granules, it can carve a channel where it can flow freely with little resistance. Water flowing through the channel does not come in contact with the filtration medium. The water continues to flow, however, so you do not realize that your filter has failed — you get water, but it is not completely filtered.
- Pockets of contaminated water can form in a loose bed of carbon granules. With changes in water pressure and flow rates, these pockets can collapse, **"dumping"** the contaminated water through the filter into the "filtered" flow.
- Since the carbon granules are fairly large (0. 1mm to 1mm in one popular pitcher filter), the **effective pore size of the filter is relatively large** (20 – 30 microns or larger). GAC filters, by themselves, cannot bacteria.
- As described above, **hot water should NEVER be run through a carbon filter**
- Also, if you think of a bed of charcoal that traps an occasional bacterium, picks up a bit of organic material, and removes the chlorine from the water, you can see how these filters might become breeding grounds for the bacteria they trap. You will see warnings about GAC filters suggesting you run water through them for a few minutes each morning to flush out any bacteria. If you are considering purchasing a GAC filter be sure to ask the representative about what they recommend you do about this potential problem (I was told by one sales person, that if the filter was not going to be used for a few days, I should remove the filtration element, place it in a plastic bag, and store it in the refrigerator until I returned).
- Unless the filter plugs up or you notice an odor in the "filtered water", it may be difficult to know when the filter has become saturated with contaminants and ineffective. That is why it is necessary to change filter cartridges according to the manufacturer's recommendation.

(b) Solid Block Activated Carbon (SBAC)

Activated carbon is the primary raw material in solid carbon block filters; but instead of carbon granules comprising the filtration medium, the carbon has been specially treated, compressed, and bonded to form a uniform matrix. The **effective pore size** can be very small (0. 5–1 micron). SBAC, like all filter cartridges, eventually become plugged or saturated by contaminants and must be changed according to manufacturer's specifications. Depending on the manufacturer, the filters can be designed to better reduce specific contaminants like arsenic, MTBE, etc.

The advantages of SBAC filters include:

- Provide a larger surface area for adsorption to take place than Granular Activated Carbon (GAC) filters for better contaminant reduction.
- Provide a longer contact time with the activated carbon for more complete contaminant reduction.
- Provide a small pore size to physically trap particulates. If the pore size is small enough, around 0. 5 micron or smaller, bacteria that become trapped in the pores do not have enough room to multiply, eliminating a problem common to GAC filters.
- Completely eliminate the channeling and dumping problems associated with GAC filters.
- SBAC filters are useful in emergency situations where water pressure and electricity might be lost. They do not require electricity to be completely effective, and water can even be siphoned through them.
- SBAC filters do not wastewater like reverse osmosis.
- Many dissolved minerals are not removed by activated carbon. In the case of calcium, magnesium, potassium, and other beneficial minerals, the taste of the water can be improved and some (usually small) nutrient value can be gained from the water.
- Simple, economical maintenance. Typically an inexpensive filter cartridge needs to be changed every few months to a year, depending on water use and the manufacturer's recommendation.
- This combination of features provides the potential for greater adsorption of many different chemicals (pesticides, herbicides, chlorine, chlorine byproducts, etc.,) and greater particulate filtration of parasitic cysts, asbestos, etc., than many other purification process available. By using other specialized materials along with specially prepared activated carbon, customized SBAC filters can be produced for specific applications or to achieve greater capacity ratings for certain contaminants like lead, mercury, arsenic, etc.

The disadvantages of SBAC filters include:

- SBAC filters, like all activated carbon filters, do not naturally reduce the levels of soluble salts (including nitrates), fluoride, and some other potentially harmful minerals like arsenic (unless specially designed) and cadmium. If these contaminants are present in your water, reverse osmosis would usually be the most economical alternative followed by distillation.
- As described above, **hot water should NEVER be run through a carbon filter**
- As SBAC filters remove contaminants from the water they gradually lose effectiveness until they are no longer able to adsorb the contaminants. There is no easy way to determine when a filter is nearing the end of its effective life except that the 'filtered' water eventually begins to taste and smell like the unfiltered water. The manufacturer's guidelines for changing filter cartridges should always be followed.

A few SBAC filtration systems have been certified for arsenic reduction:

What about the inexpensive **GAC pitcher filters** and **faucet mount filters**? What will they do for your water purification needs?

The answer to that question depends on what your water treatment goals are and the amount of filtered water you plan to use. The most popular brands are GAC filters that will sometimes also contain granules of an ion exchange medium (you can see and hear the granules rattling around).

- They do reduce the level of some contaminants of health concern and thus are better than nothing.
- **They are very limited in the type and number of contaminants they remove,** typically chlorine and perhaps lead, copper, and/or cysts are reduced.
- At about \$5 – \$10 for a filter cartridge capable of treating about 35–40 gallons, your filtered water will cost about \$0.14 – \$0.25 per gallon (\$58 – \$90 per year if you use 35 gallons per month). If you are using (or would like to use) much more than a gallon per day, and/or would like to remove more than a minimum number of contaminants, you should probably examine other options.
- Since they contain a very small amount of very loose GAC granules, **they CANNOT be considered effective treatments for most biological contaminants or chemical contaminants of health concern**. The pitcher filters which contain a microfilter are typically certified to reduce cysts. All of these filters are mostly designed to improve the aesthetics of drinking water (taste & odor improvement) and reduce levels of a limited number of harmful contaminants. They are subject to possible bacterial growth channeling, and all the other problems of larger GAC filters.
- How about the **Solid Carbon Block faucet mount filters**?
- The solid carbon block faucet mount filters are more effective than GAC filters in reducing contaminants.
- These filters, by nature, are quite small, though, and because filter effectiveness is dependent on contact time of the water with the filter media, a larger, high-quality solid carbon block filter will be more effective at reducing contaminants at the same flow rate. The difference is size can be striking - 4 ounces of activated carbon for a faucet mount filter vs. 32 ounces for a high-end filter, over 7 times more filter media.
- Like the pitcher filters, the replacement cartridges for faucet mount filters tend to be relatively more expensive (typically \$0.14 – \$0.25 per gallon) compared to a high-quality solid block activated carbon replacement filter which will filter water for about \$0.07 per gallon. Using 2 gallons of filtered water per day would cost between \$100 and \$180 to replace 4–5 ounce faucet mount cartridges and about \$50 to replace a 32 ounce high-end filter.

5. Bottled Water

Did the water in the bottle^you just purchased really come from the beautiful spring shown on the label? How can you be sure the water in the bottle is any cleaner or safer than your tap water? How does the cost, both short term and long term, compare with other water purification options?

Bottled water is simply water from some source that a company (or in the case of water vending machines, the consumer) has placed in a bottle for resale. Bottled water can have minimal (or no) processing — as in natural spring or mineral water, or it can be completely filtered and demineralized to nearly pure water and then have minerals added back in to make it taste better.

- But, how can you be sure the, water in the bottle you purchased is any cleaner or safer than your tap water? Just like with any of the other water treatment solutions, you will find reputable companies that provide a safe, quality product, and you will find companies that fill their bottles using the local, municipal water. Each company must be evaluated individually (A Review of Bottled Water and a NRDC Discussion of Bottled Water). There are 2 regulatory organizations, 1 trade association, and 1 independent certification organization which contribute to insuring the safety and quality of bottled water.

Bottled Water: Issues and Alternatives

According to the University of Illinois at Urbana-Champaign, College of Agriculture, Cooperative Extension Service, bottled water is regulated at the federal level, and in some cases, at the state level. Bottled water suppliers who are members of the International Bottled Water Association (IBWA) regulate themselves and are given unannounced inspections by NSF International. Bottled water companies can also have their product certified by NSF International.

Federal standards: Bottled water is regulated by the Food and Drug Administration (FDA), which requires manufacturers to submit samples regularly for testing. In the United States, the concentration of contaminants found in bottled water must be less than the "allowable levels" set by the FDA which are essentially the same as the maximum contaminant levels that the EPA requires municipal water supplies to meet.

State standards: Individual states must enforce the federal bottled water regulations, but they can also pass stricter standards.

International Bottled Water Association: The bottled water industry regulates itself through the International Bottled Water Association (IBWA). Bottled water FAQs. The IBWA sets manufacturing requirements, which help to ensure that bottlers meet FDA health standards. Bottled water producers who are members of IBWA are inspected annually by an independent laboratory, the National Sanitation Foundation. Through unannounced inspections, members are evaluated on compliance with the IBWA's performance requirements and FDA Quality Standards.

Not all bottled water manufacturers are members of the IBWA. Of the 475 bottling plants in the United States (in 1994), 250 are members. The label may indicate whether a bottled water comes from a member company.

You might want to check NSF International to see if the bottled water you are interested in purchasing is certified. The NSF Bottled Water Certification Program is an annual, voluntary certification process that includes both extensive product evaluations as well as on-site audits of bottling facilities. The NSF testing program provides for annual unannounced plant inspections covering every aspect of a bottler's operation, from the source of the water, through the disinfection and treatment process, and including the container closure process. The company also performs extensive product testing for over 160 chemical, inorganic, radiological, and microbiological contaminants. An interesting article from Scientific American magazine on bottled water.

The Advantages of Bottled Water include:

- An emergency source of water in the event your primary water source fails or becomes contaminated.
- A convenient source of usually safe water for drinking outside of the home.
- Bottled water, since it does not contain chlorine, and may contain a mix of minerals to enhance flavor, may taste better than untreated tap water.
- Most bottled water will contain fewer contaminants than untreated tap water.
    - The bottled water industry would like the public to believe that bottled water is more pure and safer than tap water and water produced by other treatment methods. The actual quality of bottled water depends on the bottling company, and most is usually good. High quality home water treatment methods (filtration, reverse osmosis, and distillation) can usually produce water of equal or better quality more economically, however.

The Disadvantages of Bottled Water Include:

- **Cost:** Bottled water costs anywhere from $0.25 a gallon for water from a vending machine to $0.50 for generic water jugs you purchase in a store to $1.00 – $1.50 or more for water delivered to your home. If you purchase 10 gallons of bottled water a week, you will spend $260 (at $.50/gal), $520 (at $1.00/gal), $780 (at $1.50/gal), or more every year.
    - Many of us are willing to pay extraordinary prices for the availability and convenience of certain products. That **16 ounce bottle** of "designer water" you just purchased at the gas station for $0.69 costs **$5.52 per gallon**, probably more than twice as much per gallon as your gas! Purchasing one 16 oz bottle per day (at $0.69 per bottle) will cost you **$248 per year** – that's for a total of 45 gallons. If you were to take a minute to fill your own well-cleaned bottle with filtered or distilled water, however, a 16 oz bottle of water would cost perhaps 1 - 2 cents a day, or less than **$7.30 per year!**
- **Convenience:** Using bottled water requires moving and storing jugs or bottles of water. Water weighs about eight pounds per gallon, or about 40 pounds per five gallon bottle. Failure of a small valve can lead to a big mess, as I discovered after arriving home one evening after a long day at work to find 3 gallons of water soaking into the rug.
- **Environmental Impact:** Producing bottles uses resources, and unless they are reused or recycled, they cause a waste disposal problem. Recycle or reuse the empty bottles, if at all possible. Transporting bottles of water from the bottler to stores or homes also uses resources.
    - **Keep it Clean:** According to the University of Illinois at Urbana-Champaign, College of Agriculture, Cooperative Extension Service,

If you choose bottled water after careful investigation, keep in mind that all of your hard work will go to waste if you aren't careful about keeping your bottled water clean. You have to be faithful in maintaining the hygiene of your bottled water, or you may increase your exposure to bacteria. Bacteria grow best in warm, moist areas. The wet, warm, threaded cap of an unrefrigerated bottle of water is a perfect place for bacteria to grow; they will begin to grow as soon as you break the seal. If ingested, these bacteria can cause gastrointestinal problems and other health risks. The key is to maintain the cleanliness of your bottles and store them properly. Follow these hints:

1. *Store the bottle in a refrigerator at a temperature above freezing but less than 50 degrees Fahrenheit.*
2. *Wipe the seal with a clean cloth after each use.*
3. *Avoid any type of buildup in the bottle cap.*
4. *If your bottle is refillable, make sure it is well-cleaned and rinsed before refilling. If possible, recycle the old bottle and obtain a fresh, sterile, sealed bottle."*

6. Ultra Violet Light

Water passes through a clear chamber where it is exposed to Ultra Violet (UV) Light. UV light effectively destroys bacteria and viruses. However, how well the UV system works depends on the energy dose that the organism absorbs. If the energy dose is not high enough, the organism's genetic material may only be damaged rather than disrupted.

The Advantages of using UV Include:

- No known toxic or significant nontoxic byproducts introduced
- Removes some organic contaminants.
- Leaves no smell or taste in the treated water.
- Requires very little contact time (seconds versus minutes for chemical disinfection).

- Improves the taste of water because some organic contaminants and nuisance microorganisms are destroyed.
- Many pathogenic microorganisms are killed or rendered inactive.
- Does not affect minerals in water.

The Disadvantages of using UV Include:

- UV radiation is not suitable for water with high levels of suspended solids, turbidity, color, or soluble organic matter. These materials can react with UV radiation, and reduce disinfection performance. Turbidity makes it difficult for radiation to penetrate water and pathogens can be 'shadowed', protecting them from the light.
- UV light is not effective against any non-living contaminant, lead, asbestos, many organic chemicals, chlorine, etc.
- Tough Cryptosporidia cysts are fairly resistant to UV light.
- Requires electricity to operate. In an emergency situation when the power is out, the purification will not work.

UV is typically used as a final purification stage on some filtration systems. If you are concerned about removing contaminants in addition to bacteria and viruses, you would still need to use a quality carbon filter or reverse osmosis system in addition to the UV system.

7. Water Softeners and Deionizers

Water softeners operate on the ion exchange process (specifically a cation exchange process where + ions are exchanged). In this process, water passes through a media bed, usually sulfonated polystyrene beads. The beads are supersaturated with sodium (a positive ion). The ion exchange process takes place as hard water passes through the softening material. The hardness minerals (positively charged Calcium and Magnesium ions) attach themselves to the resin beads while sodium on the resin beads is released simultaneously into the water. When the resin becomes saturated with calcium and magnesium, it must be recharged. The recharging is done by passing a concentrated salt (brine) solution through the resin. The concentrated sodium replaces the trapped calcium and magnesium ions which are discharged in the wastewater. Softened water is not recommended for watering plants, lawns, and gardens due to its elevated sodium content.

Several factors govern the efficiency of a cationic softener:

- Type & quality of resin used
- Amount of salt per cubic foot of resin for regeneration
- Brine concentration in the resin bed during regeneration
- Brine flow rate through the resin bed (contact time) during regeneration
- Raw water hardness
- Raw water temperature — softeners perform better at higher temperatures and
- Optimal flow rate of hard water through the resin bed.

Although not commonly used, potassium chloride can be used to create the salt brine for softeners designed to use KCl. In that case potassium rather than sodium is exchanged with calcium and magnesium. Before selecting an ion exchange water softener, test water for hardness and iron content. When selecting a water softener, the regeneration control system, the hardness removal capacity, and the iron limitations are three important elements to consider. More information on Hard Water and Softening.

The Advantages of Water Softeners Include:

- The nuisance factor of hard water is reduced.
- Some other other cations like barium, radium and iron may be reduced depending on the manufacturer's specifications.

The Disadvantages of Water Softeners Include:

- The process of regenerating the ion exchange bed dumps salt water into the environment.
- The elevated sodium concentration of most softened water can affect the taste and may not be good for people on low sodium diets, although sodium concentrations are typically quite low relative to sodium levels in most food.
- Cation exchange does not reduce the level of anions (like nitrates), or biological contaminants (bacteria, viruses, cysts); nor does the process reduce the levels of most organic compounds.
- Typically, approximately 50 gallons of rinse water per cubic foot of resin is required to totally remove hardness and excess salt from the resin after each regeneration.

8. Water Deionizers

Water Deionizers use both Cation and Anion Exchange to exchange both positive and negative ions with H^+ or OH^- ions respectively, leading to completely demineralized water. Deionizers do not remove uncharged compounds from water, and are often used in the final purification stages of producing completely pure water for medical, research, and industrial needs.

A potential problem with deionizers is that colonies of microorganisms can become established and proliferate on the nutrient-rich surfaces of the resin. When not regularly sanitized or regenerated, ion-exchange resins can contaminate drinking water with bacteria.

9. Ozonation

The formation of oxygen into ozone occurs with the use of energy. This process is carried out by an electric discharge field as in the CD-type ozone generators (corona discharge simulation of the lightning), or by ultraviolet radiation as in UV-type ozone generators (simulation of the ultra-violet rays from the sun). In addition to these commercial methods, ozone may also be made through electrolytic and chemical reactions.

Ozone is a naturally occurring component of fresh air. It can be produced by the ultraviolet rays of the sun reacting with the Earth's upper atmosphere (which creates a protective ozone layer), by lightning, or it can be created artificially with an ozone generator.

The ozone molecule contains three oxygen atoms whereas the normal oxygen molecule contains only two. Ozone is a very reactive and unstable gas with a short half-life before it reverts back to oxygen. Ozone is the most powerful and rapid acting oxidizer man can produce, and will oxidize all bacteria, mold and yeast spores, organic material and viruses given sufficient exposure.

The Advantages of using Ozone Include:

- Ozone is primarily a disinfectant that effectively kills biological contaminants.
- Ozone also oxidizes and precipitates iron, sulfur, and manganese so they can be filtered out of solution.

- Ozone will oxidize and breakdown many organic chemicals including many that cause odor and taste problems.
- Ozonation produces no taste or odor in the water.
- Since ozone is made of oxygen and reverts to pure oxygen, it vanishes without trace once it has been used. In the home, this does not matter much, but when water companies use ozone to disinfect the water there is no residual disinfectant, so chlorine or another disinfectant must be added to minimize microbial growth during storage and distribution.

The Disadvantages of using Ozone Include:

- Ozone treatment can create undesirable byproducts that can be harmful to health if they are not controlled (*e.g.*, formaldehyde and bromate).
- The process of creating ozone in the home requires electricity. In an emergency with loss of power, this treatment will not work.
- Ozone is not effective at removing dissolved minerals and salts.

Caution — The effectiveness of the process is dependent, on good mixing of ozone with the water, and ozone does not dissolve particularly well, so a well designed system that exposes all the water to the ozone is important.

In the home, ozone is often combined with activated carbon filtration to achieve a more complete water treatment.

10. Activated Alumina

Activated Alumina is a granulated form of aluminum oxide. In this process, water containing the contaminant is passed through a cartridge or canister of activated alumina which adsorbs the contaminant. The cartridge of activated alumina must be replaced periodically. Activated alumina devices can accumulate bacteria, so treated water may have higher bacteria counts than raw water.

The Advantages of Activated Alumina Filters Include:

- An effective way to reduce levels of fluoride, arsenic, and selenium

The Disadvantages of Activated Alumina Filters Include:

- The use of other treatment methods would be necessary to reduce levels of other contaminants of health concern.

11. 'Altered' Water

No discussion about water treatment would be complete these days without mention of what I call "altered" water that has been treated in some way to allegedly modify the physical, chemical, or 'energy' properties of water to provide some benefit to the body. These treatments fall under a wide range of categories, including: pi mag; oxygenation; hydrogenation; various 'catalytic', vortex, magnetic, & photonic treatments; microclustering; super-ionization; homeopathic successions; etc.

Note: Oxygenated water, discussed below, is just one example of the hundreds of'altered' or 'enhanced' water products promoted on the internet and in some health food stores. These products all have several characteristics in common that are discussed in more detail on the Altered Water and Drinking

Water Scams pages. It is easier to demonstrate the flaws in Oxygenated water claims than in other products that claim to use some exotic physical or chemical process, but just like oxygenated water, none of the other products are able to withstand critical scientific review.

Regardless of any alleged health benefits, these products are extremely effective at separating customers from their money – I have seen some of this bottled water sell for more than **$20 a gallon** plus shipping **(that's about 3, 000 times greater than the cost of tap water)!** Many of the 'altered' water products sell for $10 to $15 per gallon. Water 'ionizers' that generate alkaline water produce sodium hydroxide (an effective drain cleaner) and bleach and sell for up to $5, 000.

Let's look briefly at one 'altered' water example: hiOsilver oxygenated water — soon to be 02Cool Oxygen Water:

Why drink hiOsilver oxygen water? According to the marketing hype, it provides "extra energy and vitality, fresh breath, healthier gums and teeth. We do not know all of the benefits of drinking hiOsilver Oxygen Water. Many people report that they feel energized after a bottle. Studies have been done showing enhanced sports performance for athletes drinking oxygenated water. With hiOsilver, many people report that their headaches disappear, even migraine headaches. One bottle of hiOsilver Oxygen Water and you will see why we are proud of the many benefits it offers over ordinary bottled waters. . . "

hiOsilver oxygenated water water sells for $44 (for twenty four, 16 oz bottles - 3 gallons) but you do get free shipping!

If my math is correct, that's **over $14. 50 per gallon**. By comparison, tap water costs around $0, 007 per gallon (2, 000 times less than the oxygenated water). The highest quality filtered water is about $0. 07 per gallon **(that is still over 200 times less expensive than the oxygenated water)**. I suppose some people might be able to justify the exorbitant cost IF the product performed as advertised — but does it?

The **super oxygenating water cooler system** marketing materials make these 'exciting' claim — and a skeptical review:

- Extra supply of oxygen to the body for overall health improvement
- Enhancement of the brain function for clearer thinking and alertness
- More oxygen to the muscle to increase energy and performance
- More oxygen to skin cells for healthier, younger looking skin
- Enhanced metabolism and waste removal
- Enhancement of the body's ability to fight bacteria and viruses
- Better absorption of vitamins, minerals and other nutrients

Some Facts to Consider when Evaluating the Oxygenated Water Hype:

- There is less dissolved oxygen in 1 liter of 'oxygenated water' than in 1 breath of air. Taking an extra breath of air when exercising would be substantially less expensive than paying $1 to $2 for a litre of these products!
- The primary way to transport oxygen in the body is bound to hemoglobin in the red blood cells. In normal healthy exercisers, hemoglobin leaving the lungs is already 97% to 98% saturated with oxygen.

- The structure of the circulatory system ensures that any oxygen picked up in the digestive system would go through the lungs before reaching the muscles and other tissues. In the lungs any extra oxygen in the blood will reduce the amount of oxygen transferred to the blood — the final oxygen saturation of the hemoglobin would still be 97% to 98%.
- Oxygenating water has no effect on the body's ability to absorb the oxygen into the bloodstream or transfer it throughout the body.
- The concept of obtaining significant amounts of oxygen through the digestive system makes as much scientific and physiological sense as quenching your thirst by inhaling a glass of water into your lungs.

Water that you drink can take up to an hour or more to travel to the intestines where it is absorbed — the water you inhale into your lungs will absorb into your bloodstream almost immediately — so doesn't it make sense to quench your thirst by inhaling a glass of water? **(Do not try this experiment!!)** This is **not** an endorsement for inhaling water. The point is, the respiratory system is designed to absorb oxygen, not water! Oxygenated water makes sense for fish that have specialized structures (gills) for exchanging oxygen and carbon dioxide in an aquatic environment, but not for mammals. Similarly, the digestive system is designed to absorb water and nutrients, not oxygen! American Council on Exercise (ACE) Study Investigates Super Oxygenated Water Claims — Results of a study done to test health claims. Conclusion: "At this time, there is no scientific evidence or logical rationale to suggest that drinking super oxygenated water can in any way increase the amount of oxygen in the blood stream," said Porcari. "Therefore, any potential benefits of super oxygenated water would undoubtedly be attributed to the placebo effect. " A single breath of air contains more oxygen than a bottle of oxygenated water. Despite advertising claims that oxygenated water can boost sports performance, a study in the Nov. 12, 2003, Journal of the American Medical Association found that compared with tap water, it had no effect on 9 exercise-performance measures in 11 healthy men and women. Ben Goldacre of The Guardian reviews Oxygizer Oxygenated water bottled in Austria. Oxygen water: Are you being ripped off? There's very little *reliable* evidence to show that oxygenated (or oxygenized) water has any significant effect on exercise performance, energy levels, or recovery. This water isn't going to hurt you, but there's no reason to believe it's going to help you substantially.

Two Important Terms Related to Water Treatment:

- *Water Purification* — to remove specified contaminants from a water source.
- *Water Disinfection* — specifically, a purification process that kills or removes biological contaminants (cysts, bacteria, viruses, protozoans, etc.,) from a water source. *Water that has been disinfected (by UV treatment, boiling, chlorination, micro-filtration, etc.,) may still be polluted with other contaminants that are not affected by the disinfection treatment. In some cases, additional contaminants may actually be added to the water by the disinfection process.* For instance, the process of chlorination nearly always adds some disinfection by products (trihalomethanes, like chloroform), and boiling water will concentrate inorganic contaminants.

  The pore size of a filter determines how well it removes harmful particulate contaminants like asbestos, parasitic cysts, and bacteria.
- Solid Block, Activated Carbon filters have very small carbon particles bonded into a tightly packed matrix with uniform pores, typically between 0. 5 and 1. 0 micron, and can reliably remove small particulate contaminants.

- Granular Activated Charcoal (GAC) filters have relatively large, and irregular sized pores (10 microns would probably be the minimum size to expect), so it is impossible to state with any certainty what size particles would be removed. Channeling can also dump unfiltered water into the output stream.

GAC only filters should never be relied on exclusively to provide protection from small particulate contaminants.

Comparison of Long-Term Costs for Water Treatment:

- Have you considered how much it will cost to obtain pure, safe water for the next 5, 10, 20, or even 40 years?
- When you decide to protect yourself and your family from drinking water contaminants, 40 years may seem like a very long time to think about treatment costs....

However

- *If you plan to use just 10 gallons of purified water a week (less than 1. 5 gallons per day), the water treatment method you select can mean the difference between spending about $1, 700 over 40 years or spending more than $31, 000 for pure, safe water!*
- Click below on the amount of purified water you use (or would like to use) per week. This will display a table and graph comparing the average costs of some common treatment methods (bottled water, distillation, reverse osmosis, SBAC filtration, and pitcher filters).

5 gallons per week	10 gallons per week	15 gallons per week	20 gallons per week	25 gallons per week

- These cost tables (as well as the filter performance tables) convinced me to consider seriously a top—of—the line, Solid Block, Activated Carbon (SBAC) filtration system. Although the initial cost was higher than many cheaper alternatives, the cost over several years was much lower than other options - at most usage levels. The fact that SBAC filtration was simpler to use than other methods and significantly reduced all contaminants I was liable to experience in my municipal water, sold me on the technology.

EXERCISE

1. Explain 'Water Cycle'.
2. Write an Essay on 'Types of water'.
3. What is water pollution? What are its reasons?
4. Write note on different water Pollutants.
5. What are the common practices for water conservation?
6. What are the main water quality parameters? Explain.
7. Write short notes on-
   (*i*) Nitrogen Cycle
   (*ii*) Eutrophication
   (*iii*) TSS and TDS
   (*iv*) Reverse Osmosis
   (*v*) Granular Activated Carbon
   (*vi*) Disadvantages of bottled water.

5

WASTE AND WASTEWATER MANAGEMENT

5.1 WATER TREATMENT

Wastewater treatment is a process, wherein the contaminants are removed from wastewater as well as household sewage, to produce waste stream or solid waste suitable for discharge or reuse. Wastewater treatment methods are categorized into three sub-divisions, physical, chemical and biological.

Sewage treatment, or **domestic wastewater treatment**, is the process of removing contaminants from wastewater and household sewage, both runoff (effluents) and domestic. It includes physical, chemical, and biological processes to remove physical, chemical and biological contaminants. Its objective is to produce a waste stream (or treated effluent) and a solid waste or sludge suitable for discharge or reuse back into the environment. This material is often inadvertently contaminated with many toxic organic and inorganic compounds.

Process Flow Diagram for a typical large-scale treatment plant

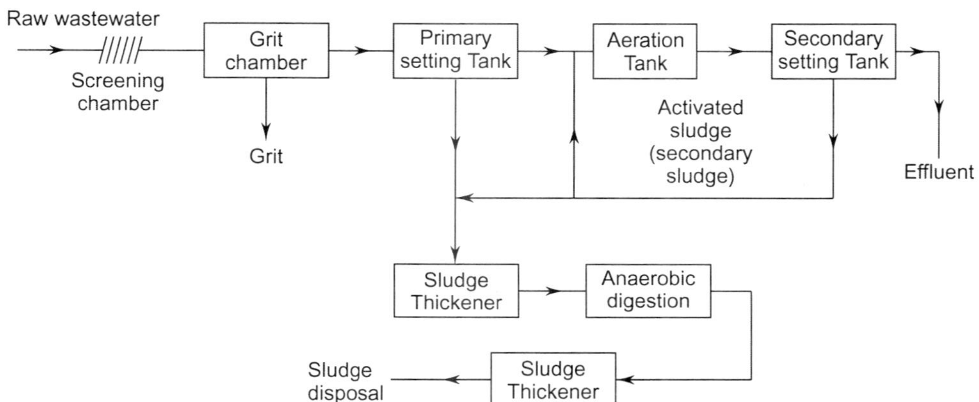

Fig. 5.1 Flow Diagram for Typical Wastewater Treatment Plant

5.2 WASTEWATER TREATMENT METHODS AND DISPOSAL

Satisfactory disposal of wastewater, whether by surface, subsurface methods or dilution, is dependent on its treatment prior to disposal. Adequate treatment is necessary to prevent contamination of receiving waters to a degree which might interfere with their best or intended use, whether it be for water supply, recreation, or any other required purpose.

Wastewater treatment consists of applying known technology to improve or upgrade the quality of a wastewater. Usually wastewater treatment will involve collecting the wastewater in a central, segregated location (the Wastewater Treatment Plant) and subjecting the wastewater to various treatment processes. Most often, since large volumes of wastewater are involved, treatment processes are carried out on continuously flowing wastewaters (continuous flow or "open" systems) rather than as "batch" or a series of periodic treatment processes in which treatment is carried out on parcels or "batches" of wastewaters. While most wastewater treatment processes are continuous flow, certain operations, such as vacuum filtration, involving as it does, storage of sludge, the addition of chemicals, filtration and removal or disposal of the treated sludge, are routinely handled as periodic batch operations.

Wastewater treatment, however, can also be organized or categorized by the nature of the treatment process operation being used; for example, physical, chemical or biological. Examples of these treatment steps are shown below. A complete treatment system may consist of the application of a number of physical, chemical and biological processes to the wastewater.

5.3 DIFFERENT TREATMENT METHODS

(*a*) **Physical:**

- Sedimentation
- Screening
- Aeration
- Filtration
- Flotation and Skimming
- Degassification
- Equalization.

(*b*) **Chemical:**

- Chlorination
- Ozonation
- Neutralization
- Coagulation
- Adsorption
- Ion Exchange.

(*c*) **Biological:**

- Aerobic
- Activated Sludge Treatment Methods
- Trickling Filtration
- Oxidation Ponds
- Lagoons

- Aerobic Digestion
- Anaerobic Digestion
- Septic Tanks
- Lagoons.

(*a*) **Physical Methods** include processes where no gross chemical or biological changes are carried out and strictly physical phenomena are used to improve or treat the wastewater.

Examples would be coarse screening to remove larger entrained objects and sedimentation (or clarification). In the process of sedimentation, physical phenomena relating to the settling of solids by gravity are allowed to operate. Usually this consists of simply holding a wastewater for a short period of time in a tank under quiescent conditions, allowing the heavier solids to settle, and removing the "clarified" effluent. Sedimentation for solids separation is a very common process operation and is routinely employed at the beginning and end of wastewater treatment operations. While sedimentation is one of the most common physical treatment processes that is used to achieve treatment, another physical treatment process consists of aeration — that is, physically adding air, usually to provide oxygen to the wastewater. Still other physical phenomena used in treatment consists of filtration. Here wastewater is passed through a filter medium to separate solids. An example would be the use of sand filters to further remove entrained solids from a treated wastewater. Certain phenomena will occur during the sedimentation process and can be advantageously used to further improve water quality. Permitting greases or oils, for example, to float to the surface and skimming or physically removing them from the wastewaters is often carried out as part of the overall treatment process.

In certain industrial wastewater treatment processes strong or undesirable wastes are sometimes produced over short periods of time. Since such "slugs" or periodic inputs of such wastes would damage a biological treatment process, these wastes are sometimes held, mixed with other wastewaters, and gradually released, thus eliminating "shocks" to the treatment plant. This is call equalization. Another type of "equalization" can be used to even out wide variations in flow rates. For example, the wet well of a pump station can receive widely varying amounts of wastewater and, in turn, pump the wastes onward at more uniform rates.

(*b*) **Chemical Treatment** consists of using some chemical reaction or reactions to improve the water quality. Probably the most commonly used chemical process is chlorination. Chlorine, a strong oxidizing chemical, is used to kill bacteria and to slow down the rate of decomposition of the wastewater. Bacterial kill is achieved when vital biological processes are affected by the chlorine. Another strong oxidizing agent that has also been used as an oxidizing disinfectant is ozone.

A chemical process commonly used in many industrial wastewater treatment operations is neutralization. Neutralization consists of the addition of acid or base to adjust *p*H levels back to neutrality. Since lime is a base it is sometimes used in the neutralization of acid wastes.

Coagulation consists of the addition of a chemical that, through a chemical reaction, forms an insoluble end product that serves to remove substances from the wastewater. Polyvalent metals are commonly used as coagulating chemicals in wastewater treatment and typical coagulants would include lime (that can also be used in neutralization), certain iron containing compounds (such as ferric chloride or ferric sulfate) and alum (aluminum sulfate).

Certain processes may actually be physical and chemical in nature. The use of activated carbon to "adsorb" or remove organics, for example, involves both chemical and physical processes. Processes such as ion exchange, which involves exchanging certain ions for others, are not used to any great extent in wastewater treatment.

(*c*) **Biological Treatment Methods** use microorganisms, mostly bacteria, in the biochemical decomposition of wastewaters to stable end products. More microorganisms, or sludges, are formed and a portion of the waste is converted to carbon dioxide, water and other end products. Generally, biological treatment methods can be divided into aerobic and anaerobic methods, based on availability of dissolved oxygen.

The purpose of wastewater treatment is generally to remove from the wastewater enough solids to permit the remainder to be discharged to a receiving water without interfering with its best or proper use. The solids which are removed are primarily organic but may also include inorganic solids. Treatment must also be provided for the solids and liquids which are removed as sludge. Finally, treatment to control odors, to retard biological activity, or destroy pathogenic organisms may also be needed.

There are generally four levels of wastewater Treatment—

(*i*) Preliminary treatment

(*ii*) Primary treatment

(*iii*) Secondary treatment

(*iv*) Tertiary advanced treatment

5.4 PRELIMINARY TREATMENT

The purpose of preliminary treatment is to protect the operation of the wastewater treatment plant. This is achieved by removing from the wastewater any constituents which can clog or damage pumps, or interfere with subsequent treatment processes. Preliminary treatment devices are, therefore, designed to:

1. Remove or to reduce in size the large, entrained, suspended or floating solids. These solids consist of pieces of wood, cloth, paper, plastics, garbage, etc., together with some fecal matter.
2. Remove heavy inorganic solids such as sand and gravel as well as metal or glass. These objects are called grit.
3. Remove excessive amounts of oils or greases.

A number of devices or types of equipment are used to obtain these objectives.

1. **Racks and Bar Screens.** These consist of bars usually spaced three-quarter inches to six inches. Those most commonly used provide clear openings of one to two inches. Although large screens are sometimes set vertically, screens are usually set at an angle of 45 to 60 degrees with the vertical. The incoming wastewater is passed through the bars or screens and periodically the accumulated material is removed. The racks or screens may be cleaned either manually or by means of automatically operated rakes. The solids removed by these units can be disposed of by burial or incineration.
2. **Comminuting Devices.** *Grinders, cutters* and *shredders*. These are devices to break or cut up solids to such size that they can be returned to the wastewater without danger of clogging pumps or piping or affecting subsequent treatment devices. They may be separate devices to grind solids removed by screens or a combination of screen and cutters installed within the wastewater flow channel in such a manner that the objective is accomplished without actually removing these larger solids from the wastewater. These latter devices are made by a

number of manufacturers under various trade names and, in most cases, consist of fixed, rotating or oscillating teeth or blades, acting together to reduce the solids to a size which will pass through fixed or rotating screens or grids having openings of about one-fourth inch. Some of these devices are even designed to operate as a low-lift pump. Unfortunately, many plants with comminuting devices develop problems within subsequent treatment units due to a build up of the shredded solids. This is usually witnessed in the aeration system of activated sludge plants. These shredded solids tend to clog diffusers and cling to the impeller blades of mechanical aerators.

3. **Grit Chambers.** Wastewater usually contains a relatively large amount of inorganic solids such as sand, cinders and gravel which are collectively called grit. The amount present in a particular wastewater depends primarily on whether the collecting sewer system is of the sanitary or combined type. Grit will damage pumps by abrasion and cause serious operation difficulties in sedimentation tanks and sludge digesters by accumulation around and plugging of outlets and pump suctions. Consequently, it is common practice to remove this material by grit chambers. Grit chambers are usually located ahead of pumps or comminuting devices, and if mechanically cleaned, should be preceded by coarse bar rack screens. Grit chambers are generally designed as long channels. In these channels the velocity is reduced sufficiently to deposit heavy inorganic solids but to retain organic material in suspension. Channel type chambers should be designed to provide controlled velocities as close as possible to 1.0 foot per second. Velocities substantially greater than 1.0 foot per second cause excessive organic materials to settle out with the grit. The detention period is usually between 20 seconds to 1.0 minute. This is attained by providing several chambers to accommodate variation in flow or by proportional weirs at the end of the chamber or other flow control devices which permit regulation of flow velocity. There are also patented devices to remove grit. One development is the injection of air several feet above the floor of a tank type unit. The rolling action of the air keeps the lighter organic matter in suspension and allows the grit relatively free from organic matter to be deposited in the quiescent zone beneath the zone of air diffusion. Excessive quantities of air can cause the roll velocity to be too high resulting in poor grit removal. Insufficient quantities of air result in low roll velocities and excessive organic matter will settle with the grit. These grit chambers are usually called aerated grit chambers.

4. **Cleaning.** Grit chambers are designed to be cleaned manually or by mechanically operated devices. If cleaned manually, storage space for the deposited grit is usually provided. Grit chambers for plants treating wastes from combined sewers should have at least two hand-cleaned units or a mechanically cleaned unit with by-pass. Mechanically cleaned grit chambers are recommended. Single, hand-cleaned chambers with by-pass, are acceptable for small wastewater treatment plants serving sanitary sewer systems. Chambers other than channel type are acceptable, if provided with adequate and flexible controls for agitation and/or air supply devices and with grit removal equipment.

   There are a number of mechanical cleaning units available which remove grit be scrapers or buckets while the grit chamber is in normal operation. These require much less grit storage space than manually operated units.

5. **Washing Grit.** Grit always contains some organic matter which decomposes and creates odors. To facilitate economical disposal of grit without causing nuisance, the organic matter is sometimes washed from the grit and returned to the wastewater. Special equipment is available to wash grit. Mechanical cleaning equipment generally provides for washing grit with wastewater as it is removed from the chamber.

6. **Quantity of Grit.** This depends on the type of sewer system, the condition of the sewer lines and other factors. Strictly domestic wastewater collected in well constructed sewers will contain little grit, while combined wastewater will carry large volumes of grit, reaching a peak at times of severe storms. In general, 1.0 to 4.0 cu.ft. of grit per million gallons of wastewater flow can be expected.
7. **Operation**. Manually cleaned grit chambers for combined wastewater should be cleaned after every large storm. Under ordinary conditions these grit chambers should be cleaned when the deposited grit has filled 50 to 60 percent of the grit storage space. This should be checked at least every ten days during dry weather.

   When mechanically cleaned grit chambers are used, they must be cleaned at regular intervals to prevent undue load on the cleaning mechanism. Recommendations of the manufacturer should be rigidly observed. This plus experience, will determine the cleaning schedule. A grit in which marked odors develop indicates that excessive organic matter is being removed in the grit chamber. Alternately, if sludge from a settling tank is excessively high in grit, or if there is excessive wear in pumps, comminutors, sludge collectors or other mechanical equipment, the reason is likely to be inefficient functioning of the grit removing process. In either case, a study of this unit should be made.
8. **Disposal of Screenings and Grit.** Screenings decompose rapidly with foul odors. They should be kept covered in cans at the screens and removed at least daily for disposal by burial or incineration. The walls and platforms of the screen chamber and screen itself should be hosed down and kept clean. Grit containing much organic matter may have to be buried to prevent odor nuisances.

Pre-Aeration Tanks

Pre-aeration of wastewater, that is aeration before primary treatment is sometimes provided for the following purposes:

1. To obtain a greater removal of suspended solids in sedimentation tanks.
2. To assist in the removal of grease and oil carried in the wastewater.
3. To freshen up septic wastewater prior to further treatment.
4. BOD reduction.

Pre-aeration is accomplished by introducing air into the wastewater for a period of 20 to 30 minutes at the design flow. This may be accomplished by forcing compressed air into the wastewater at a rate of about 0.10 cu.ft. per gallon of wastewater when 30 minutes of aeration is provided or by mechanical agitation whereby the wastewater is stirred or agitated so that new surfaces are continually brought into contact with the atmosphere for absorption of air. To insure proper agitation when compressed air is forced into the wastewater, air is usually supplied at the rate of 1.0 to 4.0 cubic feet per minute per linear foot of tank or channel. When air for mechanical agitation (either with or without the use of chemicals) is used for the additional purpose of obtaining increased reduction in BOD, the detention period should be at least 45 minutes at design flow. The agitation of wastewater in the presence of air tends to collect or flocculate lighter suspended solids into heavier masses which settle more readily in the sedimentation tanks. Pre-aeration also helps to separate grease and oil from the wastewater and wastewater solids and to carry them to the surface. By the addition of air, aerobic conditions are also restored in septic wastewater to improve subsequent treatment.

The devices and equipment for introducing the air into the wastewater are the same or similar to those used in the activated sludge process.

Pre-Chlorination

Pre-chlorination is the chlorination of a wastewater prior to primary treatment. In general, the objectives of pre-chlorination are not related to disinfection, and its use is related to either temporarily preventing further wastewater decomposition or reducing problems associated with wastewater decomposition. The objectives of pre-chlorination are:

1. Odor control
2. Protection of plant structures
3. Aid in sedimentation, and
4. Reduction or delay of Biochemical Oxygen Demand (BOD)

1. Odor Control

The decomposition of wastewater starts in sewers and becomes objectionable only after anaerobic decomposition has taken over. The degree of putrefaction that occurs is related to the time the wastewater is in the sewers which, in turn, depends on the length and grades of the sewers. Odor problems, therefore, develop where the sewers are long or where it is necessary to collect sewage in pump sumps and subsequently pump the wastewater to a treatment plant. There are few places in this state where the sewers are so long that putrefaction occurs to such a degree that offensive odors rise from the sewers before the wastewater reaches the wastewater treatment plant. If such a condition occurs, it may be possible to chlorinate the wastewater at a manhole on a trunk sewer. The amount of chlorine required varies depending on how long the decomposition of the wastewater must be delayed. It is not necessary to add sufficient chlorine to satisfy the chlorine demand, but merely sufficient to destroy odors and slow bacterial decomposition. Thus, no residual chlorine is produced. Doses of four to six mg/L are generally sufficient to control odors. Chlorine may be applied up sewer from the plant in forcemains, pump suction wells, screen chambers, grit chambers, trickling filter influent, settling tanks or wherever there is an odor problem. Normally, the practice is to start with a fairly high dose of chlorine (10 mg/L) to quickly control the odors, and gradually reduce the dose over a period of time to determine the minimum that will satisfy the local condition.

The production of offensive odors at pumping stations is a fairly common occurrence. Chlorination of the wastewater as it enters the pump sump or in the pump sump is effective as a preventative measure. The amount of chlorine required varies with the different situations but is less than that required to produce a residual. Generally, it is about the same as the chlorine demand or 25 to 50 lbs. per million gallons, but the minimum effective dose must be found by trial and error for each installation.

Another common occurrence is for wastewater to be septic, or a source of odor, as it is received at the wastewater plant. To prevent disagreeable odors during treatment, chlorination of the influent of the primary sedimentation tank is practiced which also aids in the settling properties of the sludge solids. If the purpose is only odor control and not disinfection, the chlorination need not be sufficient to produce a residual. Generally, a dose that will destroy all the reducing substances and thus slow the rate of decomposition is used. How great this dose must be depends to a large extent on how far putrefaction proceeded before the wastewater reached the plant. When putrefaction is far advanced, the chlorine dose may be equal to or greater than the dose which would produce a residual if the wastewater were fresh.

A similar situation may develop when the wastewater is received fresh but becomes septic during the treatment process. This often occurs in a new plant where the initial wastewater flow is far less than the design flow and the detention period in the primary tanks is greatly prolonged. Again pre-chlorination of the tank influent is used to delay putrefaction and resulting odors. In this case, the chlorine dose will be much less than that required if the wastewater were septic. The amount of reducing substances in the wastewater will be low and a dose of two to five mg/L of chlorine may be sufficient to prevent odors.

2. Protection of Plant Structures

Decomposition of wastewater can proceed to the point of hydrogen sulfide production, but, owing to location or low concentration, odors are not a problem. If this occurs in a pumping station, intercepting sewers or treatment plant, there may be serious corrosion. The remedy is similar to that for odor control -- chlorination sufficient to prevent hydrogen sulfide formation or to destroy hydrogen sulfide if it has been produced. The points of application are similar to those used for odor control but the quantity of chlorine may be less because only hydrogen sulfide has to be controlled. Minimum chlorine dose cannot be found without laboratory tests. In general though, this is a specific problem and the dose of chlorine can be found by trial and error. It may not be necessary to destroy all the hydrogen sulfide but only to reduce the concentration to one or two mg/L so that the amount evolved will be a minimum. Hydrogen sulfide causes structures to be damaged and weakened due to corrosion and can result in shutdown of the plant for repair. Generally, it is an economic problem, but factors other than cost must be considered. One such factor is the toxic nature of hydrogen sulfide.

3. Aid in Sedimentation

Pre-chlorination at the influent of a settling tank is sometimes practical for the benefit of improved settling. Generally, such benefits are incidental to the use of pre-chlorination for some other purpose. However, when there is a choice of the point of chlorine application, it is well to bear in mind that improved sedimentation, heavier sludge, and improved grease and oil separation are obtainable when chlorination of the primary influent is practiced.

4. Reduction or Delay of Biochemical Oxygen Demand

Chlorination of raw wastewater to produce a residual of 0.2 to 0.5 mg/L after 15 minutes contact may cause a reduction of 15 to 35 percent in the BOD of the wastewater. Generally, a reduction of at least 2 mg/L of 5 day BOD is obtained for each mg/L of chlorine applied up to the point at which a residual is produced. When units of a plant become overloaded, use can be made of chlorination to reduce the load until additional treatment facilities can be provided as the use of chlorine for BOD reduction is usually not economical. Chlorine is also used when the additional load is only temporary, such as when supernatant is returned from sludge digesters or when a plant receives intermittent discharges of industrial wastes.

Occasionally, chlorination of the plant effluent to a relatively high residual is practiced to delay or reduce the BOD load on receiving waters during short periods of extremely low stream flow. This is only an emergency procedure but does offer some aid under such conditions. Generally, the higher the residual carried the more the load is reduced, but care must be taken to prevent fish kills by chlorine.

Primary treatment is designed to remove organic and inorganic solids by the physical processes of sedimentation and flotation. Primary treatment devices reduce the velocity and disperse the flow of wastewater. In primary treatment the velocity of flow is reduced to 1 to 2 feet per minute to maintain a quiescent condition so that the material denser than water will settle out and material less dense than water will float to the surface. Approximately 40 to 60 percent of the suspended solids are removed from the waste stream (25–35% BOD reduction). The solids that remain in suspension as well as dissolved solids will usually be biochemically treated in subsequent processes for physical separation and removal in the final (secondary) settling tanks.

The size and number of primary tanks is dependent on the estimated wastewater flow and the design detention time. Generally, a detention time of 2 to 3 hours will provide a sufficient time period for most particles to settle out. Further, the settling rate of a particle depends on the strength and freshness of the wastewater being treated, the weight of the solid compared to the specific gravity of water, the size and shape of the solid and the temperature of the water. Water is more dense at lower temperatures; therefore, the required settling time increases. As the temperatures of the water increases, the required settling time decreases. Equal distribution of flow throughout the tank is critical. The greater the velocity in one area, the less the actual detention time. Solids not having sufficient time to settle out will be discharged in the effluent. Principle primary treatment devices are referred to as sedimentation tanks, primary tanks, primary clarifiers or primary settling tanks, some of which have the further function of providing an additional compartment for the decomposition of settled organic solids which is known as sludge digestion. There are several types of primary tanks in use.

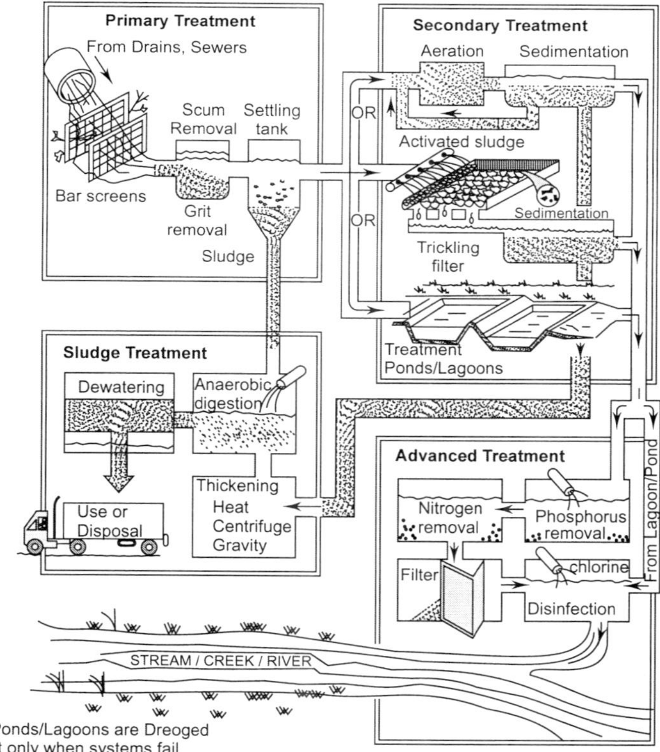

Fig. 5.2 Typical Wastewater Treatment Facility

Secondary Wastewater Treatment

Secondary treatment depends primarily upon aerobic organisms which biochemically decompose the organic solids to inorganic or stable organic solids. It is comparable to the zone of recovery in the self-purification of a stream.

The devices used in secondary treatment may be divided into four groups:

1. Trickling filters with secondary settling tanks
2. Activated sludge and modifications with final settling tanks
3. Intermittent sand filters
4. Stabilization ponds

The use of chlorine with secondary treatment is discussed under the section on Secondary Treatment.

Sludge Treatment

The solids removed from wastewater in both primary and secondary treatment units, together with the water removed with them, constitute wastewater sludge. It is generally necessary to subject sludge to some treatment to prepare or condition it for ultimate disposal. Such treatment has two objectives -- the removal of part or all of the water in the sludge to reduce its volume, and the decomposition of the putrescible organic solids to mineral solids or to relatively stable organic solids. This is accomplished by a combination of two or more of the following methods:

1. Thickening
2. Digestion with or without heat
3. Drying on sand bed -- open or covered
4. Conditioning with chemicals
5. Elutriation
6. Vacuum filtration
7. Heat drying
8. Incineration
9. Wet oxidation
10. Centrifuging

Package Units. The term "package units" is used in the field to describe equipment which has been put on the market by a number of manufacturers that is intended to provide wastewater treatment by the use of prefabricated or modular units. Package units can also refer to a complete installation, including both mechanisms and prefabricated containers. This term is also applied to installations where only the mechanisms are purchased and the containers constructed by the purchaser in accordance with plans and specifications prepared by the manufacturer.

Though specific limitations have not been established, individual package units have, in general, been small installations serving a limited population.

Package units have been adapted to practically all the treatment devices, either singly or in various combinations that have been mentioned.

Disinfection of Wastewater

Disinfection

Primary, secondary and even tertiary treatment cannot by expected to remove 100 percent of the incoming waste load and as a result, many organisms still remain in the waste stream. To prevent the spread of waterborne diseases and also to minimize public health problems, regulatory agencies may require the destruction of pathogenic organisms in wastewaters. While most of these microorganisms are not pathogens, pathogens must be assumed to be potentially present. Thus, whenever wastewater effluents are discharged to receiving waters which may be used for water supply, swimming or shellfishing, the reduction of bacterial numbers to minimize health hazards is a very desirable goal.

Disinfection is treatment of the effluent for the destruction of all pathogens. Another term that is sometimes also used in describing the destruction of microorganisms is *sterilization*. Sterilization is the destruction of *all* microorganisms. While disinfection indicates the destruction of all disease causing microorganisms, no attempt is made in wastewater treatment to obtain sterilization. However, disinfection procedures applied to wastewaters will result in a substantial reduction of all microbes so that bacterial numbers are reduced to a safe level.

In general, disinfection can be achieved by any method that destroys pathogens. A variety of physical or chemical methods are capable of destroying microorganisms under certain conditions. Physical methods might include, for example, heating to boiling or incineration or irradiation with X-rays or ultraviolet rays. Chemical methods might theoretically include the use of strong acids, alcohols, or a variety of oxidizing chemicals or surface active agents (such as special detergents). However, the treatment of wastewaters for the destruction of pathogens demands the use of practical measures that can be used economically and efficiently at all times on large quantities of wastewaters which have been treated to various degrees.

In the past, wastewater treatment practices have principally relied on the use of chlorine for disinfection. The prevalent use of chlorine has come about because chlorine is an excellent disinfecting chemical and, until recently, has been available at a reasonable cost. However, the rising cost of chlorine coupled with the fact that chlorine even at low concentrations is toxic to fish and other biota as well as the possibility that potentially harmful chlorinated hydrocarbons may be formed has made chlorination less favored as the disinfectant of choice in wastewater treatment. As a result, the increased use of ozone (ozonation) or ultraviolet light as a disinfectant in the future is a distinct possibility in wastewater disinfection. Both ozone and ultraviolet light, as well as being an effective disinfecting agent, leave no toxic residual. Ozone will additionally raise the dissolved oxygen level of the water. However, ozone must be generated and has only recently begun to compete favorably with chlorination in terms of economics. Ultraviolet light has recently undergone studies to determine its effectiveness and cost when used at large wastewater treatment plants. While the study is not yet complete, ultraviolet light now appears effective and economically competitive with chlorination as a disinfectant.

The use of both chlorine and ozone as chemical disinfectants and their disinfecting properties and actions will be considered individually. However, since chlorine continues to be used extensively as a disinfectant, we will mainly be concerned with the principles and practice of chlorination.

Solids in Wastewater

Since solids are classified in a variety of ways, they should be discussed with regard to the various categorizations that are used as well as with respect to their chemical make-up. There will, of course, be some overlap in the classification method.

(i) Organic Solids

In domestic wastewater, solids are about 50 percent organic. This fraction is generally of animal or vegetable life, dead animal matter, plant tissue or organisms, but may also include synthetic (artificial) organic compounds. These are substances which contain carbon, hydrogen and oxygen, some of which may be combined with nitrogen, sulfur or phosphorous. The principal organic compounds present in domestic wastewater are proteins, carbohydrates and fats together with the products of their decomposition. These compounds are subject to decay or decomposition through the activity of bacteria and other living organisms and are combustible, that is, they can be ignited or burned. Since the organic fraction can be driven off at high temperatures, they are sometimes called volatile solids.

(ii) Inorganic Solids

Inorganic solids are substances that are inert and not subject to decay. Exceptions to this characteristic are certain mineral compounds or salts -- such as sulfates -- which under certain conditions can be broken down. Inorganic solids are frequently called mineral substances and include sand, gravel and silt as well as the mineral salts in the water supply which produce the hardness and mineral content of the water. In general, they are non-combustible.

The amount of these solids, both organic and inorganic gives to a wastewater the characteristic frequently termed as its "strength". Actually, the amount or concentration of the organic solids and their capacity to undergo decay or decomposition is the most important part of this strength. The greater the concentration of organic or volatile solids, the stronger the wastewater. A "strong" wastewater can be defined as one containing a large amount of solids, particularly organic solids. A "weak" wastewater is one containing only a small amount of organic solids.

Solids can also be grouped depending on their physical state as suspended solids, colloidal solids and dissolved solids, each of which can include both organic and inorganic solids.

(iii) Suspended Solids

Suspended solids are those which are visible and in suspension in the water. They are the solids which can be removed from the wastewater by physical or mechanical means, such as sedimentation or filtration. More precisely, they are the solids which are retained on the filter mat or glass fiber pad in a Gooch Crucible. Suspended solids will include the larger floating particles and consist of sand, grit, clay, fecal solids, paper, pieces of wood, particles of food and garbage, and similar materials. Suspended solids are approximately 70 percent organic solids and 30 percent inorganic solids, the latter being principally sand and grit. The suspended solids portion consist of settleable solids and colloidal solids.

(iv) Settleable Solids

Settleable solids are that portion of the suspended solids which are of sufficient size and weight to settle in a given period of time, usually one hour. There are those which will settle in an Imhoff Cone in one hour. The results are reported as milliliters of settled solids per liter of wastewater. Settleable solids are approximately 75 percent organic and 25 percent inorganic.

(v) Colloidal Suspended Solids

Colloidal suspended solids are solids that are not truly dissolved and yet do not settle readily. These are somewhat loosely defined as the differences between the total suspended solids and the settleable solids. There is, at present, no simple or standard laboratory test to specifically determine colloidal matter. Most colloids will not settle out even after long quiescent periods of settling.

They constitute that portion of the total suspended solids (about 40 percent) which are not readily removed by physical or mechanical treatment facilities but may be filtered out in a Gooch Crucible. Colloidal solids are about 65 percent organic, 35 percent inorganic, subject to rapid decay, and are an important factor in the treatment and disposal of wastewater.

(vi) Dissolved Solids

The term "dissolved solids" as commonly used in discussing wastewater is not technically correct. Dissolved solids are smaller in size than suspended and colloidal solids. As used, the term means all of the solids which pass through the filter pad of a Gooch Crucible. Of the total dissolved solids, about 90 percent are in true solution and about 10 percent colloidal. Dissolved solids, as a whole, are about 40 percent organic and 60 percent inorganic.

(vii) Total Solids

Total solids, as the term implies, includes all of the solid constituents of a wastewater. Total solids are the total of the organic and inorganic solids or the total of the suspended and dissolved solids. In an average domestic wastewater, total solids are about half organic and half inorganic, and about two-thirds in solution (dissolved) and one-third in suspension. The organic solids, which are subject to decay, constitute the main problem in wastewater treatment.

(viii) Solids Determinations

The solid components of domestic wastewater can be classified in a number of ways. For example, wastewater solids can be categorized on the basis of several operational procedures used in the wastewater treatment laboratory. Total solids may be determined by driving off the water fraction, and suspended solids may be determined by filtering out the solid fraction on a porous pad and drying. Settleable solids may be determined by permitting a sample to settle in a special Imhoff Cone apparatus. The categories used most often in the wastewater treatment field are suspended solids and total solids. The colloidal fraction of domestic wastewaters comprises about 20 percent of the solid component of an "average" wastewater. As stated, this component is characterized by being non-settleable, that is, usually long periods of time would be required for them to settle by gravity alone. Colloidal components of wastewater do not fit neatly into a component that can be determined in the laboratory and colloids are found in both the suspended and dissolved solid fractions. Any estimate of wastewater composition can give only an average composition. The amounts of solids indicated cannot be applied equally to all wastewaters at all times.

5.5 TERTIARY AND ADVANCED WASTEWATER TREATMENT

The terms "primary" and "secondary" treatment have been used to generally describe a degree of treatment; for example, settling and biological wastewater treatment. Since the early 1970s "tertiary" treatment has come into use to describe additional treatment following secondary treatment. Quite often this merely indicates the use of intermittent sand filters for increased removal of suspended solids from the wastewater. In other cases, tertiary treatment has been used to describe processes which remove plant nutrients, primarily nitrogen and phosphorous, from wastewater.

Improvement and upgrading of wastewater treatment units as well as the need to minimize environmental effects has led to the increased use of tertiary treatment.

A term that is also sometimes used to indicate treatment of a wastewater by methods other than primary or biological (secondary) treatment is advanced treatment. This degree of treatment is usually achieved by chemical (for example coagulation) methods as well as physical methods (flocculation, settling and activated carbon adsorption) to produce a high quality effluent water.

Types of Advanced Wastewater Treatment

Advanced Wastewater Treatment may be broken into three major categories by the type of process flow scheme utilized:

1. Tertiary Treatment
2. Physical-Chemical Treatment
3. Combined Biological-Physical Treatment

Tertiary treatment may be defined as any treatment process in which unit operations are added to the flow scheme following conventional secondary treatment. Additions to conventional secondary treatment could be as simple as the addition of a filter for suspended solids removal or as complex as the addition of many unit processes for organic, suspended solids, nitrogen and phosphorous removal. Physical-chemical treatment is defined as a treatment process in which biological and physical-chemical processes are intermixed to achieve the desired effluent. Combined biological-physical-chemical treatment is differentiated from tertiary treatment in that in tertiary treatment any unit processes are added after conventional biological treatment, while in combined treatment, biological and physical-chemical treatment are mixed.

Another way to classify advanced wastewater treatment is to differentiate on the basis of desired treatment goals. Advanced wastewater treatment is used for:

(*i*) Additional organic and suspended solids removal

(*ii*) Removal of Nitrogenous Oxygen Demand (NOD)

(*iii*) Nutrient removal

(*iv*) Removal of toxic materials

(*v*) Advanced wastewater treatment plant effluents may be recycled directly or indirectly to increase the available domestic water supply.

(*vi*) Advanced wastewater treatment effluents may be used for industrial process or cooling water supplies.

(*vii*) Some receiving waters are not capable of withstanding the pollutional loads from the discharge of secondary effluents.

(*viii*) Secondary treatment does not remove as much of the organic pollution in wastewater as may be assumed.

The first three reasons for additional organic removal through advanced wastewater treatment are simple. The fourth requires some explanation. The performance of secondary treatment plants is almost always measured in terms of BOD and SS removals. A well designed and operated secondary plant will remove from 85 to 95% of the influent BOD and SS. However, the BOD test does not measure all of the organic material present in the wastewater. An average secondary effluent may have a BOD of 20 mg/L and a COD of 60 to 100 mg/L. The average secondary plant removes approximately 65% of the influent COD. Thus, when high quality effluents are required, additional organic removals must be accomplished. In addition to the organic materials remaining in most secondary effluents, there is an additional oxygen demand resulting from the nitrogen present in the wastewater.

In wastewaters, much of the nitrogen is found in the form of ammonia. When secondary treatment is used, a great deal of this ammonia is discharged in the effluent. Bacteria can utilize this ammonia as an energy source and convert ammonia to nitrite and nitrate.

$$NH_3 + O_2 + \text{Bacteria } NO_2 + O_2 + \text{Bacteria } NO_3$$

Another reason for advanced wastewater treatment may be to remove nutrients contained in discharges from secondary treatment plants. The effluents from secondary treatment plants contain both nitrogen (N) and phosphorous (P). N and P are ingredients in all fertilizers. When excess amounts of N and P are discharged, plant growth in the receiving waters may be accelerated. Algae growth may be stimulated causing blooms which are toxic to fish life as well as aesthetically unpleasing. Fixed plant growth may also be accelerated causing the eventual process of a lake becoming a swamp to be speeded up. Therefore, it has become necessary to remove nitrogen and phosphorous prior to discharge in some cases.

Toxic materials, both organic and inorganic are discharged into many sewage collections systems. When these materials are present in sufficient quantities to be toxic to bacteria, it will be necessary to remove them prior to biological treatment. In other cases, it is necessary to remove even small amounts of these materials prior to discharge to protect receiving waters or drinking water supplies. Thus, advanced wastewater treatment processes have been used in cases where conventional secondary treatment was not possible due to materials toxic to bacteria entering the plant as well as in cases where even trace amounts of toxic materials were unacceptable in plant effluents.

Nitrification

Biological nitrification may be used to prevent oxygen depletion from nitrogenous demand (NOD) in the receiving waters. Nitrification is simply the conversion of ammonia to nitrate in the treatment plant rather than in the receiving water. Nitrification may be carried out in the same tank as BOD removal or in a separate stage. Nitrification may be carried out either in activated sludge flocs or in fixed films.

Regardless of the particular scheme chosen, the same basic requirements for nitrification must be maintained:

- Oxygen level
- Loading rates
- Solids retention time
- Alkalinity
- pH
- Freedom from toxic materials
- Temperature

Sufficient oxygen must be available for nitrification to occur. Approximately 4.5 pounds of dissolved oxygen are required for the conversion of 1 pound of ammonia to nitrate. Dissolved oxygen sufficient to satisfy the remaining BOD is also required. In activated sludge plants the mixing requirement of the basins must also be considered. Generally, dissolved oxygen levels of approximately 2–3 mg/L are recommended for nitrification. The bacteria responsible for nitrification reproduce at a much slower rate than those responsible for BOD removal. Thus, the danger always exists for the "wash out" of the nitrifying organisms. That is, unless the nitrifying bacteria reproduce at the same or greater rate than they are removed from the system (by waste sludge) then the population of bacteria will be insufficient to carry out nitrification. For this reason, nitrification systems are operated at higher return sludge rates than conventional secondary treatment. The amount of sludge to be wasted is significantly less than from a conventional activated sludge system.

Nitrification systems are sensitive to pH variation. Optimum pH has been found to be approximately 7.8 to 9.0. Reductions in nitrification have been found outside this range. Alkalinity is also destroyed during nitrification. Theoretically, 7.2 pounds of alkalinity are destroyed in converting 1 pound of ammonia to nitrate. In low alkalinity wastewaters, Quick lime (CaO) or $Ca(OH)_2$ is often used to provide alkalinity and pH control.

Generally, the influent BOD to nitrification systems has not been found to effect performance. However, sufficient oxygen must be provided for the organic demand and organic shock loads must be avoided.

Biological Denitrification

Biological nitrification satisfies the nitrogenous oxygen demand by converting NH_3 to NO_3. In some applications, such as discharge into enclosed bodies of water or recycle to water supplies, nitrification may not be sufficient. When nitrogen removal is required, one of the available methods is to follow biological nitrification with biological denitrification.

Denitrification is accomplished under anaerobic or near anaerobic conditions by bacteria commonly found in wastewater. Nitrates are removed by two mechanisms:

1. Conversion of NO_3 to N_2 gas by bacterial metabolism and
2. Conversion of NO_3 to nitrogen contained in cell mass which may be removed by settling.

In order for denitrification to occur, a carbon source must be available. Most commonly, methanol is used. The methanol must be added in sufficient quantity to provide for cell growth and to consume any dissolved oxygen which may be carried into the denitrification reactor.

Usually 3 to 4 pounds of methanol per pound of nitrate are required. Careful control of methanol feed is necessary to prevent waste of chemicals. In addition, if excess methanol is fed to the system, unused methanol will be carried out in the effluent causing excessive BOD.

Denitrification may be carried out in either a mixed slurry reactor or in fixed bed reactors. Denitrification filters carry out both denitrification and filtration in the same unit. Mixed slurry systems consist of a denitrification reactor, reaeration basin and clarifiers. Reaeration prior to clarification is required to free the sludge from trapped bubbles of nitrogen gas.

Denitrifying bacteria grow very slowly and are extremely sensitive to temperature.

Denitrification rates have been shown to increase five-fold when the temperature is increased from 10 °C to 20 °C. Thus, operating parameters such as sludge age and retention time must be varied with temperature.

The pH in denitrification systems must be carefully controlled. The optimum pH is from 6.0 to 8.0.

Denitrification is a very sensitive and difficult process to operate. Little full scale operational experience is available. Constant monitoring of pH, methanol feed and temperature is essential to successful operation.

Filtration

Granular media filtration to remove those suspended and colloidal solids which are carried over from previous unit processes is a common unit process in advanced wastewater treatment.

Effluents of less than 10 mg/L BOD and 5 mg/L suspended solids are not uncommon for effluents from biological treatment processes after filtration.

Gravity filters similar to rapid sand filters are sometimes used. Often a combination of filter medias, such as anthracite coal and sand are used to provide coarse to fine filtration as the water passes through the filter. The water passes through the filter media and support gravel and is then collected by the underdrain system. As filtration proceeds, the headloss through the filter increases until it reaches an unacceptable level or until solids breakthrough occurs and the effluent becomes unacceptable. When either the headloss becomes excessive or solids breakthrough occurs, the filter is backwashed.

Gravity filters are generally run at 1.5 to 2.5 gpm per square foot. Pressure filters are used to obtain filter rates up to 6 gpm per square foot. Ideally, filters are designed to have the solids in the effluent and the headloss reach their allowable levels at the same time.

Summary

Advanced wastewater treatment can be used to achieve any level of treatment desired. Advanced wastewater treatment plants in Lake Tahoe, California and Windhoek, South Africa have been achieving drinking water quality since 1968. Advanced wastewater treatment plants utilize sophisticated processes and equipment. They are relatively expensive to run and operating costs as well as effluent quality are sensitive to the quality of operation.

Degrees of treatment are sometimes indicated by use of the terms primary, secondary and tertiary treatment. Tertiary treatment, properly, would be any treatment added onto or following secondary treatment.

5.6 ANAEROBIC DIGESTION

Anaerobic Digestion (AD) is a biological process similar in many ways to composting. It is a natural treatment process and, as in composting, bacteria break down organic matter and reduce its bulk or "mass".

Why are we excited about Anaerobic Digestion just now?

A remarkable combination of recent technical, political and regulatory factors is starting to propel AD forward toward the mainstream. Some of these factors are:

- **Rising oil prices:** Suddenly traditional energy sources are so much more expensive. In the past AD was a much more expensive energy source, and to be honest it remains so when all capital spend up-front is considered, but the gap is narrowing toward mainstream take-up of AD, and fast.
- **Farm waste requires better management:** As from June 30, 2006 the EU Waste Management rules apply to farming and agriculture, and their historical exemption has been removed.
- **Organic waste must be diverted from landfill:** This is the result of the EU landfill Directive and AD is an excellent technology for organic waste landfill diversion.
- The **demand for new sustainable technologies investment opportunities** powered by individuals but increasingly manifested in actions and investments by large companies and big investors. (January 2007 has seen the **big supermarkets** in the UK talking about using renewable energy technologies in their stores, including Anaerobic Digestion).

- **Rising demand for renewable fuels and bioenergy:** This includes government targets and incentives to Anaerobic Digestion which the United Kingdom and other governments provide to help them meet the targets.
- The availability of **Carbon Credits** to those who avoid Carbon Dioxide emissions: The developing world has been able to apply and obtain CER credits for nearly two years, and under the new US proposals (including the Methane to Markets project) it seems that the US may be soon paying carbon subsidies even to the developed nations where carbon emissions reductions are achieved in carbon neutral, and negative projects such as Anaerobic Digestion.

Is this a new process?

When the term "anaerobic" is used it may seem to the non-technical among us that science has conjured up a new form of digestion. Of course, this is not the case.

Methane bacteria are everywhere, and always have been. They not only occur widely in the bottom sediments of ponds, marshes, lakes and swamps, but much closer indeed in the intestinal tracts of man and many animals (especially cattle and other herbivores). Methane bacteria are usually found in association with sulphate-reducing bacteria.

Why are they all about us, you might ask? The reason is that they are scavengers, removing the decomposition products of other anaerobic bacteria and converting them to a mixture of carbon dioxide and methane gas. This mixture is that we call **"biogas"**.

What happens in an MSW Anaerobic Digestion Plant

Refuse Collection Vehicles (RCVs) deliver the collected waste to the plant and the degree of sorting then applied varies. Source separated garden and food waste often can go straight into the process, but mixed residual ('black bag') waste needs sophisticated sorting mechanically to remove the non-biodegradable contaminants. The plant in which this sorting is done is called a Materials Recycling Facility (MRF).

Sorting may involve screens, rotating drums for segregation, air classifiers, and powerful magnets. The organic waste fraction is then shredded and usually mixed with water. The waste and water slurry is then pumped into a sealed vessel where it is heated, and stirred where the it stays for up to about 3 weeks. This is known as the digestion or fermentation stage.

During this period the bacteria digest the waste and create a gas comprising of about sixty percent methane with the remainder being mostly carbon dioxide. This can be used as the source of the heat energy to warm the digestor(s), and there is usually sufficient methane left over to power an electricity generation set.

The process is normally continuous and filling and removal of the treated material takes place simultaneously. The output takes two forms. There is a solid digested material (digestate) which is often pressed to reduce the water content. The solid digestate is fibrous and can be used as a soil improver once it has been further matured usually by being placed in piles to aerobically compost, further reducing its weight, for about two weeks.

The digestate is very similar to compost once it has stood in the air for this period.

Unfortunately, even for most source segregated wastes there will be foreign matter, especially plastics etc, in the matured digestate. So, additional sorting is usually be required to remove contaminates before it can be used, and the most common is the use of a small mesh size screen.

The liquid fraction can be re-circulated back into the process, but in almost all process designs some excess water is generated. And depending on the removal of, or avoidance of, the presence of possible infectious agents from the feedstock, this can be used as a fertiliser. If the waste source was classed as contaminated (*e.g.*, food waste) and the waste is not then pasteurised within, or after, the digestion stage the resulting liquid product cannot be used on the land and has to be disposed of to sewer.

What is the Chemical Equation?

The reactions involved can be represented as follows:

$$CO_2 + 4H_2 \rightarrow CH_4 + 2H_2O$$ (*e.g.*, Methanobacterium bryantii)

$$CH_3COOH \rightarrow CH_4 + CO_2$$ (*e.g.*, Methanosarcina barkeri)

Syntrophic bacteria such as Syntrophobacter wolinii and Syntrophomonas wolfei provide the methane bacteria with their food supply via the reactions which convert propionate and other volatile acids into acetic acid and hydrogen gas.

Other supplies of food to the methanogens are other anaerobic (food decomposing) bacteria which decompose complex organic substances into volatile fatty acids.

Thus, methane bacteria provide the ultimate waste disposal system for a whole community of bacteria that grow together to provide the biological basis for what has become known as the **"anaerobic digestion process"**, which *occurs naturally in sludge digesters at sewage works and in damp putrescible deposits in landfill sites.*

Methane production also takes place in the bottom sediments of ponds, marshes, lakes and swamps, and phenomena referred to in old accounts such as the "will-o-the-wisp", no doubt originate from methane from such sources.

Effectively the **anaerobic digestion process** has been devised simply to **"engineer the inevitable"** and allow this **natural fermentation** to proceed to completion before discharging the fibrous **"digestate"** to landfill or to a **"composting plant"** further mixing and decomposition can take place to make the *digestate suitable for agricultural, and horticultural applications.*

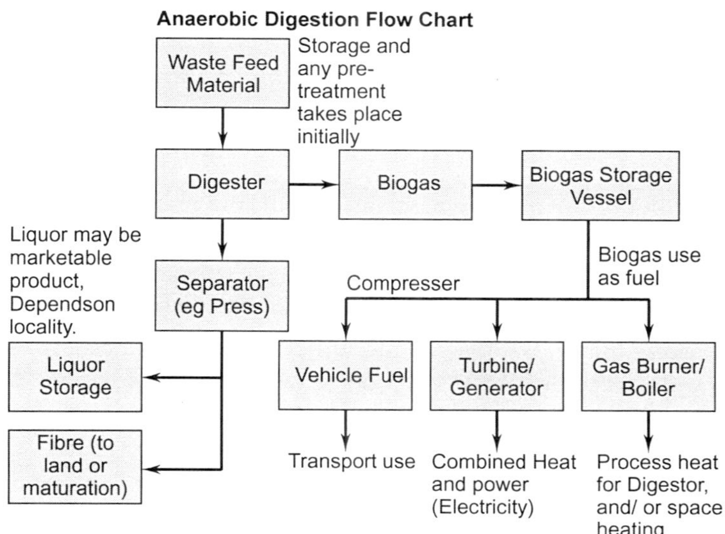

Fig. 5.3 Anaerobic Digestion Flow Chart

The Purpose of Anaerobic Digestion in MSW Treatment

The main purpose for Anaerobic Digestion in MSW Treatment has previously been to reduce the bulk (or "mass") of the material. The new generation of MSW anaerobic digesters are being promoted with the additional benefits of:

- Biogas biofuel production; which can be burnt for electricity production etc
- Sanitisation; due the AD Process' ability to operate at raised temperatures (usually using the biogas produced by the process) during the reaction, to ensure compliance with Animal By-Product Regulations (ABPR)
- Producing a digestate comprising a fibrous material, which can be spread on land, co-disposed with compost, or incinerated, and a liquor which (with care) can be used as a liquid fertiliser, but often must be further treated.

The popularity of anaerobic digestion is likely to rise as the demand for biofuel produced to replace fossil fuels increases with rising public awareness of the dangers of climate change. (See "Food for Thought", below.)

Anaerobic Digestion

Anaerobic digestion is a complex biochemical reaction carried out in a number of steps by several types of microorganisms that require little or no oxygen to live. During this process, a gas that is mainly composed of methane and carbon dioxide, also referred to as biogas, is produced. The amount of gas produced varies with the amount of organic waste fed to the digester and temperature influences the rate of decomposition and gas production.

Anaerobic digestion occurs in four steps:

- *Hydrolysis:* Complex organic matter is decomposed into simple soluble organic molecules using water to split the chemical bonds between the substances.
- *Fermentation or Acidogenesis:* The chemical decomposition of carbohydrates by enzymes, bacteria, yeasts, or molds in the absence of oxygen.
- *Acetogenesis:* The fermentation products are converted into acetate, hydrogen and carbon dioxide by what are known as acetogenic bacteria.
- *Methanogenesis:* Is formed from acetate and hydrogen/carbon dioxide by methanogenic bacteria.

The acetogenic bacteria grow in close association with the methanogenic bacteria during the fourth stage of the process. The reason for this is that the conversion of the fermentation products by the acetogens is thermodynamically only if the hydrogen concentration is kept sufficiently low. This requires a close relationship between both classes of bacteria.

The anaerobic process only takes place under strict anaerobic conditions. It requires specific adapted bio-solids and particular process conditions, which are considerably different from those needed for aerobic treatment.

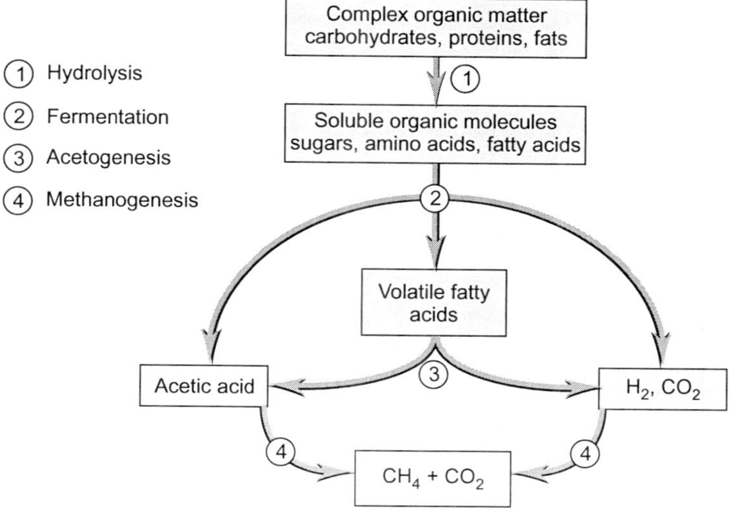

Fig. 5.4 Path of Anaerobic Digestion

Advantages of Anaerobic Digestion

Wastewater pollutants are transformed into methane, carbon dioxide and smaller amount of bio-solids. The biomass growth is much lower compared to those in the aerobic processes. They are also much more compact than the aerobic bio-solids.

Anaerobic Decomposition

A biological process, in which, decomposition of organic matter occurs without oxygen. Two processes occur during anaerobic decomposition. First, facultative acid forming bacteria use organic matter as a food source and produce volatile (organic) acids, gases such as carbon dioxide and hydrogen sulfide, stable solids and more facultative organisms. Second, anaerobic methane formers use the volatile acids as a food source and produce methane gas, stable solids and more anaerobic methane formers. The methane gas produced by the process is usable as a fuel. The methane former works slower than the acid former, therefore the pH has to stay constant consistently, slightly basic, to optimize the creation of methane. You need to constantly feed it sodium bicarbonate to keep it basic.

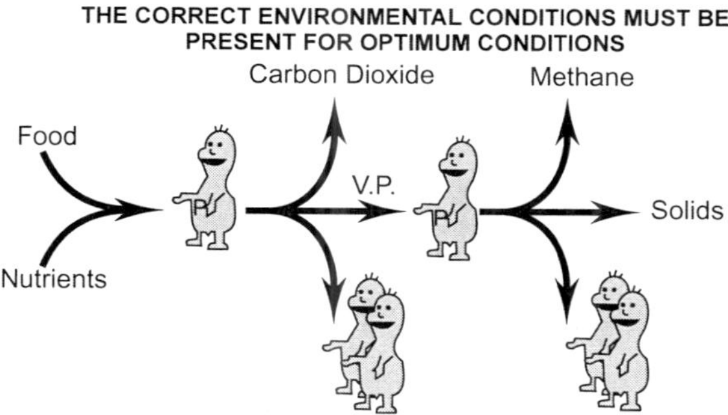

Advantages and Disadvantages of Anaerobic Treatment

The principal advantages and disadvantages of anaerobic treatment when compared with the aerobic equivalent processes are listed, and then discussed below.

Advantages of Anaerobic Treatment:

- Less energy required
- Less biological sludge produced
- Lower nutrient demand
- Methane production: Providing potential energy source with possible revenue both from sale of the energy, and benefit from government tax, and (Kyoto agreement) CDM etc., payments arising from renewable fuels/non-fossil fuel incentives
- Methane production: Anaerobic digestion contributes to reducing greenhouse gases by reducing demand for fossil fuels
- Smaller reactor volume required
- Biomass acclimatisation allows most organic compounds to be transformed
- Rapid response to substrate addition after long periods without feeding
- End product can be potentially saleable products biogas, soil conditioner and a liquid fertiliser.
- Process more effectively provides sanitisation/removal of diseases.

Disadvantages of Anaerobic Treatment:

- Longer start-up time to develop necessary biomass inventory
- May require alkalinity and/or specific ion addition
- May require further treatment with an aerobic treatment process to meet discharge requirements
- Biological nitrogen and phosphorus removal is not possible
- Much more sensitive to the adverse effect of lower temperatures on reaction rates
- May need heating (often by utilisation of process gas) to achieve adequate reaction rates
- May be more less stable after 'toxic shock' (eg after upsets due to toxic substances in the feed)
- Increased potential for production of odours and corrosive gases.
- Hazards arise from explosion. (In the EU, such additional Health & Safety Regulations as the ATEX Directive, and possibly also Gas Institute Regulations will require various compliance measures to be applied for AD.)
- Anaerobic treatment is not effective for treatment of methanogenic landfill leachate, it may (rarely) be efficacious for the early stage leachate production period while the waste is still acetogenic.

Biomethanation

Biomethanation is a general term for all the biological methods used for the production of methane biogas.

The methods available for biomethanation are:

1. Anaerobic Digestion in digestors (most of this web site is devoted to this method and it is also discussed below)
2. Methanogenesis in landfills. (For more information visit this Landfill Gas web site).
3. Upflow Anaerobic Sludge Blanket Process (UASB) (More ...)

Biomethanation by Anaerobic Digestion

There are a variety of different Anaerobic Digestion (AD) technologies available and the most well established are described in the standard water treatment textbooks. However, there is a new role for these processes as energy generators.

Historically, the most accomplished in AD Technologies have been the Chinese at the most simple low technology (mostly community and domestic) level — usually for gas supplies for cooking from sewage using AD.

Nowadays, in the large scale commercial Municipal Solid Waste (MSW) to energy market, there are an increasing number of European Companies emerging which have developed AD for MSW, over the past 10 to 20 years, who use variations of the AD process.

Potential Feed Anaerobic Digestion Plant Feed Materials

Another question posed is what is the biogas yield which can be expected from agricultural wastes, and this is also answered below.

- Dairy Farm and piggery etc., waste slurries
- Dairy Product and other Food Processing wastes
- Household Food Waste with or without Green Waste
- Brewing Industry effluents and wastes
- Supermarket Waste
- Municipal Solid Waste from Kerbside collections of Green Waste
- Municipal Solid Waste from Kerbside collections of Green Waste and Kitchen Waste
- Catering Industry Wastes
- Organic Market Wastes
- Sewage Treatment Works sludges
- Silage/ley crops (may be included and kept in storage to provide a feedstock for use to augment feed during winter seasons when the availability of other feed materials will be reduced)
- Other organic wastes available locally from other process industries.

Biogas Yield and Energy of Common Agricultural Feedstocks (Livestock Farms) and Food Waste

Feedstock	*Number of animals to produce 1 tonne/day*	*Dry Matter Content*	*Biogas Yield (M3/t)*	*Energy Value (MJ/m3) Biogas*
Cattle Slurry	20–40	12	25	23–25
Pig Slurry	250–300	9	26	21–25
Laying Hen Litter	8,000–9,000	30	90–150	23–27
Broiler Mannure	10,000–15,000	60	50–100	21–23
Food Waste	~	15	46	21–25

Anaerobic Digestion Design Biogas Digester

The terms anaerobic digestion design and biogas digester are synonymous, as is biomethanation design. However, on this web site we have chosen to use the term Anaerobic Digestion Plant design as a more generic term.

This is due to the fact that the term "biogas" suggests that the plant's primary purpose will be to produce biogas (methane), and there are many anaerobic digestion plants both in use today and historically that have not been installed primarily to produce and utilise biogas (methane for electricity generation, biofuel for transport vehicles, LPG for cooking and heating use etc,).

Anaerobic digestion design also encompasses plants which are primarily designed to:

1. **Treat an effluent** (as in industrial effluent treatment) to a quality which will allow it to be discharged to a sewer or to a watercourse according to the requirements of the site owner and the regulatory authorities;
2. **Treat the secondary (sludge) by-product** from a water treatment process to reduce volume, sanitise, and permit final disposal (*e.g.*, to land); as in the digestion of sludge created during the popular (aerobic) activated sludge sewage treatment process;
3. **Treat Solid Waste** (as in Municipal Solid Waste (MSW)) to primarily help meet local government targets (especially within the European Union (EU)) where strict targets must be achieved by member states to divert the overall tonnage of MSW away from landfill, and also to reduce still further the amount of organic/biodegradable content within the waste which is sent to landfill.
4. It is only since about 2003 in Europe, but possibly a century earlier in China, that the creation of biogas has been a viable function of anaerobic digestion plants. The reason that AD biogas has now become a valued commodity is that fossil fuel sources of methane gas have risen so much in price that (with some encouragement from government grants and tax breaks) biogas methane can be cheaper.
5. **Biogas** is, when scrubbed and pressurised, equivalent in composition and broad calorific value to the fossil fuel known as **Liquid Natural Gas (LNG)**. and at lower pressure can double for **Natural Gas** in town and city gas supply pipelines.
6. Just as the primary purpose, and scale, of Anaerobic Digestion Plants is very diverse so are the processes which have been developed to meet these challenges. Many processes are similar in principle, but do vary substantially according to the primary purpose and environmental, plus economic and political drivers.

Review

The aerobic, anoxic, and anaerobic process helps prepare the waste for decomposition by attacking the organic compounds that in wastewater. In aerobic decomposition the only microorganisms that can survive are the aerobic organisms. In anoxic decomposition the microorganisms use chemically combined oxygen that is found in nitrite and nitrate. In this process the organisms consume the organic matter to help support their life functions. In the first stage of anaerobic decomposition, acid forming bacteria use the organic matter for food which produces volatile gases, acids and facultative organisms. Second stage methane formers use the volatile acids as a food source and that produces more gases. This gas can be used as fuel.

SEPTIC TANK

For homeowners in rural areas (or places with no connection to main sewage pipes), septic tanks are the principal means of waste water disposal. If you are planning to move to a more rural setting, probably for retirement, of course, you will need to have a septic tank system.

The first known installation of a septic tank in the United States was in 1876, although Louis Mouras of Vesoul, France, was given a patent in 1881 and credited with the invention. Baffles, which regulate the flow, were added in 1905 to make the septic tank more efficient. The first baffles were made of oak boards.

At the turn of the century, there were some very large community septic tanks. In 1903, four community tanks were constructed in Saratoga, New York, with a total capacity of one million gallons.

By 1920, septic tanks began to be a common feature. After World War II, septic tanks became important to housing developments in unsewered areas.

There is nothing really complex about how a septic tank works. But, mind you, for all its simplicity, it is very efficient at what it does. The tank can be anywhere from 1000 gallons or higher and thus can accommodate large amounts of wastes. In a nutshell, septic tanks separate solid waste from liquids. The solid waste is stored in the septic tank and treated by anaerobic bacteria. The liquids, on the other hand, are dispersed throughout the soil on the property by the leaching, a mechanism which is a part of the septic tank.

The use of septic tanks may not sound sanitary but there is really nothing to worry about how they treat wastes. Well, septic tanks have been in use for many years all over the country. Nearly 25% of all homes in the country use septic tank systems.

It is important to note here that septic tanks can go wrong. People with septic tanks need to make sure that they are properly maintaining their tanks through regular cleaning. How do you clean septic tanks? To learn more about septic tanks and how to maintain them, read the rest of our pages. We have separate pages on septic tank cleaning, septic tank installation, septic tank disposal, and septic tank treatment so you will definitely find what you are looking for with regards to septic tanks here.

Septic Tank Function

Sewage or untreated household waste will quickly clog all but the most porous gravel formations. The septic tank conditions sewage to allow percolation of the liquid portion into the subsoil. The most important function of septic tanks is to protect the absorption ability of the subsoil. In doing this the septic tank does the following three things.

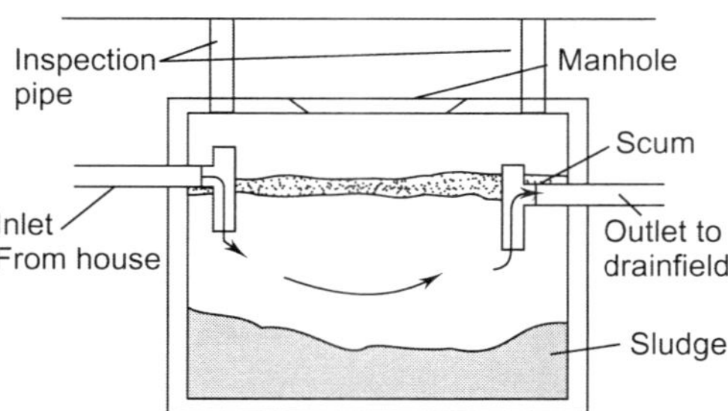

Fig. 5.5 Septic Tank

Removes solids from liquid. As sewage enters the tank, the rate of flow is reduced and heavy solids settle, forming sludge. Grease and other light solids rise to the surface, forming a scum. The sludge and scum are retained and breakdown while the clarified effluent (liquid) is discharged to the drainfield for soil absorption.

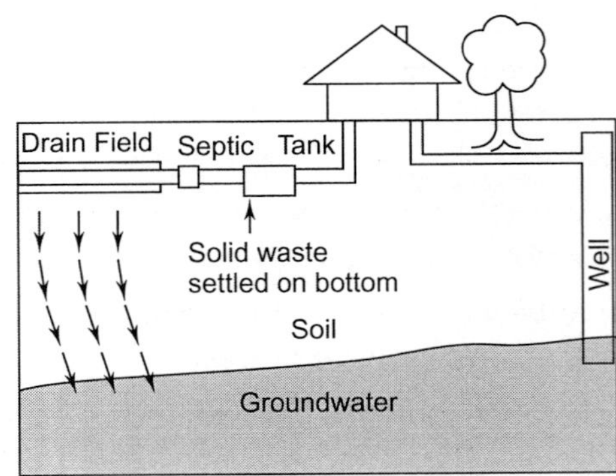

Fig. 5.6 Septic Systems can effect Groundwater

Provides biological treatment. Natural processes break down the solids and liquids by bacterial action. The breakdown occurs in the absence of oxygen (anaerobic conditions). The anaerobic conditions are referred to as "septic," giving the tank its name.

Stores scum and sludge. The solids accumulate in the bottom of the tank to form sludge. The scum is a partially submerged mat of floating solids and grease. Scum and sludge are digested over time and compact-ed into a small volume. Areas with warm climates, such as the southern United States, allow more complete breakdown of solids and scum than in the cooler climates of the North. For this reason, tanks in warm climates do not usually need to be pumped or cleaned out nearly as often as those in cold climates. Regardless of climate, a non-volatile residue of material remains in the tank. Sufficient volume for the solids must be provided in the tank between pumpings or cleanings. If the solids fill the tank and enter the drainfield, the solids can clog the soil in the drainfield.

Grease from the kitchen is detrimental to septic tank functions. Effluent from grease traps must go through septic tanks before being discharged to drainfields to prevent soil plugging. The best approach is to put kitchen grease in old milk jugs and place in the garbage rather than into the drain. Small amounts of kitchen grease can go into the septic tank without damaging the system.

Septic Tank Installation

Septic tank installation is more complicated than you think. Yes, septic tank installation is a rocket science affair. If you wish to install a new septic tank system, you must check on your property before calling for professional assistance. Does it have enough room for a septic tank? They do not do the measuring, you do it. Putting up a septic system requires you to know about water tables and soil types as well. It is imperative to know the type of soil that your property carries. Depending on the soil condition, the septic system may not be as efficient in treating waste water. If you have questions, enlist the help of the municipal office. They can help you with measuring your lot and identifying your lot's soil properties.

Septic tank installation needs the backing power of permits. A trip to the local municipal office needs to be made to be able to start with the actual installation.

If you want to save on your septic tank installation, you have to go through a lot of deciding. This is another thing that makes septic tank installation a rocket science process. Given the many options, it is really hard to pick the right septic tank installation company. Your best bet is to choose the one that offer rates that are fair enough and who you feel can work your instructions best. Think that you can do the installation yourself? Septic tank installation is not a do-it-yourself affair. It can only be done by people with experience.

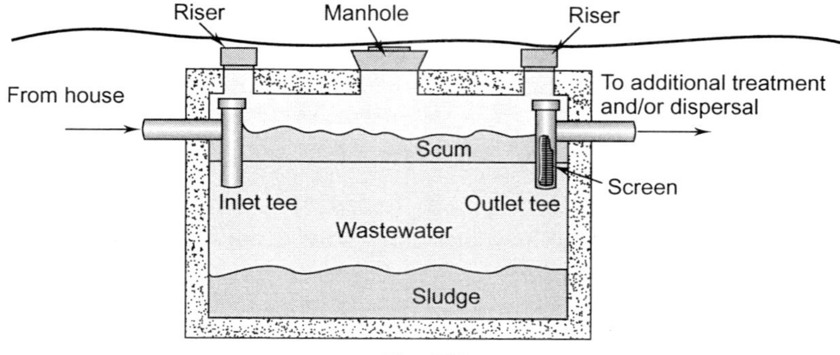

Fig. 5.7

Septic Tank Cleaning

The thing with septic tanks is, when something in them goes wrong, it can be a costly and messy affair. When it comes to septic tank maintenance, "prevention is better than cure". People with septic tanks need to properly maintain their tanks in order to avoid spending too much money for repair or treatment. Regular septic tank cleaning is the only way to prevent septic tank systems from clogging and well, causing too much financial troubles. Always remember that the more you do to keep your tank in good order, the more years it will keep working great.

So, how do you clean septic tanks? There is really only one way to clean septic tanks and that is by pumping them. Septic tank cleaning through pumping removes the solid build up which is an inevitable part of the septic process. Most septic tank system experts recommend that a typical home with a 1,000-gallon tank should be pumped every three to five years. A smaller tank, of course, needs to be pumped more often.

Septic tank maintenance is not rocket science; it is just all about you understanding the importance of regular septic tank cleaning. If your home uses a septic tank, it is your responsibility to properly maintain that tank. Failing to provide regular treatment and cleaning to your tank, will only give you a whole new set of issues to mind. Replacing "dead" septic tanks can be really costly, so being particular about septic tank cleaning is to your advantage.

Septic Tank Disposal

The toilet is not a garbage bin. Flushing cat litter, tissue, sanitary napkins, among others in the toilet will only double the amount of solids added to your septic tank. Garbage disposal, needless to say, leads to a more rapid buildup of sludge layers in the septic tank and increases the risk of clogging. You pay for septic tank cleaning and septic tank additives to prevent your septic tank from clogging. By carelessly flushing random junk in the toilet, you are countering your personal measures. In other words, you are putting your efforts to waste.

Not every from of septic tank disposal is bad, however. In some cases, it can be an advantage to helping your tank function efficiently. Dumping liquefied waste food with added enzymes, for instance can help with keeping the natural process inside the septic tank alive. By grinding waste into fine particles, septic garbage disposal devices bring down additional helpful enzymes into the waste stream and somehow help septic tanks function more efficiently.

But is it really safe to dump "unfit" waste in the septic tank? According to a lot of experts, no. If you really want a healthy septic tank, avoid septic tank garbage disposal totally. Septic tank garbage disposal devices still discharge untreated solids into the tank, which then reduces the efficiency of the tank. Reduced tank efficiency means it failing and you spending big for repairs.

Spetic Tank Treatment

It is easy to keep your septic system running clean. In order to prevent solid waste from backing up into your septic tank, aside from regularly pumping your tank, it is a clear incentive to employ septic tank treatment.

Septic tank treatment is an easy way to maintain your septic tank. The process involves the use of septic tank additives. As said, when it comes to septic tank maintenance, prevention is better than cure. The use of such chemicals will prevent your septic tank from clogging—the biggest problem concerning septic tanks. Septic tank treatment makes for a simple and cost effective way to ensure that your solid waste is being dealt with effectively. It will ensure that you have an ample amount of bacteria and enzymes in your septic tank that are necessary in keeping the system free from clog. Septic system replacement is very expensive. Adding septic tank additives is a way of guaranteeing that there are always plenty of active bacteria in your tank.

Employing septic tank treatment by using septic tank additives, although can be costly, is the only way to ensure that you will always have your tank working efficiently. By reintroducing new batches of bacteria and enzymes to your tank, you can be sure that your solid waste is always being broken down and will not back up.

Reuse of Water: What is Water Recycling?

While recycling is a term generally applied to aluminum cans, glass bottles, and newspapers, water can be recycled as well. Water recycling is reusing treated wastewater for beneficial purposes such as agricultural and landscape irrigation, industrial processes, toilet flushing, and replenishing a ground water basin (referred to as ground water recharge).

Water is sometimes recycled and reused onsite; for example, when an industrial facility recycles water used for cooling processes. A common type of recycled water is water that has been reclaimed from municipal wastewater, or sewage. The term water recycling is generally used synonymously with water reclamation and water reuse.

Through the natural water cycle, the earth has recycled and reused water for millions of years. Water recycling, though, generally refers to projects that use technology to speed up these natural processes. Water recycling is often characterized as "unplanned" or "planned." A common example of unplanned water recycling occurs when cities draw their water supplies from rivers, such as the Colorado River and the Mississippi River, that receive wastewater discharges upstream from those cities. Water from these rivers has been reused, treated, and piped into the water supply a number of times before the last downstream user withdraws the water. Planned projects are those that are developed with the goal of beneficially reusing a recycled water supply.

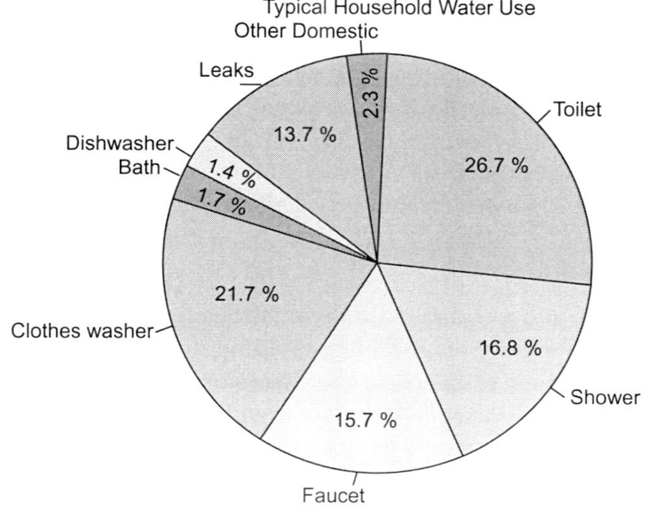

American Water Works Association, 1999

How Can Recycled Water Benefit Us?

Recycled water can satisfy most water demands, as long as it is adequately treated to ensure water quality appropriate for the use. Figure 1 shows types of treatment processes and suggested uses at each level of treatment. In uses where there is a greater chance of human exposure to the water, more treatment is required. As for any water source that is not properly treated, health problems could arise from drinking or being exposed to recycled water if it contains disease-causing organisms or other contaminants.

The US Environmental Protection Agency regulates many aspects of wastewater treatment and drinking water quality, and the majority of states in the US have established criteria or guidelines for the beneficial use of recycled water. In addition, in 2004, EPA developed a technical document entitled "Guidelines for Water Reuse," which contains such information as a summary of state requirements, and guidelines for the treatment and uses of recycled water. State and Federal regulatory oversight has successfully provided a framework to ensure the safety of the many water recycling projects that have been developed in the United States.

Recycled water is most commonly used for nonpotable (not for drinking) purposes, such as agriculture, landscape, public parks, and golf course irrigation. Other nonpotable applications include cooling water for power plants and oil refineries, industrial process water for such facilities as paper mills and carpet dyers, toilet flushing, dust control, construction activities, concrete mixing, and artificial lakes.

Although most water recycling projects have been developed to meet nonpotable water demands, a number of projects use recycled water indirectly[1] for potable purposes. These projects include recharging ground water aquifers and augmenting surface water reservoirs with recycled water. In ground water recharge projects, recycled water can be spread or injected into ground water aquifers to augment ground water supplies, and to prevent salt water intrusion in coastal areas. For example, since 1976, the Water Factory 21 Direct Injection Project, located in Orange County, California, has been injecting highly treated recycled water into the aquifer to prevent salt water intrusion, while augmenting the potable ground water supply.

While numerous successful ground water recharge projects have been operated for many years, planned augmentation of surface water reservoirs has been less common. However, there are some existing projects and others in the planning stages. For example, since 1978, the upper Occoquan Sewage Authority has been discharging recycled water into a stream above Occoquan Reservoir, a potable water supply source for Fairfax County, Virginia. In San Diego, California, the Water Repurification Project is currently being planned to augment a drinking water reservoir with 20,000 acre-feet per year of advanced treated recycled water.

What are the Environmental Benefits of Water Recycling?

In addition to providing a dependable, locally-controlled water supply, water recycling provides tremendous environmental benefits. By providing an additional source of water, water recycling can help us find ways to decrease the diversion of water from sensitive ecosystems. Other benefits include decreasing wastewater discharges and reducing and preventing pollution. Recycled water can also be used to create or enhance wetlands and riparian habitats.

Water Recycling Can Decrease Diversion of Freshwater from Sensitive Ecosystems

Plants, wildlife, and fish depend on sufficient water flows to their habitats to live and reproduce. The lack of adequate flow, as a result of diversion for agricultural, urban, and industrial purposes, can cause deterioration of water quality and ecosystem health. Water users can supplement their demands by using recycled water, which can free considerable amounts of water for the environment and increase flows to vital ecosystems.

Water Recycling Decreases Discharge to Sensitive Water Bodies

In some cases, the impetus for water recycling comes not from a water supply need, but from a need to eliminate or decrease wastewater discharge to the ocean, an estuary, or a stream. For example, high volumes of treated wastewater discharged from the San Jose/Santa Clara Water Pollution Control Plant into the south San Francisco Bay threatened the area's natural salt water marsh. In response, a $140 million recycling project was completed in 1997. The South Bay Water Recycling Program has the capacity to provide 21 million gallons per day of recycled water for use in irrigation and industry. By avoiding the conversion of salt water marsh to brackish marsh, the habitat for two endangered species can be protected.

Recycled Water May Be Used to Create or Enhance Wetlands and Riparian (Stream) Habitats

Wetlands provide many benefits, which include wildlife and wildfowl habitat, water quality improvement, flood diminishment, and fisheries breeding grounds. For streams that have been impaired or dried from water diversion, water flow can be augmented with recycled water to sustain and improve the aquatic and wildlife habitat.

Water Recycling Can Reduce and PreventPollution. When pollutant discharges to oceans, rivers, and other water bodies are curtailed, the pollutant loadings to these bodies are decreased. Moreover, in some cases, substances that can be pollutants when discharged to a body of water can be beneficially reused for irrigation. For example, recycled water may contain higher levels of nutrients, such as nitrogen, than potable water. Application of recycled water for agricultural and landscape irrigation can provide an additional source of nutrients and lessen the need to apply synthetic fertilizers.

What is the Future of Water Recycling?

Water recycling has proven to be effective and successful in creating a new and reliable water supply, while not compromising public health. Nonpotable reuse is a widely accepted practice that will continue to grow. However, in many parts of the United States, the uses of recycled water are expanding in order to accommodate the needs of the environment and growing water supply demands. Advances in wastewater treatment technology and health studies of indirect potable reuse have led many to predict that planned indirect potable reuse will soon become more common.

While water recycling is a sustainable approach and can be cost-effective in the long term, the treatment of wastewater for reuse and the installation of distribution systems can be initially expensive compared to such water supply alternatives as imported water or ground water. Institutional barriers, as well as varying agency priorities, can make it difficult to implement water recycling projects. Finally, early in the planning process, agencies must implement public outreach to address any concerns and to keep the public involved in the planning process.

As water demands and environmental needs grow, water recycling will play a greater role in our overall water supply. By working together to overcome obstacles, water recycling, along with water conservation, can help us to conserve and sustainably manage our vital water resources.

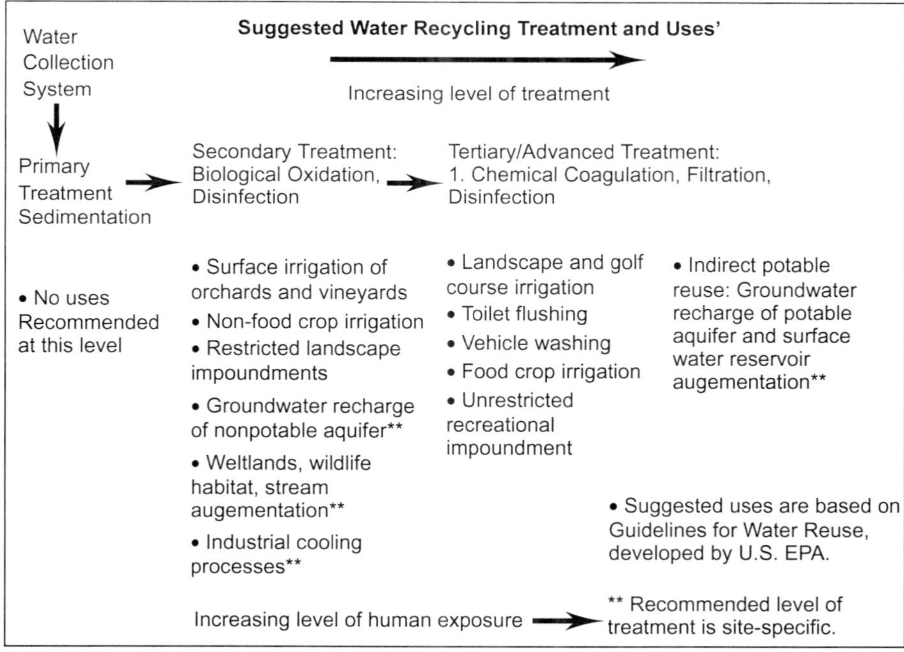

Water Saving Practice: Rain Water Harvesting

What is Rain Water Harvesting?

The principle of collecting and using precipitation from a catchments surface.

An old technology is gaining popularity in a new way. Rain water harvesting is enjoying a renaissance of sorts in the world, but it traces its history to biblical times. Extensive rain water harvesting apparatus existed 4000 years ago in the Palestine and Greece. In ancient Rome, residences were built with individual cisterns and paved courtyards to capture rain water to augment water from city's aqueducts. As early as the third millennium BC, farming communities in Baluchistan and Kutch impounded rain water and used it for irrigation dams.

Artificial Recharge to Ground Water

Artificial recharge to ground water is a process by which the ground water reservoir is augmented at a rate exceeding that obtaining under natural conditions or replenishment. Any man-made scheme or facility that adds water to an aquifer may be considered to be an artificial recharge system.

Why Rain Water Harvesting?

Rain water harvesting is essential because:

- Surface water is inadequate to meet our demand and we have to depend on ground water.
- Due to rapid urbanization, infiltration of rain water into the sub-soil has decreased drastically and recharging of ground water has diminished.

Rain Water Harvesting Techniques

There are two main techniques of rain water harvestings:

1. Storage of rainwater on surface for future use.
2. Recharge to ground water.

The storage of rain water on surface is a traditional techniques and structures used were underground tanks, ponds, check dams, weirs etc. Recharge to ground water is a new concept of rain water harvesting and the structures generally used are:

Pits: Recharge pits are constructed for recharging the shallow aquifer. These are constructed 1 to 2 m, wide and to 3 m. deep which are back filled with boulders, gravels, coarse sand.

Trenches: These are constructed when the permeable stram is available at shallow depth. Trench may be 0.5 to 1 m. wide, 1 to 1.5 m. deep and 10 to 20 m. long depending up availability of water. These are back filled with filter. materials.

Dug Wells: Existing dug wells may be utilised as recharge structure and water should pass through filter media before putting into dug well.

Hand Pumps: The existing hand pumps may be used for recharging the shallow/deep aquifers, if the availability of water is limited. Water should pass through filter media before diverting it into hand pumps.

Recharge Wells: Recharge wells of 100 to 300 mm. diameter are generally constructed for recharging the deeper aquifers and water is passed through filter media to avoid choking of recharge wells.

Recharge Shafts: For recharging the shallow aquifer which are located below clayey surface, recharge shafts of 0.5 to 3 m. diameter and 10 to 15 m. deep are constructed and back filled with boulders, gravels & coarse sand.

Lateral Shafts with Bore Wells: For recharging the upper as well as deeper aquifers lateral shafts of 1.5 to 2 m. wide and 10 to 30 m. long depending upon availability of water with one or two bore wells are constructed. The lateral shafts is back filled with boulders, gravels and coarse sand.

Spreading Techniques: When permeable strata starts from top then this technique is used. Spread the water in streams/Nalas by making check dams, nala bunds, cement plugs, gabion structures or a percolation pond may be constructed.

5.7 DIVERSION OF RUN OFF INTO EXISTING SURFACE WATER BODIES

Construction activity in and around the city is resulting in the drying up of water bodies and reclamation of these tanks for conversion into plots for houses.

Free flow of storm run off into these tanks and water bodies must be ensured. The storm run off may be diverted into the nearest tanks or depression, which will create additional recharge.

Urbanisation effects on Groundwater Hydrology:

(*i*) Increase in water demand
(*ii*) More dependence on ground water use
(*iii*) Over exploitation of ground water
(*iv*) Increase in run-off, decline in well yields and fall in water levels
(*v*) Reduction in open soil surface area
(*vi*) Reduction in infiltration and deterioration in water quality.

Methods of Artificial Recharge in Urban Areas:

(*i*) Water spreading
(*ii*) Recharge through pits, trenches, wells, shafts
(*iii*) Rooftop collection of rainwater
(*iv*) Roadtop collection of rainwater
(*v*) Induced recharge from surface water bodies.

Computation of Artificial Recharge from Roof Top Rainwater Collection

Factors taken for computation:

Roof top area 100 sq.m. for individual house and 500 sq.m. for multi-storied building.

Average annual monsoon rainfall – 780 mm.

Effective annual rainfall contributing to recharge 70% – 550 mm.

	Individual Houses	*Multistoried Building*
Roof top area	100 sq. m.	500 sq. m.
Total quantity available for recharge per annum	55 cu. M.	275 cu. M.
Water available for 5 member Family	100 days	500 days

Benefits of Artificial Recharge in Urban Areas:

- Improvement in infiltration and reduction in run-off.
- Improvement in groundwater levels and yields.
- Reduces strain on Special Village Panchayats/Municipal/ Municipal Corporation water supply
- Improvement in groundwater quality
- Estimated quantity of additional recharge from 100 sq. m. roof top area is 55.000 liters.

5.8 HARVESTING RAINWATER HARNESSING LIFE

A Noble Goal — A Common Responsibility

Groundwater exploitation is inevitable is Urban areas. But the groundwater potential is getting reduced due to urbanisation resulting in over exploitation. Hence, a strategy to implement the groundwater recharge, in a major way need to be launched with concerted efforts by various Governmental and Non-Governmental Agencies and Public at large to build up the water table and make the groundwater resource, a reliable and sustainable source for supplementing water supply needs of the urban dwellers.

Recharge of groundwater through storm run off and roof top water collection, diversion and collection of run off into dry tanks, play grounds, parks and other vacant places are to be implemented by Special Village Panchayats/Municipalities/Municipal Corporations and other Government Establishments with special efforts.

The Special Village Panchayats/Municipalities/Municipal Corporations will help the citizens and builders to adopt suitable recharge method in one's own house or building through demonstration and offering subsidies for materials and incentives, if possible.

5.9 ATTRIBUTES OF GROUNDWATER

There is more groundwater than surface water:

- Groundwater is less expensive and economic resource.
- Groundwater is sustainable and reliable source of water supply.
- Groundwater is relatively less vulnerable to pollution.
- Groundwater is usually of high bacteriological purity.
- Groundwater is free of pathogenic organisms.
- Groundwater needs little treatment before use.
- Groundwater has no turbidity and colour.
- Groundwater has distinct health advantage as art alternative for lower sanitary quality surface water.
- Groundwater is usually universally available.
- Groundwater resource can be instantly developed and used.
- There is no conveyance losses in groundwater based supplies.
- Groundwater has low vulnerability to drought.
- Groundwater is key to life in arid and semi-arid regions.
- Groundwater is source of dry weather flow in rivers and streams.

ONSITE SANITATION:

There are different technologies available for dealing with waste water management and excreta. Some are suitable for dealing with off-site sanitation and others can be applied to on-site sanitation.

In this document the concept of on-site sanitation will be described, followed by the description of different types of latrines.

What is on-site sanitation?

1. *What is sanitation?*

   Sanitation refers to excreta and waste water management as well as runoff water and solid and industrial waste.

   It is the whole range of strategies used to solve problems raised by excreta, solid and industrial waste and runoff water, excluding production and distribution of drinking water. Two main types can be distinguished:

   - off-site sanitation which is appropriate for large scale exploitation, based on technical and economic feasibility studies (sewer networks, runoff water drains, etc)

   - and on-site sanitation.

2. *The concept of on-site sanitation*

   (*a*) *Definition.* On-site sanitation is the whole of actions related to the treatment and disposal of domestic waste water that cannot be carried away by an off-site sanitation system because of low density of population.

   (*b*) *Characteristics of on-site sanitation.* The autonomous part of on-site sanitation refers to the techniques used design, as well as its financing, implementation and maintenance.

   Because the autonomous part of on-site sanitation is defined as such it is often regarded as the opposite of off-site sanitation, although the boundary between the two systems is not always as clear and well-identified.

   In this context, we talk about:

   - individual on-site sanitation, when a house (plot) makes use of the soil as a treatment medium (example of soak-away, latrines, etc.,)
   - grouped on-site sanitation (or semi off-site sanitation), when many individual houses are linked to a network leading to a treatment system, or small communities: grouped sanitation doesn't always use the soil as treatment medium (filtration beds, activated sludge are examples of purification systems at the end of the chain).

   This classification is itself characterized by boundaries that are not clearly defined.

On-site Sanitation Latrines

1. *Definition of a latrine.* It is a site or a structure, located normally outside the house or building, destined to receive and store excreta and sometimes, to process it. The latrine is the most used sanitation system in the world. (Different types of appropriate latrines are used in on-site sanitation. This type of sanitation protects sensitive species of aquatic fauna by avoiding discharge concentrations in small waterways. Besides, it is cheaper than off-site sanitation, since the construction of a treatment plant is not required. When 1 gram of faecal matter can contain 10 million viruses, 1 million bacteria, 1 thousand cyst parasites and about a hundred worm eggs, it goes without saying, that a system that takes care of the storage and the disposal of excreta must be found.

2. *Description of some types of latrines.* Latrines can improve living conditions of populations and solve sanitation needs in developing countries. Conventional water born sewerage systems have been incapable of meeting populations' needs in these countries. Sanitation technologies most commonly used nowadays are pit toilets and pour-flush toilets.

   (*a*) *Simple pit latrine or traditional latrine.* It is a simple wooden or concrete slab installed over a pit of 2 m or more in depth. This support should stand on a sufficiently waterproof edge of the pit to avoid surface water (runoff and grey water) entering and destroying the facility. The pit should be lined in case of unstable soil where there is a risk of walls collapsing. However, in the case of lining, openings have to be provided in the walls to allow infiltration of liquid in the soil.

   Excreta fall directly into the pit through a drop hole or a seat. This type of latrine is a simple pit covered by a slab with a drop hole. A superstructure is necessary to provide privacy and protection.

   Advantages:

   - Relatively cheap
   - Can be constructed by the user (particularly in rural areas)
   - Does not need water to function
   - Easy to maintain.

   Disadvantages:

   - Considerable nuisance because of flies, insects (mosquitoes if pit is wet)
   - Bad smell.

   (*b*) *Ventilated improved pit latrines* (VIP). These are also called self-ventilated improved pit latrines. The principle is to cancel or to reduce harmful side-effects (smells and flies) related to traditional latrines by providing a vent pipe higher than the superstructure. This vent-pipe is fitted at the end by a mesh that stops flies from entering. Wind that blows at the top of the pipe creates an ascending air current between the pit and the outside atmosphere, and a descending air current between the superstructure and the pit through a drophole. Smells are evacuated through the vent pipe.

   Advantages:

   - Relatively cheap
   - Can be constructed by the user
   - Does not need water to function
   - Easy to maintain
   - No smell and no flies
   - It can take different sorts of cleansing materials (solid and liquid).

   Disadvantages:

   - Darkness is indispensable within the superstructure to fight off flies
   - Only functions properly when conveniently oriented towards the wind
   - No surrounding obstacle (trees and buildings) should be higher than the vent pipe.

   There are two types of VIP latrines: one pit latrines and double or multiple pit latrines.

   (*c*) *Pour-flush latrines or toilets.* A pour-flush latrine is composed of a pan with a water-seal installed in a superstructure. The water-seal is connected to a pit by a pipe. In certain cases, the water-seal can be installed on the slab of the latrine. The water-seal flushes out excreta with just enough water to drain off the solids and to restore the water level in the water-seal. Another advantage of the water-seal is that it stops flies and smells from coming out of the pit. The pit can be dug far from the superstructure and connected with a pipe or covered drains. The pan is laid down on the floor on top of the water-seal for urine reception, under minimal hygienic conditions.

Advantages:

- Relatively cheap
- No smells, no flies and mosquitoes
- Agreeable to use
- Can be improved by a connection to a sewer network at the right moment
- Low water consumption 2–3 litres of water for each flush.

Disadvantages:

- A source of water is needed
- The use of solid cleansing material is not advisable (except paper)
- Skilled labour (mason) is required.

There are two kinds of VIP latrines: pour-flush latrines with one pit and pour-flush latrines with double or multiple pits

(*d*) *Eecological sanitation.* Ecological latrines are latrines that ensure the recovery of waste by separating urine and faeces in view of their reuse in the fertilization of soil for agricultural purposes. (reintroducing nutrients into the soil such as: nitrogen, phosphorous and potassium).

The variety of available eco-sanitation systems helps, in most cases, to choose a system that is culturally acceptable. The issue of cost is relative, some systems are complicated and expensive, others are simple and cheap. There is always a compromise to be made between cost and functionality: cheap solutions imply more handling and maintenance of sanitation systems. In the case of more expensive systems, there could be less handling and maintenance. These latrines have been set up in order to preserve the environment and to recover waste products. Pits are completely lined and the bottom is waterproof in order to avoid infiltration into the soil.

An eco-latrine is composed of a pit constructed above ground. The pit is covered by a slab on which a superstructure is built. The pit is ventilated like the VIP latrines but its volume is less important, because the filling time calculated is less than a year. The storage of ash inside the superstructure (in a bowl) is advisable so that it can be mixed with the excreta, after every use, in order to control the moisture content. A tank installed outside the superstructure is used for the storage of the urine.

Advantages:

- Recovery of urine and waste products
- No pollution of ground water
- Very hygienic
- Easy to maintain
- Does not need water to function.

Disadvantages:

- Containers for urine storage are needed
- Water is not recommended for cleansing

- Addition of ash is required after every use
- Overhead pits require climbing stairs to access the facility
- Skilled labour (mason) is required.

Apart from these latrines described above, there are other facilities which also have advantages and disadvantages. They are: cat latrines, composting latrines, septic tank latrines, aqua privies, etc.

It is important to know that the choice of the right latrine is based on a series of studies that go into all the relevant factors which need to be considered for an efficiently functioning latrine. CREPA can be of assistance in this process.

ENVIRONMENTAL IMPACT ASSESSMENT

An **Environmental Impact Assessment** (**EIA**) is an assessment of the possible impact—positive or negative—that a proposed project may have on the environment, together consisting of the natural, social and economic aspects. It was made mandatory in 1994 for certain type of projects.

The purpose of the assessment is to ensure that decision makers consider the ensuing environmental impacts to decide whether to proceed with the project. The International Association for Impact Assessment (IAIA) defines an environmental impact assessment as "the process of identifying, predicting, evaluating and mitigating the biophysical, social, and other relevant effects of development proposals prior to major decisions being taken and commitments made."

After an EIA, the precautionary and polluter pays principles may be applied to prevent, limit, or require strict liability or insurance coverage to a project, based on its likely harms. Environmental impact assessments are sometimes controversial.

The city of today is a major performer for the secondary sector of the economy. Manufacturing and processing facilities not only provide the city with the essential commodity inputs but also provide the means of living for the populace. Of late, ad-hoc decisions for industrial development have led to an adverse impact on the local environment at costs which are much higher than the benefits actually accrued. In view of the deteriorating environmental conditions in and around industrial townships, it has become necessary to account for the environment while planning for such areas.

Environmental Impact Assessment (EIA) is a relatively new planning and decision making tool first enshrined in the United States in the National Environmental Policy Act of 1969. It is a formal study process used to predict the environmental consequences of any development project. EIA thus ensures that the potential problems are foreseen and addressed at an early stage in project planning and design.

Environmental Assessment is taken up in this exercise as a rapid assessment technique for determining the current status of the environment and identifying impact of critical activities on environmental parameters. Based on this analysis we can draw up an Environmental Management Plan that would ensure impact monitoring and mitigation planning.

Environmental Assessment enables us in carrying out Environmental Cost-Benefit Analysis of projects at an initial stage. It is thus a pre-cursor to detailed analysis of environmental impacts, which are taken up only if a need for the same is established. It gives a view of the actors involved in the 'development-environment linkages. This is required in view of the fact that the community at large is always at a loss in terms of deterioration of living environment that accompanies industrial development. Based on Environmental Assessment, the regulatory measures can be identified and the roles of concerned agencies defined for achieving more efficient environmental management.

In view of the fact that development is an ever growing process, its impact on the environment is also ever increasing, leading to rapid deterioration in environmental conditions. As such Environmental Assessment provides a rational approach to sustainable development.

Extensively developed rapid assessment techniques often avoid carrying out of detailed studies which need more resources in terms of time and money. This exercise is an attempt in developing an approach to Environmental Assessment technique, primarily for industrial townships.

Environmental Impact Assessment (EIA) may be defined as a formal process used to predict the environmental consequences of any development project. EIA thus ensures that the potential problems are foreseen and addressed at an early stage in the projects planning and design.

The phrase 'Environmental Impact Assessment comes from Sec. 102 (2) of the National Environmental Policy Act (NEPA), 1969, USA. Some rudiments of EIA are implicit even in early examples of environmental legislation. Napoleon in 1910 issued a decree which divided noxious occupations into categories: those which must be far removed from habitations, those which may be permitted on the outskirts of towns, and those which can be tolerated even close to habitations, having regard to the importance of the work and the importance of the surrounding dwellings.

In India, the environmental action formally started with the participation of late Smt. Indira Gandhi in the UN Conference on Human Environment in Stockholm in 1972. A National Committee on Environmental Planning & Coordination (NCEPC) was established to be the apex body in the Department of Science and Technology. The term 'Environment figured for the first time in the Fourth Five Year Plan (1969–74) which recorded that 'harmonious development is possible only on the basis of a comprehensive appraisal of environmental issues. The Tiwari Committee (Committee on Review of Legislative Measures and Administrative Measures), in its report in 1980, recommended creation of a Department of Environment as a nodal agency to ensure environmental protection, to carry out environmental impact studies of proposed development projects, and to have administrative responsibility for pollution monitoring and control. The department came into being in 1980 within the Ministry of Science and Technology under the charge of the then Prime Minister. In 1989 the subjects of wildlife and forestry were added to the list and a new Ministry of Environment and Forests was created with the Prime Minister holding its charge. Since its inception the Department (under the Ministry) has issued various guidelines on EIA for various projects.

EIA, in brief, extrapolates from scientific knowledge to assess the problem consequences of some human interventions on nature. Although EIA uses the techniques of science, it differs from ordinary scientific inquiry, because it is dealing with events which have not yet occurred, may not occur, and whose chances of occurrence may be changed by the very statement that they may occur.

Some measures are required to be taken in the future to reduce the anticipated environmental degradation. Before starting a major project, it is essential to assess the present environment without the project, and the likely impact of the project on the environment, when it is completed. Therefore, an Environment Impact Assessment has to be made before starting a project. For analysis of environmental impacts, many professions and disciplines have to be involved. Like economic and engineering feasibility studies, Environmental Impact Assessment is a management tool for officials and managers who make important decisions about major development projects. The Environmental Impact Assessment should have the following objectives:

(*i*) Predict environmental impact of projects
(*ii*) Find ways and means to reduce adverse impacts
(*iii*) Shape project to suit local environment
(*iv*) Present the predictions and options to the decision-makers.

The EIA statement should cover brief description of project, brief description of existing environment, likely impact of project, the mitigation and protection measures, consideration of alternatives, and summary with conclusions.

Why EIAs?

EIAs have two roles—legal and educational. The legal one is quite straight forward: to ensure that development projects such as a housing estate, a road/bridge or some such construction project has a minimal impact on the environment in its entire 'lifecycle' — *i.e.*, during design, construction, use, maintenance, and demolition. Many countries now have laws stipulating that unless an EIA study is carried out (particularly for large infrastructure projects), permission for construction will not be granted by the local authority. But countries in Africa will possibly see EIA processes as a 'hinderance' to development as environment is not yet a priority! The educational one is equally important and probably a forerunner to the legal role — to educate everyone one involved - professionals and users included, of the potential environmental impacts of anything we do. We need to look at all our daily actions as eventually and cumulatively affecting the environment. This includes our daily choices, where a delicate balance between financial and environmental considerations need to be made automatically - without thinking!

Methodology Environmental Impact Assessment

Environmental Assessment Method

This page describes the purpose of our Environmental Assessment method, the assessment stages, and details of the methods used. The method has been transferred to a spreadsheet.

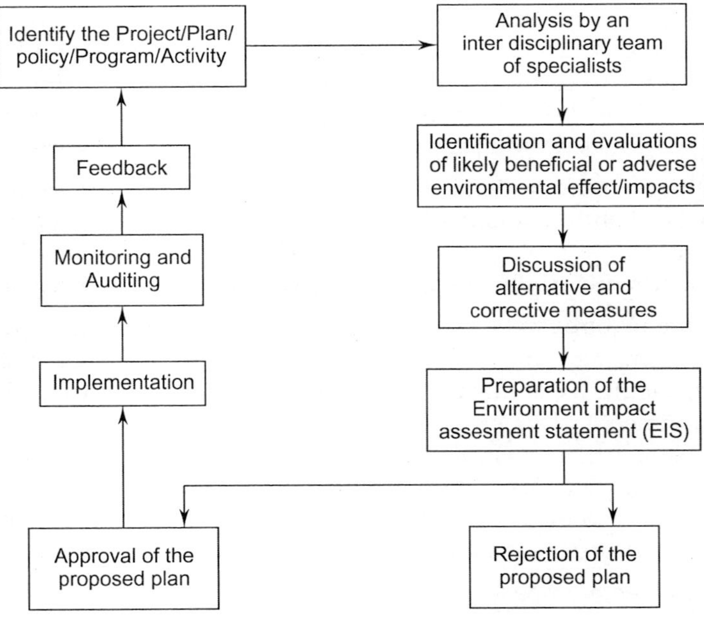

Fig. 5.8 Environmental Impact Assessment Process

Assessment Stages

Identify environmental aspects and agree significance

A check list of the sustainability issues and environmental aspects is provided, together with a participative method for weighting the significance of each aspect. The significance of the aspect is that to the environment, and is considered independent of the technical solution proposed in the scheme.

Assess the Environmental Performance

The environmental performance of the technical solution or scheme is assessed under each aspect, using detailed assessment methods such as resource use planning and life cycle assessment, or risk assessment, depending upon the purpose of the assessment.

The most important environmental indicators for energy projects are identified, and general data provided for comparisons with alternative energy schemes.

Evaluate the Environmental Impact

The environmental impact is evaluated against each aspect by multiplying the environmental significance and the scheme performance.

Our method assesses the overall environmental impact by calculating a cumulative score from the specific impacts.

Details of the Assessment Method

Agreeing the environmental aspects and their significance

The recommended approach is to agree the aspects and their significance between the stakeholders in the project. This is achieved by involving representatives of all the stakeholders in a participative group process in the early stages of the method.

Identifying the Environmental Aspects

Environmental aspects are identified under the following headings:

- Social inclusion
- Economic development
- Resource use
- Transport
- Ecological protection
- Environmental protection.

The list given in the spreadsheet is intended to cover the generally significant aspects for any energy project, but additional aspects may be added if required.

EXERCISE

1. What do you mean by treatment of water? Draw a flow diagram for the treatment of surface water and briefly explain the methods used for the treatment of surface water.

   [*Raj. Tech. Univ; Sem II, 2009*]

2. Discuss the various methods of treatment of wastewater. Justify the need for the treatment of wastewater. *[Raj. Tech. Univ; Sem II, 2009]*
3. What are the various methods of disposal of wastewater? Explain then and justify the need for disposal of wastewater. *[Raj. Tech. Univ; Sem II, 2009]*
4. Describe various parameters to determine the quality of drinking water?
5. Write different sources of water and describe them. *[RTU, June 2009]*
6. What do you mean by treatment of water? Draw a flow diagram for the treatment of surface water and briefly explain the methods used for the treatment of surface water. *[RTU, August 2009]*
7. Describe the filtration and disinfection methods for treatment of water. *[RTU, June 2008]*
8. Enlist the chemical, physical and bacteriological standards of drinking water with their acceptable limits. *[RTU, June 2008]*
9. Discuss different ways of disposal of wastewater after treatment. *[RTU, June 2008]*
10. Discuss the rural and urban wastewater collection system. *[RTU, June 2008]*
11. What do you understand by primary, secondary and tertiary treatment of water, explain in detail.
12. Discuss different ways of disposal of wastewater after treatment. *[RTU, June 2008]*
13. What are different treatments for wastewater. Describe in detail.
14. Describe the filtration and disinfection methods for the treatment of water. *[RTU, June 2008]*
15. What is septic tank. How can you care and feeding of a septic tank.
16. Describe various constructive suggestions for reuse and saving of water.
17. What is onsite sanitation. Define various impacts of onsite sanitation.
18. What is EIA. Discuss the important aspects in the design of an EIA process.
19. Discuss the advantages and limitations of environmental impact assessment.
20. Justify the need for wastewater treatment briefly. Outline the different steps of domestic wastewater treatment.
21. With the help of suitable sketches differentiate between activated sludge process and trickling filter process of wastewater treatment.
22. List various factors affecting ground water quality. What are the desirable characteristics of potable water.
23. What are the two main parts of a septic system. Explain it and write the necessary steps which should be taken for the maintenance of the septic system properly.
24. When a septic system must be inspected. How do I know if my septic system has failed. Write the effects and impacts of failing septic system to human health and water bodies.
25. What is the need of on-site sanitation. Mention the operation and maintenance of some major types of on-site sanitation facility.
26. Comment upon the statement that disposal of excrete is of fundamental importance, not only for the health of the community but also because of the social and environmental benefits it brings.
27. What do you understand by primary, secondary and tertiary treatment of water? Explain in detail.
28. Write short notes on
    (*i*) On-site sanitation
    (*ii*) Wastewater management

6

AIR POLLUTION

6.1 INTRODUCTION

Dust, dirt, harmful gases and industrial emissions, all form part of what is called air pollution. These pollutants cause a deterioration in air quality and are capable of harming life forms. Air pollution is a serious problem and can be eased with some simple solutions.

When was the last time you stepped out of your house and inhaled clean and fresh air that really invigorated all your senses? Here's a look at some of the causes of air pollution that has resulted in the thick smog which most city dwellers are subjected to.

Air pollution is a condition triggered by the presence of air-borne pollutants that affect the quality of air we inhale. These pollutants could either be the result of chemical emissions or the particulate material from biological waste. The condition has reached alarming proportion in the modern world, with a large scale industrialization and vehicle-emissions being the primary culprits. The pollutants that are air borne cause a lot of harm to humans and animals, other than permanent damage to the natural environment.

Air pollution is something that we cannot really ignore now-a-days. This is evident from the moment we step out of our house and are greeted with black colored smog that hits us directly reminding us that breathing clean air is more of a distant dream. It is so easy for us to endlessly rant and rave about the causes of air pollution and its ill effects, but little do we realize that each person is responsible for all the causes of air pollution and the situation that we face today. Take a look around you at the dismal state of affairs. The thick smog that is seen in the morning hours is not really due to somebody else but rather due to each and every one of us. Here is a look at the causes of air pollution and how it can affect us if the matter is not taken care of at this stage itself.

Facts about Air Pollution

Here are a few reports that have come out recently about the adverse effects of air pollution which will make you sit up and sniff the air around you apprehensively:

- According to a study, living in a major city places people at a higher risk than living in the radioactive zone in Chernobyl.

- According to the World Health Organization (WHO), about 2 million premature deaths are caused each year due to air pollution in cities across the world.
- A recent study has revealed that exposure to fine particle matter in polluted air increases the risk of hospitalization due to respiratory and cardiovascular diseases.
- According to a German study, people who breathe in the fumes of heavy traffic regularly have higher chances of getting hardening of the arteries, which is associated with the risk of heart attack.
- A Scottish study has shown that jogging with traffic around results in reduced blood flow to the heart. This is particularly dangerous for people with stable heart disease, because it can trigger off cardiac arrhythmia or even a heart attack.
- According to a study conducted on eight-year-olds in Mexico City, being exposed for a long time to high levels of polluted air reduces lung function and growth in children.
- According to a 20-year study conducted on the residents of Los Angeles, it has been shown that the adverse effects of air pollution have been grossly underestimated. The researchers opine that chronic health problems due to particulate matter in the air may be double or even triple times greater than the estimates available currently.

So what is air pollution and why is it so harmful? Air pollution is caused when it gets filled with too much gases, particulate matter, and droplets of liquid. In cities, the air gets polluted by the exhaust fumes of vehicles, along with the pollutants given off by construction work and industry. In the country, the dust given off by tractors working on fields, vehicles being driven on gravel or dirt tracks, smoke given off by crop and wood being burnt, and work carried out in rock quarries, are some of the causes of air pollution.

Another major air pollutant in cities is ozone that occurs at ground level. Ozone forms when nitrous oxides and hydrocarbons react with sunlight. However, not everything about ozone is bad. In fact, its presence in the upper atmosphere is beneficial because it keeps out harmful ultra-violet rays, which is one of the major causes of skin cancer. Ozone only becomes problematic when it occurs near the ground where it can be inhaled. When inhaled, ozone causes reduced lung capacity, choking, and coughing.

When the air is polluted, it causes irritation of the throat, lungs and eyes. Some of the common symptoms are: a burning sensation in the eyes, tightness in the chest, and coughing. It exacerbates respiratory conditions like emphysema and asthma, and reduces the body's capacity to fight off infections of the respiratory system. Also, people afflicted with heart disease, like angina, are usually very sensitive to air pollution. People who exercise outdoors are also susceptible to the symptoms of air pollution, because it involves deeper and faster breathing.

In fact, polluted air is particularly detrimental to those who have lung and/or heart disease. When the pollution levels become very high, it can lead to them having to curtail their activities and even result in hospitalization. Severe air pollution has even been known to cause death in the recent past. However, such high levels of pollution are now not as common in the US.

Another group that is susceptible to the effects of air pollution is children. When they live in areas that have high levels of pollution, children tend to be prone to illnesses like earaches and bronchitis.

Although some groups of people feel the effects of air pollution more acutely, one of the positive factors is that when there is an improvement in the quality of air, the symptoms caused by air pollution are quickly alleviated for most people who are healthy.

The long-term effects of being exposed to low air pollution levels are still being studied.

Here are some of the ways you can prevent the harmful effects of air pollution affecting you and your family:

- Try staying indoors as much as possible in the daytime. Usually the air indoors is less polluted than outdoors.
- If you cannot avoid going outside, try to do it in the early morning or after sunset. This is particularly important when there are high levels of ozone, which is usually the case in many big cities, because sunshine triggers off the creation of ozone.
- When the pollution levels are high, try not to exert yourself. The harder you breathe, the more polluted air you inhale into your lungs.

The above steps should be enough to protect you against air pollution if you are healthy. However, in case you work or live near some source of pollution, or if you are afflicted with chronic lung or heart problem, it is best to seek medical advice on the best ways to deal with it.

Causes of Air Pollution

Carbon dioxide is one the main pollutants that causes air pollution. This is because, although living beings do exhale carbon dioxide, this gas is harmful when emitted from other sources, which are caused due to human activity. An additional release of carbon dioxide happens due to various such activities. Carbon dioxide gas is used in various industries such as the oil industry and the chemical industry. The manufacturing process of most products would require the use of this gas. There are various human activities that add to the increased proportions of carbon dioxide in the atmosphere. The combustion of fossil fuels and the harmful effects of deforestation have all contributed towards the same. show that amongst the various gases emitted during a volcanic eruption, carbon dioxide remains to be at least 40% of the emission. Scientists have now therefore identified carbon dioxide as one of those elements that have contributed to global warming.

Causes of air pollution are not limited to this. The combustion of fuels in automobiles, jet planes etc all cause the release of several primary pollutants into the air. The burning of fossil fuels in big cities which is seen at most factories, offices and even a large number of homes, it is no wonder that air pollution is increasing at an alarming rate. The release of other harmful gases all adds to the state that we see today. Although carbon dioxide plays an important role in various other processes like photosynthesis, breathing an excess of the same also causes harmful effects towards one's health.

The various causes of air pollution that releases harmful gases into the atmosphere are caused due to the increasing number of power plants and manufacturing units or industries that mostly have activities related to the burning of fuels. Besides, as mentioned earlier, most automobiles, marine vessels, activities that involve the burning of wood, fumes that are released from aerosol sprays, military activities that involve the use of nuclear weapons, all are the numerous causes of air pollution.

Carbon monoxide is another such gas which, although was present in the atmosphere earlier, is now considered to be a major pollutant. An excess of the same has a harmful effect on our system. There are many reasons why carbon monoxide can be released into the atmosphere as a result of human activities. This is also produced due to any fuel burning appliance and appliances such as gas water heaters, fireplaces, woodstoves, gas stoves, gas dryers, yard equipments as well as automobiles, which add to the increased proportion of this gas into the atmosphere.

Sulfur dioxide is yet another harmful pollutant that causes air pollution. Sulfur dioxide is emitted largely to the excessive burning of fossil fuels, petroleum refineries, chemical and coal burning power plants etc. Nitrogen dioxide when combined with sulfur dioxide can even cause a harmful reaction in the atmosphere that can cause acid rain.

Nitrogen dioxide is one more gas that is emitted into the atmosphere as a result of various human activities. An excess of nitrogen dioxide mainly happens due to most power plants seen in major cities, the burning of fuels due to various motor vehicles and other such sources, whether industrial or commercial that cause the increase in the levels of nitrogen dioxide.

These and a number of other hazardous air pollutants are emitted with the various numbers of activities that we carry out during the day which are the main causes of air pollution.

6.2 CLASSIFICATION OF FACTORS CAUSING AIR POLLUTION

1. Classification on the Basis of Physical States of Air Pollutant

(*a*) *Gases:* Which freely mix with air without settling down. These are SO_2, SO_3, NO_x, H_2S, CO_2, O_3, HF, aerosols, hydrocarbons and photochemical oxidants.

(*b*) *Particulates:* Small solid particles and liquid droplets are collectively known as particulates, *e.g.*, dust, smoke, smog, asbestos, lead, Hg and cadmium, mist, oil and grease etc.

(*c*) *Deforestation.*

(*d*) *Internal combustion engines.*

2. Classification on the Basis of Origin

(*i*) *Primary air pollutants* which are directly emitted into the atmosphere and are found as such *e.g.*, CO, NO_2, SO_2 and hydrocarbons.

(*ii*) *Secondary air pollutants* which are derived from the primary pollutants due to chemical or photochemical reaction in the atmosphere *e.g.*, O_3, peroxyactylnitrate (PAN) and photochemical smog etc.

3. Classification on the Basis of Chemical Composition

(*i*) *Organic pollutants, e.g.*, hydrocarbons, aldehydes, ketones, amines and alcohols.

(*ii*) *Inorganic pollutants:*

(*a*) Carbon compounds, (CO, CO_2 and carbonates).

(*b*) Nitrogen compounds (NO and NH_3)

(*c*) Sulphur compounds (SO_2, H_2S, SO_3 and H_2SO_4)

(*d*) Halogen compounds (HF, HCl, MF_X where M is a metal and x is the valency of M).

(*e*) Oxidizing agents (O_3).

4. Classification on the Basis of Major Sources of Air Pollution

(*i*) *Natural sources:* The natural sources of air pollution are volcanic eruptions releasing poisonous gases, forest fires, natural organic and inorganic decay or vegetative decay, marsh gases, deflation of sand and dust, extra terrestrial bodies, cosmic dust, pollen grains of flowers, soil debris, and fugitive spores.

(*ii*) *Man made sources:* Man made sources such as increase in pollution, deforestation, burning of fossil fuels, emissions from vehicles, rapid industrialization, agricultural activities and wars, are the major causes of air pollution.

(iii) *Emissions of major industrial air pollutants:* The major air polluting sources are sulphuric acid industry (emits SO_3 and SO_2), nitric acid plants (oxides of nitrogen), fluoride industry (HF and cryolite), chloroalkali plants (Cl_2 amd H_2 gas, CO and CO_2), hydrochloric industry (toxic chlorine and chlorine monoxide), iron and steel industries, radio active natural sources, phosphoric acid plants, petrochemical refineries (kerosene, gasoline, formaldehyde and fuel oil), power plants.

(iv) *Aerosol pollutants:* These air pollutants remain suspended in air and consist of fine particles of different organic and inorganic compounds having diameter less than 100 μ.

(v) *Metallic contaminants:* A number of toxic and non-toxic metals occur in the atmosphere. Most of the metals are destructible poisons to living beings, *e.g.*, Cd, Pb, Cr, Be, Ba, Mn etc., which are most toxic.

(vi) *Carcinogens:* A group of carcinogens, such as benzidine, β-naphthylamine, vinyl chloride, α-naphthylamine, ethylene dichloride etc. present in air cause cancer in man and animal affecting DNA and cell growth.

(vii) *Biological contaminants:* There exits various air borne microorganisms, pathogens, bacteria, viruses and parasites, which are added as air pollutants in the atmosphere.

Other inorganic particles, *e.g.*, flyash, silica, asbestos and dust from transport, mining, metallurgical and other industrial process.

6.3 AIR POLLUTANTS GASES

These pollution causing factors are gaseous in physical state. These gases mix freely with air without settling down. These are mainly as follows—

(a) **Oxides of Nitrogen(NO_2 and NO):** These oxides are produced from acid manufacturing plants, explosive industries, power plants and acid-packing industries. There are eight possible oxides of nitrogen but only NO_2 and NO are more significant from air pollution point of view. The formation of NO from nitrogen is favoured at high temperatures (~1210 to 1765° C), which are usually obtained in the combustion process involving air. Rapid cooling of combustion product prevents the dissociation of NO. Further the oxidation of NO to NO_2 is also favoured at high temperature (~1100 °C) but the amount of NO_2 formed does not exceed from 0.5% of the total nitrogen oxides. NO_2 is also formed by photolytic reaction in the atmosphere. Some of the reactions involved are as follows–

$$N_2 + O_2 \xrightarrow{1210-1765\,°C} 2NO$$
$$2NO + O_2 \longrightarrow 2NO_2$$
$$NO_2 + h\nu \xrightarrow{398\text{ mm}} NO + O^{\cdot}$$
$$O_2 + h\nu \xrightarrow{242\text{ mm}} O^{\cdot} + O^{\cdot}$$
$$O + O_2 \longrightarrow O_3$$
$$NO + O_3 \longrightarrow NO_2 + O_2$$
$$2NO_2 + O_3 \longrightarrow N_2O_5 + O_2$$
$$4NO_2 + 2H_2O + O_2 \longrightarrow 4HNO_3$$

The average lifetime of NO and NO_2 is four days and three days respectively in the atmosphere. They undergo various photochemical reactions leading to the formation of HNO_3 which gets precipitated as nitrates during rainfall or other natural process. The chemical and photochemical reactions involving NO_2 and hydrocarbons induced by sunlight are responsible for the formation of photochemical smog. The nitrogen oxides formed react with organic compounds such as hydrocarbons present in gasoline, in the presence of sunlight, producing peroxyacetylnitrate (PAN), which has formula—

$$\begin{array}{c} R-\underset{\underset{O}{\|}}{C}-O-O-\underset{\underset{O}{\downarrow}}{N}=O \end{array}$$

PAN was identified as the major visible toxicant in the photochemical smog. The formation of smog as described earlier causes following effects (*i*) Limits the visibility on roads, (*ii*) Causes eye irritation, (*iii*) Gauses difficulty in breathing.

(*b*) **Sulphur Oxides:** The most common type of air pollution from the 14th century to the early twentieth century is from the smoke and gases released by the burning of coal. The lower grades coal contains sulphur in its reduced forms, much of it as iron pyrite, FeS_2.Coal (and fuel oils) also contains reduced sulphur in organic forms such as RSH or R–S–S–R, where R represents a complex of carbon. The sulphur is oxidized to the primary pollutant sulphur dioxide when the fuel is burned–

$$4FeS_2 + 11O_2 \longrightarrow 2Fe_2O_3 + 8SO_2$$

The SO_2 formed is eventually oxidized to sulphur trioxide (SO_3) by a number of possible reactions whose net result is

$$2SO_2 + O_2 \longrightarrow 2SO_3$$

SO_3 combines with water vapours and produces a mist of sulphuric acid—

$$SO_3 + H_2O \longrightarrow H_2SO_4$$

Reduced sulphuric acid mists are responsible for the bluish haze in atmosphere. The sulphuric acid mist can react with other materials present in the air, such as metal oxide dusts (*e.g.*, CaO), NH_3 and salt crystals (NaCl), to form sodium sulphate particulates—

$$H_2SO_4 + CaO \longrightarrow CaSO_4 + H_2O$$

$$H_2SO_4 + 2NH_3 \longrightarrow (NH_4)_2SO_4$$

$$H_2SO_4 + 2NaCl \longrightarrow Na_2SO_4 + 2HCl$$

Effects:

(*i*) The presence of the gas causes cardiac and respiratory diseases(*e.g*. asthma and bronchitis).

(*ii*) Eye irritation.

(*iii*) Damage to agriculture.

(*iv*) Throat troubles and corrosion of metals.

(*v*) It also affects plant cells in membrane damaging; chlorophyll destruction, metabolism inhibition and growth yield reduction.

(*vi*) SO_3 and SO_2 cause a very dangerous effect on living being on earth surface causing acid rain.

(*c*) **Hydrogen Sulphide:** It is produced by the decomposition of sewage wastage or organic matter and as a by-product from various industries. H_2S is more poisonous than even carbon monoxide and causes an adverse effect on living as well as on non-living world.

(*d*) **Carbon Monoxide:** It is colourless, odourless and a toxic gas which is slightly water soluble and extremely dangerous due to great affinity for haemoglobin. Carbon monoxide is released by the partial combustion of fuel in automobile industries and oil refineries. Domestic heat appliances, cigarette and bidi smoke are other sources of CO. It is estimated that the amount of CO librated every year is about 290 million litres in atmosphere. It is corrosive and a very toxic air pollutant. It causes headache, visual irritability, paralysis and even death in human beings. It has 250 times greater affinity for haemoglobin than O_2. It forms a stable compound with haemoglobin namely carboxy haemoglobin. It can be represented as—

$$O_2 + Hb \longrightarrow HbO_2 \quad \text{(Under normal conditions)}$$

(Oxyhaemoglobin)

$$HbO_2 + CO \longrightarrow HbCO + O_2 \quad \text{(In presence of CO)}$$

(Carboxyhaemoglobin)

The immediate effect of CO poisoning is loss of judgement, which is responsible for many automobile accidents. Further exposure to higher levels of CO leads to various metabolic disorders such as asphyxiation and causes death.

In the study conducted by the **National Environment Engineering Research Institute (NEERI),** the concentration of carbon monoxide at street level was found to be in the range of 10 ppm to 35 ppm, a value much higher than would be expected from the prevailing traffic density.

(*e*) **Carbon Dioxide:** CO_2 is also colourless and odourless in nature. It is discharged into atmosphere in the form of smoke, which is produced by combustion of fuels such as coal, wood, petroleum products and gaseous fuels. It is also released during respiration of plants and animals. The increase in concentration of CO_2 in atmosphere causes respiratory disorder and suffocation. CO_2 also plays an important role in heating up of environment due to trapping of infra-red rays from sun, which is called **Greenhouse effect.**

(*f*) **Hydrogen Fluoride:** It is released from the industries of phosphate fertilizers, aluminium metallurgical process, brick-kilns, pottery kilns and coal burning industries. It is also produced by automobiles. Due to the presence of this pollutant there are many types of disorders in human body *e.g.*, irritation, skeleton disorders and respiratory diseases. It also causes fluorosis in cattle breeding by using plants affected with HF.

(*g*) **Hydrocarbons:** These are made up of hydrogen and carbon. Hydrocarbons, which are volatile in nature, takes part in atmospheric reaction with other organic compounds producing ozone. These hydrocarbons affect human body adversely due to very reactive in nature. Even carcinogenic hydrocarbons like benzopyrenes can react with DNA thus causing mutation and cancer. Inhalation of aromatic hydrocarbon vapour produces more serious ill effects like irritation to muscles, membranes, problems in respiratory system, lungs, cancer and disorders in nervous system and may even lead to death.

Methane is a major naturally occurring hydrocarbon emitted in the atmosphere. It is produced by bacteria during anaerobic decomposition of organic matter in soil, water and sediments—

$$2CH_2O \xrightarrow{\text{Bacteria}} CH_4 + CO_2$$

The evaporation of organic solvents like C_6H_6, $C_6H_5CH_3$, $CC1_4$, $CHCl_3$ etc. emit about 10% hydrocarbons in the air during industrial operation. Hydrocarbons enter the atmosphere directly due to petroleum refineries or by evaporation from the fuel tanks of automobiles. Some olefins produced by cracking of saturated hydrocarbons are very harmful-

$$C_8H_{18}(g) \longrightarrow 2C_2H_4(g) + C_4H_8(g) + H_2(g)$$

Unsaturated hydrocarbons such as ethylene and butene can react with species such as O, NO or O_3 to produce some of the most toxic components of photochemical smog.

(h) **Internal Combustion engines:** When fuel is combusted in an engine different type of air polluting gases and particulate pollutants *e.g.,* Pb etc., are discharged in atmospheric air. The fuels used are the mixture of various hydrocarbons of different molecular masses and formulae. When these are combusted in the internal combustion engine of an automobile in the presence of air, the following reaction takes place-

$$\text{Hydrocarbons} \longrightarrow x\,CO_2 + y\,H_2O + \text{Heat}$$

Vapours of unburnt hydrocarbons produce a number of petrochemical oxidants and photochemical smog with O_2 and N_2 which cause adverse effects on physiological activities of living beings.

6.4 PARTICULATE POLLUTANTS

The pollutants which are particulate in nature are called particulate pollutants. They exist in particle form in atmosphere, thus causing air pollution. They are mainly as follows-

(a) **Dust:** Dust is composed of fine solid particles and their size ranges from 1 micron to 100 microns. Even dust particles of the size 0.1 micron have been observed in the atmosphere. They are liberated mainly from mines, speeded winds, industrial furnaces, traffics, house cleaning dust, combustion operations, metallurgical processes and many other activities taking place in the era of the industrialization. Dust makes the atmospheric air unfit for respiration.

Atmospheric dust causes allergic and respiratory diseases; if it possesses silica then it causes silicosis.

(b) **Radioactive Gases and Dust:** The radioactive gases are liberated from natural and artificial radioisotopes. These get mix with dust. They cause leukemia, cataract, cancer and affects human being genetically such as mutation in human gametes.

(c) **Lead (Pb):** It is discharged from automobile emission; burning of coal, lead arsenate pesticides and lead smelters. It causes respiratory, excretory disorders, lever and kidney damage and mental retardation in children. Lead content in human blood exceeding 40 ppm is considered to be dangerous. Tetra Ethyl Lead (TEL) is more harmful than inorganic lead. According to **Vincent J. Schaefer,** Head of the Atmospheric Science Research Centre of New York State University, over seeding of clouds is caused by extremely tiny crystals of lead iodide, formed from tiny particles of lead released from tetraethyl and traces of iodine in the atmosphere. The decline of rain and snowfall would be serious enough to upset the equilibrium that exists between air, water and the earth, if automobiles continue to use leaded gasoline.

(d) **Mercury (Hg):** The main sources of Hg particles are mining and refining of mercury. Pure mercury is not particularly toxic and inhalation of a very small amount of Hg does not produce any ill effect. Hg vapours are very dangerous, it cause irritation, destruction of lung tissues, protoplasmic poisoning. Organomercuric compounds are highly toxic and may cause irreversible damage to nervous system and brain.

(e) **Cadmium (Cd) and Arsenic (As):** They are discharged in atmosphere as by products of Pb, Zn and Cu metallurgy, nuclear fission plants, Cd-Ni batteries, Cd producing industries. Inhalation of Cd fumes causes kidney damage, bronchitis, gastric disorders, cancer and heart diseases. Arsenic is librated from fugicides, pesticides, byproduct of mining activities and chemical grades.

(*f*) **Smoke and Fumes:** Smoke particulates consist of solid and liquid particles ranging from 0.05 to 1.0 microns which are formed during incomplete combustion of carbon containing materials. It includes smoke of gaseous pollutants like oxides of nitrogen, carbon monoxide and hydrocarbons etc.

Fumes are generally obtained by the condensation of vapours by sublimation, distillation, boiling, calcinations and by several other chemical reactions. Generally organic solvents, metals and metallic oxides form fume particles having a size less than 1 micron.

(*g*) **Mist and Spray:** Mist or liquid particles are formed by condensation of vapour, having a size less than 10 μ, e.g.

$$SO_2\,(gas) \xrightarrow{22\,^\circ C} SO_2\,(liquid)$$

Spray constitutes liquid particles obtained from the parent liquid by the process of mechanical disintegration.

Effects of Air Pollution on Humans

The effects of air pollution on humans are fatal and life threatening. WHO statistics report that over 2 million people succumb to the fatalities attributed to air pollution. Consistent exposure to the pollutants leads to the development of:

- Cardiopulmonary disease
- Pneumonia
- Premature mortality
- Heart attack
- Asthma
- Difficulty in breathing
- Wheezing and coughing
- Acute vascular dysfunction
- Thrombus formation
- Cystic fibrosis
- Chronic obstructive pulmonary disease
- Chronic bronchitis
- Emphysema

There are a number of ways, by which these emissions could be controlled. Particulate control is possible with the use of mechanical collectors, electrostatic precipitators, bag-houses and scrubbers like the Baffle, Cyclonic, Ejector venturi and Mechanical scrubbers. Nitrogen dioxide control is possible with the help of low consumption burners and scrubbers, selective catalytic and non-catalytic reduction and even catalytic converters. Sulfur dioxide or acid gas can be effectively controlled with dedicated use of wet and dry scrubbers and the latest introduction of flue gas desulfurization.

The world's most polluted cities include Australia, America, the UK, China and India as forerunners. The greenhouse effect is a life threatening global phenomenon that is the creation of air pollutants. This phenomenon is the result of the trapped greenhouse gases in the upper atmosphere. Accumulation of carbon dioxide gas, methane, nitrogen oxides, hydrofluorocarbons, perfluorocarbons and chlorofluorocarbons h ave brought on a major climate change, increased acidity of ocean waters and major modifications in marine ecosystems.

How you can help to prevent Air Pollution:

- Carpool— This will help to reduce the number of vehicles on the already congested roads.
- Always keep your car tuned properly so that it remains in a good condition.
- Save energy— Try to use minimum amounts of natural gas and even electricity. Whenever possible, avoid the use of air conditioner and use a fan instead.
- Always buy recycled products.
- Reuse things such as paper and plastic bags, paper etc. This will contribute a lot towards reducing the effects of air pollution and global warming.
- Avoid the use of firecrackers. You don't really need it to express your feeling of happiness.
- Go in for water-based paints instead of varnishes.
- If you really cannot avoid using your car, plan your work systematically to reduce air pollution.

6.5 PREVENTION OF AIR POLLUTION

Air pollution being a serious problem, should be checked at any cost. There are a number of methods by which we can control and prevent air pollution up to a great extent. These are as follows:

1. The minimization of automobile emission should be carried out by cleaning exhaust gases with the help of catalysts.
2. The chimney in various industries causing air pollution as in power plants, brick manufacturing plants etc. should be very high so that pollutant gases may reach to the upper level of atmosphere far from the ground level. The pollutant minimizing devices should be applied in these chimneys or stacks so that the amount of polluting gases e.g. SO_2 etc. can be reduced within the chimney.
3. Deforestation should be checked strictly and programmes of plantation should be encouraged. Besides absorbing CO_2, many plants are capable of absorbing significant amount of hydrogen sulphide and nitric acid. In the absence of trees and green vegetation air becomes richer with harmful pollutants like N_xO_y, H_2S and O_3, thus causing air pollution.
4. SO_2 formation by combustion processes can be controlled by desulphurization of fuel. SO_2, and acid fumes can be removed by passing the stack or fuel gases through a bed of alkaline alumina, activated carbon or limestone powder where SO_2 etc. are absorbed.
5. The use of power generating devices such as coal and wood combustion should be reduced in domestic as well as industrial activities. In place of these, modern power generating devices should be used which are electricity, solar energy, nuclear power plants which will help in reducing air pollution.
6. Industries should be established at a sufficient distance from populated area so that air—polluting gases may get diluted in the atmosphere before reaching to populated area. This process is known as **zoning or buffer zoning.**
7. **Clause Process:** It is a method used for controlling the pollution caused by H_2S and SO_x, which are liberated by refineries. The result of this process is the formation of elemental sulphur as a byproduct. It is a two-stage process, which occurs, in the following two steps-

$$2H_2S + 3O_2(g) \longrightarrow 2H_2O + 2SO_2$$
$$2H_2S + SO_2 \longrightarrow 2H_2O + 3S$$

8. Acids and chemical fumes are removed by passing gases and vapours through a tower filled with coke and a water counter current is applied against it. By this method we can recover the useful volatile substances simultaneously, which have commercial values.
9. Smoke concentration can be reduced by maintaining high temperature during the combustion of fuel and feeding of fuel continuously. Installation of **Cottrell electrostatic precipitator** minimizes the smoke quantity.
10. Extraction ventilations are more frequently applied for the removal of dust from air. In this process firstly a stream of air carrying suspended dust particles is maintained at a sufficient velocity to keep the dust in suspension and secondly the rate of the stream is reduced suddenly to some extent to cause the dust particles to settle down in a settling chamber. Thus air is made dust free by the process.
11. Air pollution can be controlled by using **cyclone collector.** This equipment is usually used as a precleaner of air. The principle of this equipment is that the particulate matter present in air possesses greater moment of inertia than that of gases and fumes. In cyclone collector, moment of inertia of particulate is less than that of gases and fumes. In cyclone collector, the smoke or gas containing particulate is allowed to flow into a tight circular spiral fitted chamber. Inside the cyclone there is a device to rotate the fume at a desired speed. The particulates in air feel greater inertial effect, being heavier particle, due to centrifugal force. Consequently these particles tend to move away from the centre towards the wall of chamber from where they start to settle down due to force of gravity. Thus particulate so gathered at the bottom of chamber are removed periodically.
12. Air pollution can be controlled by fuel selection and better utilization. Windmills and other such sources of power and fuel should be used. The aim is to generate power without burning fossil fuels which emit CO_2.

Air Pollution Control Equipment

Here are a few air pollution control systems that are being used by vehicles and industries. They help to either remove pollutants from a stream of exhaust before they are emitted into the air or destroy them.

Air Pollution Control Systems To Reduce Particulate Matter

Wet Scrubbers: These include a number of devices that remove pollutants from furnace flue gas as well as other gas streams. The pollutants are removed by the polluted gas stream being forced through a scrubbing liquid or by using some other method of bringing it into contact with the liquid. Wet scrubbers are used in a number of industries like large power plants, asphalt plants, steel plants, fertilizer plants, and acid plants.

Electrostatic Precipitator (ESP): Also known as Electrostatic Air Cleaners, this air pollution control system is a particulate collecting device which uses the force created by an induced electrostatic charge to remove particulate matter from any flowing gas, *e.g.*, air. These filtration devices are highly efficient and are very effective in removing fine particles like smoke and dust from the air stream. ESPs are used for controlling particulate emissions in various industries like oil refineries, pulp mills, and oil and coal fired utilities that generate electricity which produce smoke.

Dust Cyclones: These are used to remove particulate matter from a gas or air stream, without using filters, using vortex separation instead. Mixtures of fluids and solids are separated by using gravity and rotational effects. There is large scale use of cyclones in oil refineries as well as the cement industry wherein they form a part of the kiln preheaters.

Air Pollution Control Systems To Reduce NOx (Nitrogen Dioxide and Nitrogen Oxide)

Exhaust Gas Recirculation (EGR): This is a technique used for reducing NOx that is used in most diesel and gasoline engines. A part of the exhaust of an engine is recirculated back into its cylinders. When the incoming air is intermixed with the recirculated exhaust gas, it results in diluting the mixture with inert gas, reducing the adiabatic flame temperature and also lowering the excessive oxygen in diesel engines. The peak combustion temperature is also lowered because the specific heat capacity of the mix is increased by the exhaust gas. Since high temperatures cause Nox to form much faster, EGR helps in limiting NO_x from being generated. Nox is produced when a mixture of oxygen and nitrogen is subjected to high temperatures.

Catalytic Converter: This is a device that is used to diminish the toxicity of emissions that are produced by internal combustion engines. First introduced in 1975 in the US in order to comply with the tightening regulations by the Environmental Protection Agency, catalytic converters are still used most commonly in the exhaust systems of motor vehicles. Some of the other places they are use in are: trains, mining equipment, forklifts, generator sets, and other machines equipped with engines.

Air Pollution Control Systems To Decrease Volatile Organic Compounds (VOC)

Gas Flare: Also called a flare stack, this is a chimney that is erected on oil rigs or oil wells, as well as landfills, chemical plants, and refineries. When flammable gas or unusable waste gas plus liquids are discharged by pressure relief valves, this device is used to burn them off. This device is also used in landfills to burn and/or vent the waste gas that is produced by the decomposing materials.

Biofilters: This is a technique for pollution control which uses living matter to trap and biologically degrade pollutants. In air pollution control, the pollutants in the air are subjected to microbiotic oxidation. In other words, when it is applied in the filtration and purification of air, microorganisms, such as fungi and bacteria that are embedded in a biofilm, are used to degrade the air pollutant.

Acid Rain

Acid rain was first reported by Scottish scientist Robert Angus Smith in 1852. However the term 'acid rain' came into existence in 1972. Acid rain is the manifestation of serious air pollution. The gases emitted from various sources mix with the rain to form an acidic precipitation, called *acid rain*. This acidic precipitation is capable of threatening the balance of entire ecological cycle. Acid rain affects all the living beings directly or indirectly. The future implications of acid rains can be severe, if proper measures are not taken on time.

Any form of precipitation which is acidic in nature is called acid rain. Acid rain is the result of excessive emissions of sulfur and nitrogen caused by human activity, which reacts with other compounds to form acids. Acid rain has detrimental effects on animals, plants and infrastructure.

In its purest state, rain water is like distilled water. It does not have carbon dioxide dissolved in it. It is neutral, with a pH level of 7. pH is the concentration of hydrogen ions in an aqueous solution. If the pH level is above 7, it is said to be basic, and if it is below 7, it is said to be acidic in nature.

As rain water falls through the atmosphere, particles suspended in the air are dissolved in it. These substances are generally dust, pollen grains and carbon dioxide (CO_2). Emissions of volcanoes and lightning tend to decrease the pH level of acid rain, making it even more acidic. CO_2 combines with water to form carbonic acid (H_2CO_3).

$$H_2O(l) + CO_2(g) = H_2CO_3(aq)$$

Carbonic acid ionizes in water to form low concentrations of carbonate and hydronium ions.

$$2H_2O(l) + H_2CO_3(aq) = CO_3^{2-}(aq) + 2H_3O^+(aq)$$

Carbonic acid is a weak acid. It brings down the pH of the rain water to 6.0–5.2. With pH levels ranging between 6.0–5.2, rain water is acidic, but still not dangerous. This is a reversible reaction.

The problem occurs when rain water combines with gaseous oxides of sulphur, nitrogen, and phosphoric and hydrochloric acid mists. The latter two and sulphur are released into the atmosphere from automobile exhausts, industries and electric power plants. Nitrogen forms a major part of atmospheric composition. These chemicals bring down the acid rain pH level to 5.6–3.5. Sometimes, the pH level can even become as low as 2. This phenomenon of acidic rain water precipitation, is called acid rain. Rain, snow, sleet, freezing rain, hail, fog and dew are other forms of precipitation.

Chemistry of Acid Rain

Sulphuric acid and nitric acid are the main acids present in acid rain. Sulfuric acid is formed as follows:

- Sulphur released into the atmosphere combines with atmospheric oxygen to form sulfur dioxide (SO_2)
- Sulphur dioxide reacts with atmospheric water to form sulphurous acid – $SO_2(g) + H_2O(l) = H_2SO_3(aq)$
- Sulphurous acid is also present in acid rain.
- Sulphur dioxide gradually oxidizes to form sulphur trioxide (SO_3) – $2SO_2(g) = O_2(g) = 2SO_3(g)$
- Sulphur trioxide reacts with water to form Sulphuric acid (H_2SO_4) – $SO_3(g) + H_2O(l) = H_2SO_4(aq)$

  Nitrogen dioxide(NO_2) is formed as follows:
- Nitrogen combines with atmospheric oxygen to form nitrogen dioxide (NO_2). Nitrogen dioxide reacts with water to form nitrous acid (HNO_2) and nitric acid (HNO_3) – $2NO_2(l) + H_2O(l) = HNO_2(aq) + HNO_3(aq)$

Acid rain is a mild combination of mainly sulfuric and nitric acid. Sulphurous acid and nitrous acid are less stable and are present only in very low amounts. Following are the various adverse effects of acid rain on living organisms and infrastructure.

Wet Acid Rain

Every source of energy (coal, fuel, wood or petroleum products) has sulphur and nitrogen. These two elements, when burnt in atmospheric oxygen, are converted into their respective oxides (SO_2 and NO_2), which are highly soluble in water.

$$S + O_2 \longrightarrow SO_2$$
$$2SO_2 + O_2 \longrightarrow 2SO_3$$

In case of nitrogen, following reactions takes place—

$$NO + O_3 \longrightarrow NO_2 + O_2$$
$$NO_2 + O_3 \longrightarrow NO_3 + O_2$$
$$NO_3 + NO_2 \longrightarrow N_2O_5$$

Under the humid conditions of air, N_2O_5 invariably reacts with water vapours and form droplets of HNO_3—

$$N_2O_5 + H_2O \longrightarrow 2HNO_3$$

Some HNO_2 is also formed.

$$N_2O_3 + H_2O \longrightarrow 2HNO_2$$

HNO_3 and HNO_2 then return to the earth's surface. However, HNO_3 can be removed as a particulate or as particulate nitrate after reaction with bases such as NH_3.

SO_2 in humid atmosphere forms droplets of H_2SO_4.

$$SO_2 + 1/2\,O_2 + H_2 \xrightarrow{NO} H_2SO_4$$

Presence of hydrocarbon andNO_x steps up the oxidation rate of reaction. In water droplets, ions such as Mn(II), Fe(II), Ni(II) and Cu(II) catalyze the oxidation reaction.

HNO_3 and H_2SO_4 thus formed combine with HCl (emitted from natural and man made sources) to generate precipitation, which is commonly referred to as **acid rain.** Normally unpolluted rain is weakly acidic and has a pH of 5.6

Oxidation of Sulphur Dioxide: The foremost chemical reaction of SO_2 in air is its oxidation to SO_3 which reacts with the moisture to give H_2SO_4. The H_2SO_4 or its sulphates occur as aerosols—

$$SO_2 \xrightarrow{NO} SO_3 \xrightarrow{H_2O} H_2SO_4 \xrightarrow{NH_3} \underbrace{(NH_4)_2SO_4 \text{ and } NH_4HSO_4}_{\text{Aerosols}}$$

Oxidation step proceed in three ways—

1. Catalytic oxidation.
2. Photochemical oxidation.
3. Oxidation by radicals.

Areas of Acid Rains

In highly populated cities, internal combustion engines, used by masses, emit pollutants like CO_2, CO, SO_2, NO, NO_2, etc. on road around them in atmosphere, which are absorbed by moisture present in air forming various acids, *e.g.*, H_2SO_4, HCl, and HNO_3 etc. Thus more acid rains are likely to occur in such areas where there is tendency of formation of acids due to absorption of various gases by the moisture present in the environment—

Combustion Engines $\longrightarrow$ Pollutants (CO, CO_2, SO_2, NO_x etc.) $\longrightarrow$ Moisture + Pollutants (In atmosphere) $\longrightarrow$ Acids (Causing rain)

A typical sample of acid rain contain following cations and anions.

Cation	Conc. (mole/l * 10^6)
H^+	56
NH_4	10
Ca_2	7
Na^+	5
Mg_2	3
K^+	2
Total	**83**

Anion	Conc. (mole/l * 10^6)
SO_4^{2-}	51
NO_3^-	20
Cl^-	12
Total	83

Adverse Effects of Acid Rains

Acid rain exerts both direct and indirect effect on the organisms and materials. The direct effects of acid rain are determined by concentration of pollutants in the air. They are mainly of local nature with a geographical extent of few kms. They decline rapidly with distance from the emission source. They affect organisms and materials and cause more harm at a distance up to hundreds and sometimes thousands of kms.

The dry deposition has several direct effects on the environment. It attacks building materials, mainly, limestone, sandstone, marble, steel, nickel and other metals causing a loss of millions of rupees. When deposited in gaseous form it causes direct damage to plants and trees, the visible injury being gradual yellowing or depigmentation of leaf tissues, *i.e.,* chlorosis.

The wet deposition has direct as well as indirect effects. It increases the acidity of lakes and rivers which is made worse by inflow of acids and metals from nearby soils. Wet deposition also affects aquatic as well as terrestrial ecosystems. Wet deposition can transport metals such as Al, Cd, Hg and Cr into soil water, ground water, lakes and streams, depleting the stocks of nutrients in the soil, thereby causing harm to various ecosystems. Acidity kills fish, algae, bacteria and aquatic system gets collapsed into the sterility leaving a crystal clear but ultimately a dead lock.

The increased acidity of rainwater causes damage of fresh water life. In the form of mist it causes the damage of plant leaves, changes the rate of metabolism of organisms. These also produce irritation to eyes and mucous membrane. This acid rain accelerates the rate of corrosion. It dissolves salts in the soil (*e.g.,* $CaCO_3$) and metals like aluminium, which passes in rivers where they cause toxic effects to aquatic life forms.

(*a*) **Effects of Acid Rain on Aquatic Biota**:

1. A significant reduction in fish population accompanied by decrease in the variety of species in food chain has been observed.
2. Many bacteria and blue green algae are killed due to the acidification, disrupting the whole ecological balance. Acid rain killed fishes in lakes and destroyed trees in a wide swathe across Europe.

(*b*) **Effect of Acid Rain on Terrestrial Ecosystem** :

1. In West Germany, about 13% of the forests died and nearly 20 million acres of forests are affected by acid rains. Forests in Switzerland, Netherlands and Czechoslovakia have been damaged by such rains, Actually nutrients like calcium, potassium, iron and magnesium have been away from soil by acids. These nutrients are most essential for the plant growth.

2. The effect of acid precipitation on terrestrial vegetation indicates reduced rate of photosynthesis and growth and increased sensitivity to drought and disease.
   (*i*) Acid deposition weakens the trees like pine, which can be easily attacked by pathogens and drought.
   (*ii*) Acidification of soil changes biology and chemistry. Plants can easily absorb Cd from the acidified soil. High levels of Cd in plants are injurious for animals and human beings.
   (*iii*) Root systems are damaged by the uptake of aluminium released from the soil. Nitrates may be released from the soil by acid run–off waters.
   (*iv*) Trees pull heavy metal ions, like Cd, Cr and Al into their systems.
   These ions are imunobile in many soils, but acidic water mobilizes them and cause tree damage.
   (*v*) Acid rain has severely retarded the growth of crop such as pea, beans, potato, broccoli and carrots etc.

(*c*) **Effects of Acid Rain on Lake Ecosystem:**

1. The activity of the bacteria and other microscopic animals is reduced in acidic water. So the dead materials and other accumulated substances lying on the bottom of lakes are not rapidly decomposed. Thus essential nutrients as nitrogen and phosphorus stay locked up in plant and animal remains. Biomass production is reduced and fish population declines.
2. Acid rains cause a number of complications in ponds, rivers and lakes where it accumulate as **acid snow.** In summer rapid snowmelt gives a jolt of acid water to lakes. This acid fell is most damaging to young fish, algae, and insects and to the food chain.
3. Black flies, mosquitoes, deer flies and other aquatic worms occur abundantly where fishes are eliminated. So they appear to thrive in acid conditions. Dragon fly larvae and water boatmen also flourish in acidified lakes.
4. Aquatic plants such as broad-leafed pondweeds do not grow in acidic water. This could affect the feeding and breeding habits of aquatic species.

(*d*) **Effects of Acid Rain on Human Beings:**

1. Acid rain has been found to be very dangerous to the living organisms as it can destroy life. Acidification can play havoc with human nervous system, respiratory system and digestive system by making the person an easy prey to neurological diseases. This happens because these acids produce highly toxic compounds which contaminate the potable water and enter man's body.
2. Acidic rains containing air pollutants contribute to variety of safety hazards, associated with reduced visibility due to smog etc. These contaminants can be nuisance in several aspects and cause adverse health effects.

(*e*) **Effects of Acid Rain on Buildings:**

1. **Stone laprosy**— Acid rain causes extensive damage to buildings and structural materials of marble, limestone, slate and mortar etc. Limestone is attacked rapidly.
   $$CaCO_3 + H_2SO_4 \longrightarrow CaSO_4 + H_2O + CO_2$$
   The attack on marble is termed as **stone laprosy.**
2. The Taj mahal in Agra is also suffering at present due to SO_2 and H_2SO_4 acid fumes or air pollutants released from Mathura refinery.

3. Due to acidity, levels of heavy metals as Al, Mn, Pb, Cd, Cr and Cu in water increases beyond the safe limit which indirectly affects the buildings.
4. H_2S tarnishes silver and blackens leaded house paints. Ozone produces cracks in rubber. Trace of radio active elements present in radio active rain severely damages the buildings.

Control of Acid Rain

There is an urgent need for proper and regular monitoring to provide timely warnings about acidification of our environment. Short-terms control of acid deposition problem can be achieved by using lime.

With air pollution a serious cause of concern in European countries, the European Union (EU) Environment Ministers have agreed in early March 1996 to reduce diesel fuel emissions all over Europe, because diesel emissions contain nitrogen oxides, which are known to cause acid rain and also diesel particles, some of which are carcinogenic in nature. Most of the machinery covered under the proposals account for about 30% of total nitrogen oxide emissions from diesel vehicles. The proposal was to reduce nitrogen oxide emission by 42% and diesel particles by 67% by the end of the year 2003.

Various chemical industries and automobiles release acidic oxides (like SO_2, NO_2, HCl etc.) into the atmosphere. These oxides and HCl dissolve in moisture present in atmosphere etc. form corresponding acids, which then fall snowing on earth as acid rain.

$$SO_2 + H_2O \longrightarrow H_2SO_3 \text{ Sulphurous acid}$$
$$2SO_2 + O_2 + 2H_2O \longrightarrow 2H_2SO_4 \text{ Sulphuric acid}$$
$$4NO_2 + 2H_2O + O_2 \longrightarrow 4HNO_3 \text{ Nitric acid}$$
$$HCl + H_2O \longrightarrow HCl(aq.) \text{ Hydrochloric acid}$$

Automobiles (consisting of internal combustion engines) exit pollutants like CO_2, CO, hydrocarbons etc; while large number of industries exit pollutants like CO_2, CO, SO_2, SO_2, NO, NO_2 etc. in the atmospheric air, which are absorbed by moisture, present in air, forming corresponding acids.

Some mechanical or chemical methods should be adopted by the industries and automobiles to prevent the formation of harmful pollutants and gases, which may cause acid rains. Combustion engines should be so manufactured that they do not release harmful pollutants in the atmosphere, which may be absorbed by the moisture present in the environment causing acid rain.

Effects of Acid Rain

Acid rain can induce long lasting effects on the life on earth. The various ecological components affected by acid rain are:

Human Beings

It is not the acid rain that directly affects the human beings, but the pollution that causes it. Air pollutions leads to several health problems like bronchitis, asthma or damage to lungs. The sulfur dioxide and nitrogen oxides may lead to irritation of eyes, ears and nose and poor visibility. These pollutants can also cause throat infections and dry coughs. In addition, several problems may arise due to consumption of intoxicated plants or animals.

Plants

Acid rain can deplete the nutrients in the soil, thereby depriving the plants of their food. Nutrients in soil dissolve in acid rain and get washed away before the plants can utilize them. The poor quality acidic soil is not only incapable of buffering but it also becomes intoxicated with harmful substances like mercury and aluminum, which are absorbed by the plants. Also, the acidic fog that covers the trees, deprives the leaves of sunlight and makes them weak. As a result, the leaves are unable to perform photosynthesis, rendering the trees weak and susceptible to various diseases and infections. Crops and other plantation are less likely to be affected by the acid rain, as the soil is enriched by the fertilizers. But applying excessive fertilizers is not beneficial for soil quality in the long run.

Aquatic Life

The acid rain interferes with the natural p^H of water, which is between 6 to 8. When acid rain mixes with water in lakes and streams, the p^H of water is reduced to 5. p^H level below 5.0 is extremely detrimental to aquatic life. The young ones of most species are particularly sensitive to the acidic water. They find it even more difficult to survive in such water than adults. This endangers the entire species. Many species of fishes, frogs and other marine organisms have already become extinct while several others are on the verge of extinction.

Animals and Birds

Animals and birds are not directly affected by the acid rain, but as they are the part of food chain, they invariably have to face the effects. Consuming plants that are intoxicated by the acid rain can produce certain problems in animals and birds. Reduction in forest also deprives them of their habitat. This may result in substantial decrease in their population.

Other Effects of Acid Rain

Other effects of acid rain include damage to several monuments and places of historic importance. The major pollutants of acid rain, sulfur dioxide and nitrogen oxide erode the marble and limestone which form the major part of construction materials for most buildings. Fading and peeling of color of buildings is the common effect of acid rain. Many great structures in the world like Taj Mahal in India, Washington Monument, Renaissance buildings in Italy, Acropolis in Greece have all been affected by the acid rain.

Buildings and Monuments: Acid rain causes severe damage to buildings and marble statues. Acid rain reacts with the calcium carbonate ($CaCO_3$) to form soluble calcium hydrogen carbonate or calcium bicarbonate, $Ca(HCO_3)_2 - CaCO_3 + \text{Acid rain} = Ca(HCO_3)_2(aq)$

Calcium bicarbonate is a powdery substance, which is easily washed away with water or more specifically, rain water. This is the way acid rain has partly eroded many world-famous monuments and buildings like the Taj Mahal in India, St. Paul's Cathedral in London, and the Statue of Liberty in New York.

Acid rain can destroy stained glass windows in churches, bridges made of steel, and railway tracks. It corrodes metal, ruins the paint color, weakens leather and forms a crust on glass surfaces.

Soil: Soil contains many harmful minerals such as mercury and aluminum. These elements can't be absorbed by plants and trees and are thus harmless. Upon contact with acid rain, these chemicals undergo chemical reactions with the acids. As a result, compounds of aluminum, lead and mercury are formed. Plants and trees can easily absorb these compounds. Such elements, which are extremely harmful to living forms, ultimately affect the entire food chain. These chemicals not only harm the flora, but also the animals that feed on them.

Measures

Though the havoc played by acid rain cannot be reversed, certain actions can be taken to avoid further damage. General awareness among common people and pressure from various environmental groups have forced the Governments to take measures against the acid rain. Lakes and streams can regain their PH balance by the process of liming. Liming means dissolving large quantities of alkaline substances like quicklime into the water bodies. In addition, measures like reducing power consumption, cutting down on fuel and afforestation can also bring down the pollution level to a great extent.

Solutions to Acid Rain

Acid rain is potent enough to destroy life on Earth. It damages anything it comes in contact with. It ruins forests, water bodies, soil, infrastructure, and the health of living beings. It's high time we take all possible measures to control it. There is a solution to every problem, and acid rain is no exception. Take a look at the various measures we can take.

- Human beings should reduce the use of fossil fuels. This would lessen the emission of nitrogen and sulfur in the environment.
- Public transportation, car pools and walking can reduce nitrogen, sulfur and lead emissions into the atmosphere.
- Sulfur and nitrogen are mostly released into the atmosphere from the burning of fossil fuels (*e.g.,* Coal). Switching over to alternative forms of energy such as geothermal, water, wind, and solar power would help to a great extent.

All of the above measures are simple steps that can easily be adopted on an individual level. Acid rain has become an international issue because of its serious and definite danger to life on earth. Many international treaties have been signed *i.e.,* the Sulfur Emissions Reduction Protocol and the Convention on Long-Range Transboundary Air Pollution. Installing Flue Gas Desulfurization *e.g.,* wet scrubber in coal-burning power plants to remove sulfur-containing gases - is one of the steps taken by the USA, and is followed by a number of developed and developing countries.

Global Warming and Greenhouse Effect

Global warming results from the effect of what is called the 'greenhouse effect'. Carbon dioxide (CO_2), methane, water vapor, nitrous oxide and ozone are the main greenhouse gases. The main purpose of a greenhouse is to allow sun's radiation (heat) to enter the enclosure easily, however, does not allow the heat to be radiated back to the atmosphere. Similar to a greenhouse, our atmosphere allows the sun's radiation to heat the Earth and slows down the radiated heat to go back into the space. This greenhouse effect is essential to sustain life on Earth.

The greenhouse effect is a natural warming process of the earth. The theory is that when the sun's energy reaches the earth, some of it is reflected back to space while the rest is absorbed. The absorbed energy warms the earth's surface which then emits heat energy back toward space in the form of long wave radiation. This outgoing radiation is partially trapped by greenhouse gases such as carbon dioxide, methane and water vapor which then radiate the energy in all directions, warming the earth's surface and atmosphere.

Troubling facts about the greenhouse effect include that the increasing amounts of greenhouse gases intensify the greenhouse effect. Higher concentrations of CO_2 and other greenhouse gases trap more infrared energy in the atmosphere than what occurs naturally. The additional heat further warms the atmosphere and Earth's surface. Climate models suggest this natural warming is being enhanced by human activities that increase concentrations of greenhouse gases in the atmosphere and thus, in turn, intensify the greenhouse effect. The increase in atmospheric concentration of greenhouse gases, accentuating the natural greenhouse effect globally, results in global warming.

Now, due to the increased level of greenhouse gases, the atmosphere's ability to trap and hold the heat increases more than required. As a result, the average temperature of the Earth increases and this natural phenomenon is known as 'global warming'. It is also defined as a sustained increase in the atmosphere that causes abnormal climate change. In the following description about global warming and carbon dioxide, a brief explanation regarding effects of global warming and CO_2 emission in the atmosphere, has been provided.

Global Warming and Carbon Dioxide Emissions

Power Plants

Power plants are the major contributors to the increased level of carbon dioxide emissions in the atmosphere. The plants work for generating electricity by burning fossil fuels in a massive scale, which produces large amounts of CO_2. Coal is known to be responsible for about 93% of the emissions in the power plants. Natural gas produces 80% less carbon per unit of energy than coal, and hence so much of impact on the environment results in the ever-increasing pace of global warming.

Vehicles

Gasoline-burning engines produce about 20% of carbon dioxide emitted in the atmosphere. In the United States, 33% of emissions is from vehicles. Sports bikes and vehicles result in more emission than general vehicles designed for normal roads.

CO_2 Due to Deforestation

The ever-increasing deforestation is also one of the main culprits of global warming. As we are all aware of the fact that, trees take in huge amount of CO_2 from the atmosphere and releases oxygen. So, if there are no trees, there will be no absorption of this global warming gas and thus, the situation worsens further. This fact is important when it comes to the reason as to why one of the global warming causes is carbon dioxide? As the world is progressing its way towards development, forests are vanishing away from the face of the Earth. More urbanization results in more deforestation for land and timber requirements, and all these factors boil down to one 'dead end', called global warming.

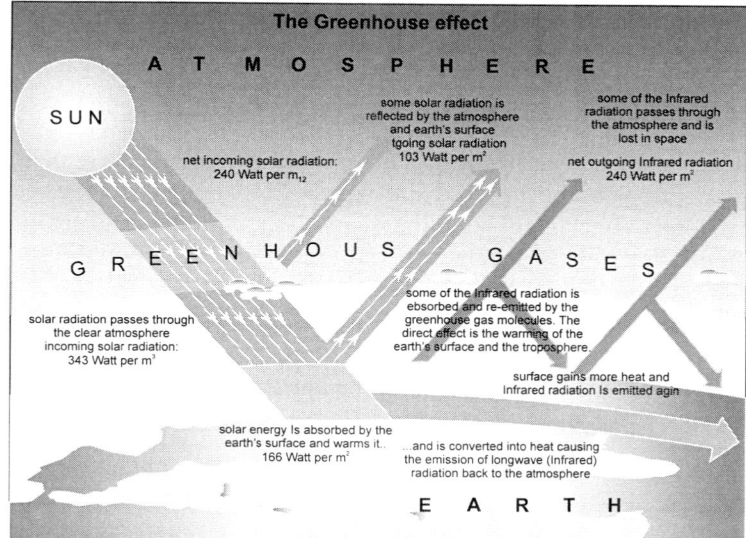

Fig. 6.1 Greenhouse Effect

Some Major Negative Effects of Global Warming

Increase in Sea Level

As the Earth warms further, polar ice caps would melt and increase the water levels in seas and oceans. It is estimated that melting down of all glaciers will cause the water level to rise to about 230 feet! So you can imagine what it would be like? Further, ice caps are fresh water and their melting and mixing with the and seas and oceans will greatly imbalance the eco-system. This is because, the ice caps will desalinate the water and make it less salty. This process will disturb the ocean currents and hence their regulation of temperature. Ice caps are known to reflect much of sunlight back into space, further cooling the Earth. Melting of these reflectors will leave oceans and seas to be totally exposed to sun's radiation. These water bodies cannot reflect light but absorb it as they are dark in color. This will further contribute to the warming of the Earth.

Diseases and Drought

Places of the Earth that will drastically warm up due to global warming, will face more incidents of diseases and drought. Such areas will have increased infestation by disease-carrying insects and heat waves.

Hurricanes

Due to the melting of ice caps, oceans will absorb more heat resulting in warmer waters. This will favor stronger and more frequent inc idents of hurricanes.

Increasing emissions of CO_2 into the atmosphere is making matters worse than ever when it comes to dealing with 'global warming and carbon dioxide'. We can also form an idea about the 'most probable' aftermaths of the always-increasing temperature of out planet Earth, and the idea is definitely not a pleasing one. All other factors, along with those responsible for increased amount of carbon dioxide causing global warming, are perhaps impossible to do away with. These factors are arising only due to human needs; development being an apt example, and so they can be regarded as 'inevitable'.

Effects of Global Warming

Global warming affects many different facets of life on earth. There will be winners and losers, even within a single region. But globally, the losses are expected to far outweigh the benefits. The regions that will get most severely affected are often the regions that emit the least green house gases. This is one of the challenges that policy makers face in finding fair international responses to the problem. Some of the likely effects of global warming include:

Rise in Sea Level: It is a serious worry that rising sea levels from the melting of the polar ice caps could severely flood many countries. A rise in sea levels of one meter, which many experts are predicting by the year 2100 (and some as soon as 2030), would flood 15 percent of Egypt, and 12 percent of Bangladesh. The Maldives in the Indian Ocean would almost completely disappear.

Water Resources: Many of the major rivers in Europe and Asia emanate from the glaciers in the mountains. For example, the whole of the grain producing northern belt of India is fed by rivers which originate from glaciers in the great Himalayas; electricity is also produced wherever the rivers are dammed. Over the years it has been seen that because of global warming the glaciers are receding at an alarming rate, impacting the flow of the rivers, causing a reduction in production of foodgrain and electricity.

Health and Disease: It is a known fact that people today are inflicted by many new diseases unheard of in the last century. Increase in CO_2 levels in the cities is the cause of pollution, and the effects of pollution are well documented. Global warming is also responsible for the introduction of some new diseases. Bacteria are known to be more effective and multiply much faster in warmer temperatures compared to cold temperatures. The increase in temperature has led to increase in the microbes that cause diseases.

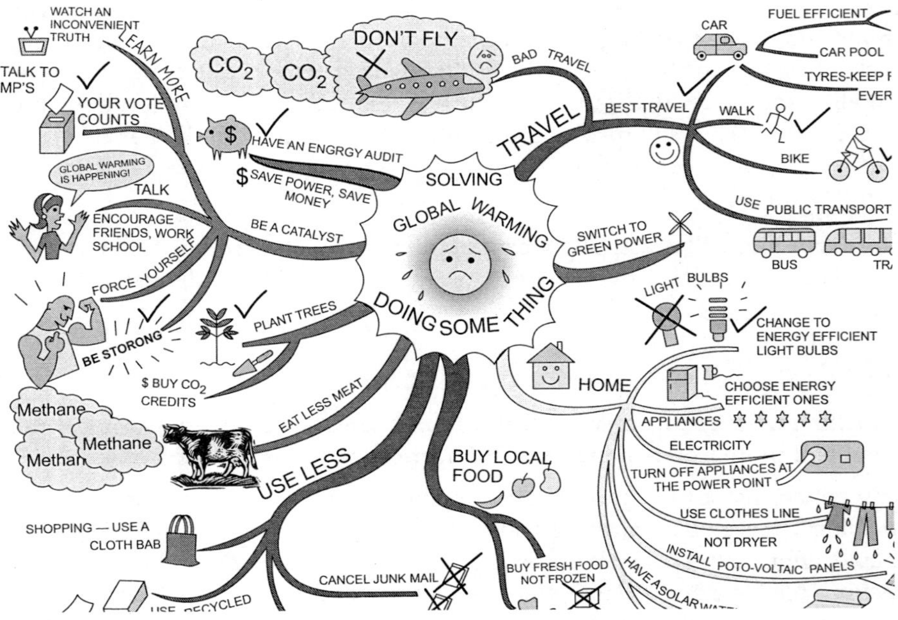

Fig. 6.2 Impacts of Global Warming

Agriculture: It is said that should there be a few degree heating up of the earth it would have a potentially negative effect on the production of corn in North America, where much of the world's foodgrain comes from. This would result in higher prices of foodgrain causing starvation in third world countries. On the other hand, it may so happen that colder regions further north would be able to grow crops that have never been cultivable before. However, it is a known fact that the availability of cultivable land decreases as you go north.

Ecosystems: We are aware of the fact that the ecosystem of the world is fragile and delicately balanced. Any change brought about by global warming could result in disastrous consequences. Some of these the world has experienced in the last few decades by way of floods, droughts, frequent hurricanes, tornadoes, and severe cold waves. There have also been an increased number of forest fires in the recent past.

There is, however, a very divergent view on global warming facts. Many skeptics are of the view that global warming is a good and natural phenomenon. There are various benefits of global warming, and as per them it will increase humidity in tropical deserts, melting of snow bound areas will result in vast tracks of land getting available for agriculture, the increase in carbon dioxide in the atmosphere will trigger plant growth. Another argument of global warming is that the earth has been gradually warming up through history, and nature will continue to adapt to these changes.

Despite the views of some skeptics mentioned above, there is a recorded negative impact of global warming. Due to the direct link between global warming and the greenhouse effect, the only way to turn the situation around is for everyone to do their bit to reduce the emission of greenhouse gases. Reduction of use of fossil fuels, use of alternate sources of energy, restricting the emission of pollutants by industries, reducing forest degradation, eliminating the use of CFCs and planting more trees will help us overcome this seemingly insurmountable problem.

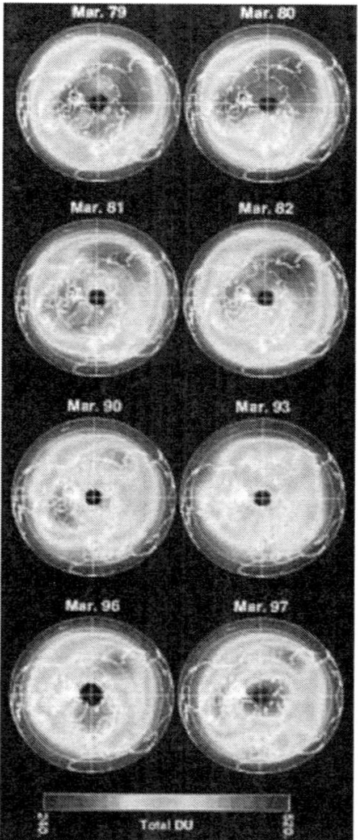

Fig. 6.3 The Ozone Layer Over Time

Ozone Depletion

The ozone layer protects the Earth from the ultraviolet rays sent down by the sun. If the ozone layer is depleted by human action, the effects on the planet could be catastrophic.

Ozone is present in the stratosphere. The stratosphere reaches 30 miles above the Earth, and at the very top it contains ozone. The suns rays are absorbed by the ozone in the stratosphere and thus do not reach the Earth.

Ozone is a bluish gas that is formed by three atoms of oxygen. The form of oxygen that humans breathe in consists of two oxygen atoms, O_2. When found on the surface of the planet, ozone is considered a dangerous pollutant and is one substance responsible for producing the greenhouse effect.

The highest regions of the stratosphere contain about 90% of all ozone.

In recent years, the ozone layer has been the subject of much discussion. And rightly so, because the ozone layer protects both plant and animal life on the planet.

The fact that the ozone layer was being depleted was discovered in the mid-1980s. The main cause of this is the release of CFCs, chlorofluorocarbons.

Antarctica was an early victim of ozone destruction. A massive hole in the ozone layer right above Antarctica now threatens not only that continent, but many others that could be the victims of Antarctica's melting icecaps. In the future, the ozone problem will have to be solved so that the protective layer can be conserved.

Causes of Ozone Layer Depletion

Only a few factors combine to create the problem of ozone layer depletion. The production and emission of CFCs, chlorofluorocarbons, is by far the leading cause.

Many countries have called for the end of CFC production because only a few produce the chemical. However, those industries that do use CFCs do not want to discontinue usage of this highly valuable industrial chemical.

CFCs are used in industry in a variety of ways and have been amazingly useful in many products. Discovered in the 1930s by American chemist Thomas Midgley, CFCs came to be used in refrigerators, home insulation, plastic foam, and throwaway food containers.

Only later did people realize the disaster CFCs caused in the stratosphere. There, the chlorine atom is removed from the CFC and attracts one of the three oxygen atoms in the ozone molecule. The process continues, and a single chlorine atom can destroy over 100,000 molecules of ozone.

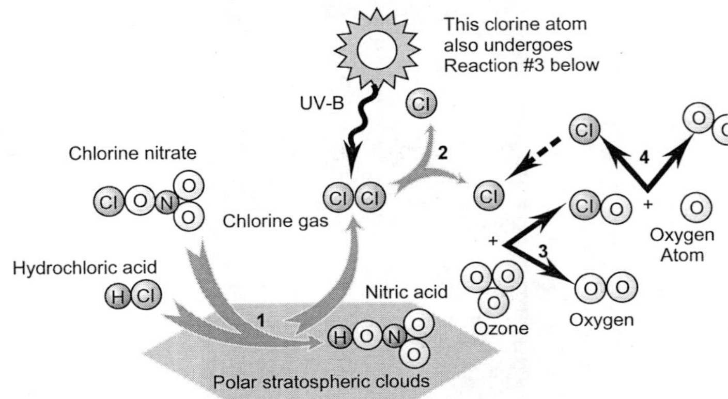

Fig. 6.4 Mechanism of Ozone Layer Depletion

In 1974, Sherwood Rowland and Mario Molina followed the path of CFCs. Their research proved that CFCs were entering the atmosphere, and they concluded that 99% of all CFC molecules would end up in the stratosphere.

Only in 1984, when the ozone layer hole was discovered over Antarctica, was the proof truly conclusive. At that point, it was hard to question the destructive capabilities of CFCs.

Even if CFCs were banned, problems would remain. There would still be no way to remove the CFCs that are now present in the environment. Clearly though, something must be done to limit this international problem in the future.

Mechanism of Ozone Layer Depletion

It includes, 1. **The natural process.** 2. **The anthropogenic process.**

1. **The natural process:** Atmospheric oxygen absorbs ultraviolet radiation shorter than 240 nm and photodissociates into two oxygen atoms. These unite with other O_2 molecules to form ozone. During the process surplus energy of nascent O_3 is often transferred to the nearby molecules as kinetic energy which slightly raises the surrounding atmospheric temperature.

   Ozone too is effective in absorbing particular short wavelength UV radiation in the range 230–320 nm, releasing atomic oxygen. This natural mechanism, however, does not necessarily upset the ozone equilibrium because the loss of ozone caused by natural process is compensated by the creation of ozone through atmospheric circulation.

   Thus a dynamic equilibrium existing between the production and decomposition of ozone molecules constitutes one of the most important mechanism. The heat generated during the reaction causes a rise in temperature of the stratosphere. Secondly, the photochemical process absorbs most of the harmful solar ultraviolet radiations of wavelength from 2000 Å to 3500 Å.

   Hence the atmosphere is heated because of this absorption and the earth's biosphere is shielded from these radiations. Without it, life on the earth would be completely destroyed. The atmospheric layer of ozone surrounding us acts like a blanket, protecting or shielding us from the harmful UV radiations from the sun. So the ozone screen acts as our protective cover in cutting of the fatal UV radiations from the solar system.

   Ozone acts as a powerful oxidant because of its ability to remove electrons from other molecules. It occurs as a natural component of air. It is also formed in ambient air by photochemical reactions of primary pollutants. Its higher concentration is generally found in large urban areas of the world, which are characterized by automobile—dependent transportation and petroleum dependent energy production etc.

   Surface ozone is produced as a result of chemical reactions in the atmosphere driven by the action of UV radiation in the sunlight, involving both nitrogen oxides (NO_x) and volatile organic matter (VOM). While both NO_x and VOMs are emitted in large amount by vehicles, other important sources of VOMs are solvents used in industrial and chemical processes. According to British scientists, this surface ozone is responsible for the substantial reduction in the yields of major crops like wheat and rice. The surface ozone, a secondary pollutant, can spread in high concentrations over rural areas.

   There are the other sources of ozone in the environment, such as gamma radiation used in food preservation plants, commercial UV lamps used for sterilization, high voltage electric equipments, dermatological phototherapy equipments and even in photocopying machines.

Ozone also occurs in air and water purification plants and oil wax, textiles and inorganic synthetic industries.

2. **The anthropogenic process:** We know that there are natural chemical and photochemical processes which lead to the continuous formation and destruction of ozone in the stratosphere. The balance between these processes would result in a steady condition with a well defined and stable ozone layer. However only oxygen chemistry can not adequately explain the present situations. Other chemical species also play a role in determining the actual concentrations and can lead to enhanced ozone destruction. Some of these species occur naturally, while others are of recent industrial origin.

   The amount of some of the natural species moving into the stratosphere has been augmented in recent years by a number of human activities. Many of the processes, which are responsible for the depletion of ozone layer, share a general mechanism of the type,

$$X + O_3 \longrightarrow XO + O_2$$
$$XO + O \longrightarrow X + O_2$$

Net reaction $$O + O_3 \longrightarrow O_2 + O_2$$

Here X is the reacting species. The most common of these species have been identified to be free radicals, such as $HO_x(\bullet H, \bullet OH, HOO\bullet)$, $NO_x(\bullet NO, \bullet NO_2)$ and ClO_x ($\bullet Cl$, $\bullet ClO$). Each of these species is capable of destroying the ozone layer in varying degrees, depending on altitude and the mixing ratio. For example, near the stratosphere the HO_x radicals are responsible for about 70% of the total mechanism of ozone destruction, including the oxygen only process.

Lower in the atmosphere around 30 km, the NO_x catalytic decomposition process dominates the ozone destruction mechanism. The NO_x cycle in the region just above the troposphere has been found to be responsible for about 70% of the ozone destruction, but recent studies show that this figure may be even more.

(*a*) **The hydroxyl radical (HO_x) cycle:** The HO_x radicals are generated by either of the following two photochemical reactions.

$$O^* + H_2O \longrightarrow 2\bullet OH$$
$$H_2O + h\upsilon \longrightarrow \bullet H + \bullet OH$$

where O is excited state oxygen. These reactive free radical species ($\bullet$ H or $\bullet$ OH) decompose ozone in the following manner—

$$\bullet OH + O_3 \longrightarrow O_2 + HOO\bullet$$
$$HOO\bullet + O \longrightarrow \bullet OH + O_2$$

$$O + O_3 \longrightarrow O_2 + O_2$$

Production of $\bullet$ OH radical is largely due to the natural processes and their relative importance increases with increasing altitude. The hydrogen radical decomposes the ozone in the following way—

$$\bullet H + O_3 \longrightarrow O_2 + \bullet OH$$
$$\bullet OH + O \longrightarrow \bullet H + O_2$$

$$O + O_3 \longrightarrow O_2 + O_2$$

(*b*) **The nitric oxide (NO_x) cycle:** The nitric oxide cycle contributes in a significant way in the natural destruction of stratospheric ozone in the lower region of the stratosphere—

$$NO + O_3 \longrightarrow NO_2 + O_2$$
$$NO_2 + O \longrightarrow NO + O_2$$
$$\overline{O + O_3 \longrightarrow O_2 + O_2}$$

In the upper atmosphere, NO is produced by reactions involving atomic and ionic species of nitrogen and oxygen. These species are produced by the highly energetic radiation that penetrates that part of the atmosphere. Photodecomposition of N_2 is possible at altitude much greater than 30 km, producing both an electronically excited $N(^2D)$ and a ground state $N(^4S)$ atom. The ground state nitrogen subsequently reacts with O_2, producing another natural source of NO—

$$N_2 + h\upsilon\ (\lambda < 126\ \text{mm}) \longrightarrow N(^2D) + N(^4S)$$
$$N(^4S) + O_2 \longrightarrow NO + O$$

At altitudes of less than 30 km, additional NO is also generated via a reaction,

$$N_2O + O(^1D) \longrightarrow 2NO$$

Where $O(^1D)$ is excited state oxygen. The principal source of nitrous oxide is the process of denitrification in soils having high concentration of nitrate ion.

$$2\ NO_3^-\ (aq) + 2[CH_2O] + 2H_3O^+(aq) \longrightarrow N_2O + 2CO_2 + 5H_2O$$

The increase in atmospheric nitrous oxide (N_2O) mixing rations is due (to some extent) to enhanced rates of denitrification resulting from greater use of nitrogenous fertilizers worldwide. Although the mixing ratio of nitrous oxide in the troposphere is steadily increasing, the production of stratospheric nitric oxide through various processes can, in a large part, be attributed to natural causes. Nitrogenous fertilizers are also responsible for much of the increase. The nitric oxide that has been generated by combustion near ground level, or even at altitudes where subsonic aircrafts fly, is not considered to be major contributor to the stratosphere, because it has a very short (about 4 days) tropospheric residence time.

(*c*) **Chlorine radical (ClO_x) cycle:** The CFCs and halons are entirely manmade with wide applications in air conditioning, refrigeration, aerosols, electric and metal clearing, foam blowing and modern fire fighting. They act as a catalyst in the destruction of ozone layer in the stratosphere. The rate of destruction of ozone is further enhanced by the oxides of nitrogen (NO_x) released from the exhausts of large fleet of supersonic aircrafts (SST). A wide range of chlorofluorocarbons (CFCs) are known, but widely used are $CFC1_3$ and CF_2C1_2, which are in use since 1930.

CFCs and halons remain inactive in the troposphere and it takes about 20–40 years for these chemicals to travel and reach the stratosphere, but after that their intermediate product (chlorine radical) remains active for more than 100 years. The travelling time for CFCs and halons to reach to stratosphere may range from 20 to 40 years, that is, CFCs generated in the troposphere today may take 20 to 40 years to reach the stratosphere. Once these CFCs and halons reach to stratosphere, the chlorine and bromine atoms present in these chemicals are released as a result of interaction with UV radiations in the stratosphere.

Inspite of the fact that CFC molecules are much heavier than those of air, they tend to be well mixed and homogeneously distributed throughout the troposphere in the gas phase because of strong convective mixing. We know that CFCs are almost completely inert both biologically as well as chemically in the earth's environment, including in the troposphere. Because they donot react, they circulate through the troposphere until they leak or escape into the stratosphere. At higher altitudes they are capable of undergoing UV photolytic decomposition, as a result of being exposed to the intense flux of energetic UV radiation—

$$CFCl_3 + h\nu\ (\lambda < 290\ nm) \longrightarrow \bullet CFCl_2 + \bullet Cl$$

The released chlorine radical is capable of taking part in the catalytic decomposition of ozone as follows—

$$\bullet Cl + O_3 \longrightarrow ClO\bullet + O_2$$
$$ClO\bullet + O \longrightarrow \bullet Cl + O_2$$
$$\overline{O + O_3 \longrightarrow O_2 + O_2}$$

A second and perhaps third chlorine radical could also be produced by further decomposition of $\bullet CFCl_2$ radical, generating additional $\bullet$ Cl radical responsible for ozone depletion—

$$\bullet CFCl_2 + h\nu \longrightarrow \bullet CFCl + Cl\bullet$$
$$\bullet CFCl + h\nu \longrightarrow \bullet CF + Cl\bullet$$

Chlorine atoms catalyse the destruction of ozone. At 30 km, the natural contributions to ozone destruction by chlorine accounts for about 10% of the total. A significant natural precursor of atomic chlorine is CH_3Cl, which is produced biogenically from the oceans, from the burning of vegetation and from volcanic emissions. Once in the atmosphere, CH_3Cl undergoes photochemical reaction,

$$CH_3Cl + h\nu \longrightarrow \bullet CH_3 + \bullet Cl$$

CH_3Cl is partially removed by reaction with $\bullet$ OH.

$$CH_3Cl + \bullet OH \longrightarrow \bullet CH_2Cl + H_2O$$

Other natural sources of chlorine are various forms of chlorine. Such sources as HCl released by volcanoes or Cl^- ions released from sea–salt spray are shortlived in the troposphere and most of them are washed out through the rainfall before they reach the stratosphere. There are other catalytic cycles also involving chlorine. For example,

$$HOO\bullet + \bullet ClO \longrightarrow HOCl + O_2$$
$$HOCl + h\nu \longrightarrow \bullet OH + \bullet Cl$$
$$\bullet Cl + O_3 \longrightarrow \bullet ClO + O_2$$
$$\bullet OH + O_3 \longrightarrow HOO\bullet + O_2$$
$$\overline{2O_3 \longrightarrow 3O_2}$$

In the case of CFCs, the following mechanism has also been proposed,

$$CFCl_3 + h\nu \longrightarrow \bullet CFCl_2 + \bullet Cl$$
$$\bullet Cl + O_3 \longrightarrow \bullet ClO + O_2$$
$$\bullet ClO + O_3 \longrightarrow \bullet Cl + 2O_2$$

Similarly, another mechanism for ozone depletion may be,

$$\bullet BrO + \bullet ClO \longrightarrow \bullet Br + \bullet Cl + O_2$$
$$\bullet Br + O_3 \longrightarrow \bullet BrO + O_2$$
$$\bullet Cl + O_3 \longrightarrow \bullet ClO + O_2$$
$$\overline{2O_3 \longrightarrow 3O_2}$$

Bromine (from, say halons) can also replace chlorine in the cycle and together these reactions account for about 60% of the routes by which ozone is destroyed by halogens. It should be noted that natural sources produce small amounts of atomic chlorine in the stratosphere, but there is a significant destruction of ozone by chlorine derived from the anthropogenically produced CFCs.

Taking into consideration the ozone depleting CFCs and halons, scientists are trying to modify the relative amounts of fluorine, chlorine and hydrogen in new compounds. Thus a common approach is to incorporate hydrogen in the structure, as in **hydrochlorofluorocarbons (HCFCs)** or replace chlorine and produce hydrofluorocarbons(HFCs). Increasing the content of hydrogen reduces the inertness of the compound and in this way the lifetime of the compound in troposphere may be made short. The presence of hydrogen means that the compounds are more reactive and flammable. But the higher reactivity is a disadvantage in many applications. If the proportion of fluorine is increased at the expense of chlorine, then a highly stable compound will be formed because the bond energy of C–F bond is larger than C–Cl bond. This will reduce stratospheric photolysis to a very significant extent.

Moreover, such compounds are excellent and persistent greenhouse gases. For compounds containing no chlorine, obviously no Cl atoms can be released, so there is no potential to deplete ozone. Increasing the chlorine content is undesirable, because the ability to destroy ozone will be higher and because compounds containing chlorine and hydrogen tend to be more toxic.

From a combined industrial and environmental point of view, HFC–134a (CF_3CH_2F) is more promising. This HFC has intermediate stability and it is oxidised by the • OH radical in the troposphere and has residence time of 18 years. It is neither very flammable nor have chlorine. So it is not expected to have ozone depleting potential. HCFC–123 (CF_3CHCl_2) is another replacement for CFC. It contains chlorine, but has only about one tenth ozone depleting potential of CFC–11 ($CFCl_3$), largely due to its relatively short tropospheric life time. **Halons,** the bromine analogues of CFCs have widely been used in fire extinguishers because of the following important reasons.

(*a*) Their high density, due to which they settle over and smother a fire at ground level.

(*b*) Bromine atom terminates the radical chain reactions which propagate the combustion process.

Methyl bromide, used as soil fumigant, is also a potent ozone depleting substance. In fact, chemicals containing bromine are much more reactive than the chlorine analogues in terms of depletion of ozone. Because of this, ODP values for bromine compounds are very high. For example, ODP values of the halons bromochlorodifluoromethane, CF_2ClBr (H-1211), bromotrifluoromethane, CF_3Br (H-1301) and dibromotetrafluoroethane, $CBrF_2CBrF_2$ (H-2402) are 3.0, 10.0 and 6.0 respectively.

Antarctic and Arctic Ozone Hole Formation

The **ozone hole** is actually an area of thinner normal ozone. Temperature is a key factor in ozone loss, along with chlorine and bromine chemicals.

In 2001, the Antarctic ozone hole reached a maximum size of more than 26.5 million sq.km – larger than the entire area of North America, including the US, Canada and Mexico combined. In the year 2000, it briefly approached 11.30 million sq. km. In 2002, ozone hole over Antarctica was not only much smaller than what it was in 2000 and 2001, but it has split into two. It was measured at 15.6 million sq. km. In September 2002, a figure well below the more than 24 million sq. km spread seen consecutively during the last six years. The hole's total column ozone was measured below 220 **dobson units.**

Ozone depletion is an annual phenomenon over the South Pole during August and September when coldest temperatures are prevalent. Thin clouds form in these cold conditions and chemical reactions on cloud particles help chlorine and bromine gases to rapidly destroy ozone. By early October, temperatures typically start to warm and the ozone layer starts to recover.

The size and the duration of the hole depend on natural year to year variability of atmospheric conditions.

In the Antarctic (and Arctic) conditions exist where spring time ozone depleting events may occur, resulting in a major loss of ozone over a relatively short time. The conditions include a complex combination of climate factors as well as accumulation of reservoir species which react vigourously to bring about a major loss of ozone, especially at the onset of the first day light of the polar spring.

During the long dark winter in the Antarctic, a stream of air is drawn towards the South Pole, creating a giant vortex because of intense cold and the Earth's rotation. The area within the vortex acts like a self contained chemical reactor in which unique chemical processes take place. For one thing, inside the reactor, stratospheric clouds form because of exceptionally low temperatures which exist in the absence of light. The clouds are classified into two types—

(*a*) The polar stratospheric clouds (PSC) are formed at about 193 K and consist of about 1 μm diameter particles of HNO_3 and H_2O in the ratio of 1 : 3.

(*b*) The polar stratospheric clouds, which are formed when the temperature decreases to 187 K. These clouds are made up of relatively pure particles of water–ice upto 10 μm in size.

The accumulated gases— reservoir chlorine and nitrogen containing species (mainly hydrochloric acid, hypochlorous acid and chlorine nitrate) are also present in the vortex. During the winter, these species take part in heterogeneous reaction on the surface of PSCs and consequently release chlorine molecules and HOCl—

$$HCl + ClONO_2 \xrightarrow{PSCs} Cl_2 + HNO_3$$

$$H_2O + ClONO_2 \xrightarrow{PSCs} HOCl + HNO_3$$

The relatively stable situation remains until after the onset of sunlight. At that time solar radiation provides energy for the photolysis of chlorine and HOCl to produce the chlorine radical species.

$$Cl_2 + h\nu \longrightarrow \bullet Cl + \bullet Cl$$

$$HOCl + h\nu \longrightarrow \bullet Cl + \bullet OH$$

The chlorine radicals so formed are then available for ozone depletion by catalytic cycles or according to following reactions—

$$2 \bullet Cl + 2O_3 \longrightarrow 2 \bullet ClO + 2O_2$$

$$\bullet ClO + \bullet ClO \longrightarrow ClOOCl$$

$$ClOOCl + h\nu \longrightarrow \bullet ClOO + \bullet Cl$$

$$\bullet ClOO \longrightarrow \bullet Cl + O_2$$

$$\overline{2O_3 + h\nu \longrightarrow 3O_2}$$

Depletion of ozone occurs rapidly so that even in some days, ozone levels decrease dramatically to half or even less of their winter value. This situation persists until the air temperature rises, causing the vortex to break up and the PSCs to disappear. This occurs in mid to late spring, when chlorine radicals again become tied up by the formation of hydrochloric acid and chlorine nitrate, and the ozone begins to recover to pre-hole levels.

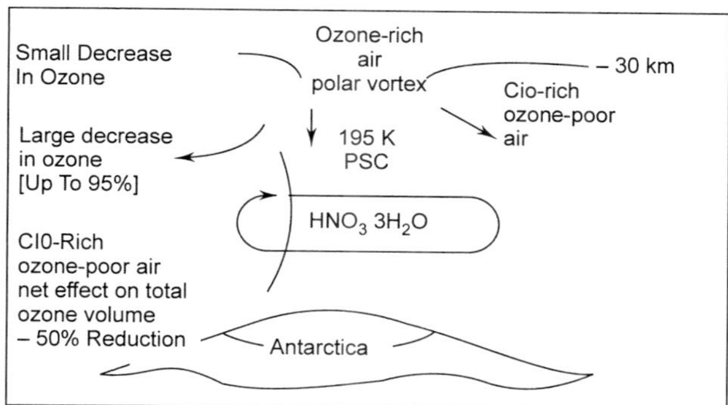

Fig. 6.5 The Antarctic ozone hole showing the polar vortex and the location and relative amounts of ozone loss during the polar sunrise.

In the Arctic atmosphere, as a result of the eruption of Mt Pinatubo in the Phillippines during 1991, sulphate was found to be present at higher concentrations than usual. Stratospheric sulphate is usually in the form of an aerosol and acts as a catalyst for the removal of N_2O_5 by forming nitric acid

$$N_2O_5 + H_2O \xrightarrow{\text{Sulphate aerosol}} 2HNO_3$$

This reaction is a means by which NO_x species are removed from the polar stratosphere, thus eliminating one of the species which is associated with holding cycles that tie up chlorine monoxide radicals ($\bullet\, ClO + \bullet\, NO_2 + M \rightarrow ClONO_2 + M$). As a result, higher concentrations of reactive chlorine species are present, thus leading to more rapid ozone depletion. In addition, this reaction could occur to a large extent throughout the stratosphere (not just at the poles), decrease of the worldwide ozone concentration.

Effects of Ozone Layer Depletion

Even minor problems of ozone depletion can have major effects. Every time even a small amount of the ozone layer is lost, more ultraviolet light from the sun can reach the Earth.

Every time 1% of the ozone layer is depleted, 2% more UV-B is able to reach the surface of the planet. UV-B increase is one of the most harmful consequences of ozone depletion because it can cause skin cancer.

The increased cancer levels caused by exposure to this ultraviolet light could be enormous. The EPA estimates that 60 million Americans born by the year 2075 will get skin cancer because of ozone depletion. About one million of these people will die.

In addition to cancer, some research shows that a decreased ozone layer will increase rates of malaria and other infectious diseases. According to the EPA, 17 million more cases of cataracts can also be expected.

The environment will also be negatively affected by ozone depletion. The life cycles of plants will change, disrupting the food chain. Effects on animals will also be severe, and are very difficult to foresee.

Oceans will be hit hard as well. The most basic microscopic organisms such as plankton may not be able to survive. If that happened, it would mean that all of the other animals that are above plankton in the food chain would also die out. Other ecosystems such as forests and deserts will also be harmed.

The planet's climate could also be affected by depletion of the ozone layer. Wind patterns could change, resulting in climatic changes throughout the world.

Solution

The discovery of the ozone depletion problem came as a great surprise. Now, action must be taken to ensure that the ozone layer is not destroyed.

Because CFCs are so widespread and used in such a great variety of products, limiting their use is hard. Also, since many products already contain components that use CFCs, it would be difficult if not impossible to eliminate those CFCs already in existence.

The CFC problem may be hard to solve because there are already great quantities of CFCs in the environment. CFCs would remain in the stratosphere for another 100 years even if none were ever produced again.

Despite the difficulties, international action has been taken to limit CFCs. In the Montreal Protocol, 30 nations worldwide agreed to reduce usage of CFCs and encouraged other countries to do so as well.

However, many environmentalists felt the treaty did "too little, too late", as the Natural Resources Defense Council put it. The treaty asked for CFC makers to only eliminate half of their CFC production, making some people feel it was inadequate.

By the year 2000, the US and twelve nations in Europe have agreed to ban all use and production of CFCs. This will be highly significant, because these countries produce three quarters of the CFCs in the world.

Many other countries have signed treaties and written laws restricting the use of CFCs. Companies are finding substitutes for CFCs, and people in general are becoming more aware of the dangers of ozone depletion.

6.6 AIR QUALITY STANDARDS

Definition

Air Quality Standards are legal limits placed on levels of air pollutants in the ambient air during a given period of time. As such, they characterize the allowable level of a pollutant or a class of pollutants in the atmosphere and thus define the amount of exposure permitted to the population and/or to ecological system.

Air Quality Standards are expressions of public policy and thereby requirements for action. Thus, they are not based solely on air quality criteria but are also based on a broad range or economic, social technical and political considerations. Air quality standards have evolved differently in different countries depending on exposure conditions, the socio-economic situation, and the importance of other health related problems.

Single or Multiple Standards

A major decision that must be taken in the adoption of air quality standards is whether there should be one standard for an entire jurisdiction of different standards for different areas within the jurisdiction. All political jurisdictions within an air quality region would have the same air quality standards. In general, air quality regions would not be contiguous but would be separated by non-urban hinterland. If two separated regions were to expand in area in the future until they become contiguous, they would presumably then be merged into one larger air quality region. Since each air quality region would adopt its own air quality standards, they could differ among the several regions. However, there should be a datum below which no regin's air quality standards should drop.

6.7 KINDS OF AIR QUALITY STANDARDS

Ambient Air Quality Standards

These are the legal limits placed on the concentration of air pollutants in a community where people and things are exposed. Enforcement of ambient air quality standards is difficult. To know if the actual concentrations are in compliance requires an air monitoring programme. Ambient air quality standards need to be distinguished from threshold limit values (TLV) for work-room atmospheres, e.g., those published by the American Conference of Government Industrial Hygienists. These latter values are the permissible exposure levels for healthy adult worker for eight hours per day, five days per week. In contrast, air quality standards are permissible exposures of all living and non-living things for 24 hours per day, 7 days per week.

Other Air Quality Standards:

(*a*) Quasi-emission standards called 'point of impingement standards' which are the limit on specific pollutants of the ambient air at ground level required by national, state, or local regulations to be used in diffusion computations to determine limits of emission from specific sources.

(*b*) Soiling index, which is the measurement of transmitted or reflected light through or from a spot of particulate matter collected on a filter for a prescribed period of time. (The darkness of stains produced by drawing polluted air through filter paper has been used to estimate suspended particulate matter concentration.

(*c*) Odour standards.

(*d*) Visibility standards.

(*e*) Standards for particulate matter deposited, etc.

6.8 AIR QUALITY STANDARDS : A COMPARISON

The cost to a community is the paramount consideration in deciding which of the devices or techniques available to control air pollutants will provide an acceptable atomsphere. To eliminate air pollution entirely would be excessively expensive. Then, the economics involved demands that some level of pollutants be permitted in the atmosphere. This permissible level as mentioned earlier is called 'air quality standards'.

Thus air quality standards imply acceptable pollutants concentration in the atmosphere enveloping a community.

Several countries in the world which are advanced industrially and which have an air pollution problem have developed standards for air quality which they strive to attain by adopting suitable air pollution control measures. Table 6.1 gives the air quality standards adopted by the Environmental Protection Agency, USA. Table 6.2 gives the standards for suspended particulate matter for various countries. Table 6.3 gives the standards for Sulphur dioxide for various countries. Table 6.4 gives the standards for dust fall for various countries. Table 6.5 gives the standards for sulphation rate in Canada.

Original standards expessed in different units in original data have been converted to suitable units for easy understanding. Due to this change, the original value may match to the nearest expressed values as mentioned in the various tables.

Table 6.1 Air Quality Standards Adopted by EPA

No.	Parameter	Standard concentration		Remarks
		$\mu g/m^3$	ppm	
1.	Suspended particulate matter	75 260		Annual geometric mean 24 h
2.	Sulphur dioxide	80 365	0.83 0.14	Annual mean 24 h
3.	Carbon monoxide	10000	9	8 h. Not more than once per year.
		40000	35	1 h. Not more than once per year.
4.	Nitrogen oxides	100	0.05	Annual mean
5.	Ozone	235	0.12	1 h daily maximum. Not more than once per year.
6.	Non-methane hydrocarbons	160	0.24	6–9 h. Annual mean Not more than once
7.	Lead	1.5		3-month average.

In India, on the basis of land use and other factors, the various areas of a state may be classified into three categories by concerned **State Pollution Control Boards:**

(*a*) Industrial and mixed-use areas,

(*b*) Residential and rural areas,

(*c*) Sensitive areas.

Table 6.2 Standards for Suspended Particulate Matter

Country	$\mu g/m^3$	Averaging Time
Bulgaria	150	24 hours
Czechoslovakia	150	"
East Germany	150	"
Finland	150	"
Poland	200	"
Israel	150	"
USA		
Delaware-rural	60	"
Delaware-residential	75	"
Delaware-commercial	95	"
Delaware-industrial	125	"
South Carolina	25	"
Texas-residential	125	"
Texas- commercial	150	"
Texas-industrial	175	"

Texas-others	200	"
USSR	150	"
Canada		
Ontario-residential-rural	60	1 Year
Ontario-industrial-cum commercial	110	"
France	60	"
USA		
Missouri	75	"
Montana	75	"

Table 6.3 Standards for Sulphur Dioxide

Country	*μg/m³*	*Averaging Time*
Belgium	150	24 hours
Bulgaria	50	"
Canada		
Alberta	150	"
Newfoundland	150	"
Ontario	290	"
Czechoslovakia	150	"
East Germany	150	"
Finland	250	"
France		
Paris	75	"
Italy	380	"
Netherlands	125	"
Poland	75	"
Rumania	250	"
Sweden	250	"
Switzerland	75	"
Turkey	150	"
USA		
Colorado	260	"
Delaware-rural	50	"
Delaware-residential	80	"
Delaware-commerical	100	"
Delaware-industrial	160	"
Missouri	120	"

USSR	150	"
Canada		
Alberta	30	1 Year
Manitoba	60	"
Ontario-industrial-cum commercial	130	"
Ontario-residential-rural	50	"
Spain	150	"

Table 6.4 Standards for Dust Fall

Country	*Dust fall rate MT/km²/Month*
Canada	
Ontario-residential-rural	8
Ontario-industrial-commercial	15
Federal Republic of Germany-industrial	13
Poland	3
Canada	
Alberta-residential	6
Alberta-industrial-commercial	17
USA	
Missouri-non-residential	4
Missouri-heavy industrial	10
Montana-residential	6
Montana-heavy industrial	12
Oregon-residential-commercial	6
Oregon-industrial	12

Table 6.5 Standards for Sulphation Rate

Country	*Sulphation rate mg SO_3/100 cm^2/day*
Canada	
Manitoba	1.00
Ontario	0.40
Saskatchewan	0.80

Category (A) becomes self evident by the intensity of industrial activity in an area and is bound to have somewhat inferior quality of air compared to other categories. The category (C) covers hill stations, tourist resorts, sanctuaries, national parks, national monuments, health resorts and other such areas where the nation conserves its clean environment even if that implies some curbs on economic activity. All areas not specifically declared by the concerned State Pollution Control Board to be classified in category (A) or category (C) will be automatically deemed to fall in category (B). *Table 6.6 gives the Indian Ambient Air Quality Standards.*

When monitored uniformly over the 12 months of an year with a frequency of not less than once in a week with a sampling time of eight hours for any sample and analysed according to procedures specified by the Central Board, the concentrations for the pollutants shall be 95% of the time within the limits prescribed.

Whenever and wherever three consecutive measurements spaced at least one week apart, or any three out of ten consecutive measurements spaced by at least one week apart found to exceed limits spacified in Table 6.6 for the respective category it would be considered adequate reason to institute regular weekly monitoring and further investigation.

Table 6.6 Indian Air Quality Standards

Category		*Concentration in µg/m³*			
	Area	*SPM*	SO_2	NO_x	*CO*
A.	Industrial and mixed use	500	120	120	5000
B.	Residential and rural	200	80	80	2000
C.	Sensitive	100	30	30	1000

The World Health Organization (WHO) Technical Report discusses the effects of sulphur dioxide and particulate matter, carbon monoxide, photo-chemical oxidants and nitrogen dioxide. It was the opinion of the committee that exposure to these pollutants should be kept as low as possible. Taking into consideration all the evidences available to it, the committee reached a consensus on long-term goals for these pollutants (Table 6.7)

Table 6.7 Recommended Long-Term Goals

Pollutant	*Measurement method*	*Limiting value*
Sulphur dioxide*	British standard procedure**	Annual mean 60 µg/m³ (0.02 ppm) 98% observations below 200 µg/m³ (0.060 ppm)
Suspended particles*	British standard procedure**	Annual mean 40 µg/m³ (0.016 ppm) 98% observation below 120 µg/m³ (0.048 ppm)
Carbon monoxide	Non-dispersive infrered	8 h average 10 mg/m³ (9 ppm) 1 h maximum 40 mg/m³ (30 ppm)
Photochemical oxidant	Measured by neutral buffered KI method expressed as Ozone	8 h average 60 µg/m³ (0.03 ppm) 1 h maximum 120 µg/m³ (0.06 ppm)

* Values for sulphur oxides and suspended particulates apply only in conjunction with one another.

** Methods are not necessarily those recommended but indicate those on which these units have been based. Where other methods are used, an appropriate adjustment may be necessary.

EXERCISE

1. (*a*) What do you mean by air pollution? Discuss various air pollutants and their sources.

   (*b*) Describe ozone depletion, its harmful effects and causes of ozone depletion. How the ozone depletion can be prevented. [*RTU August 2009*]

2. Write short notes on the following:

   (*i*) Greenhouse effect

   (*ii*) Depletion of Ozone layer

   (*iii*) Discuss the effect of various air pollutants [*RTU June 2008*]

3. What are primary and secondary air pollutants? Tabulate the allowable limit of important air pollutants in air. [*RTU June 2008*]

4. What are the harmful impacts of air pollution on human being? [*RTU June 2008*]

5. (*a*) Define the term air pollution, pollutants and contaminants, particulates, smog, primary and secondary pollutants. [*RTU June 2008*]

   (*b*) Write a short note on "Chlorofluorocarbon"

6. (*a*) Discuss the mechanism of depletion of ozone layer. What are the harmful effects due to the formation of 'ozone hole'? Discuss the substitutes of chlorofluorocarbons. (CFCs).

   (*b*) What is greenhouse effect? How is it caused? What are its adverse effects on the environment.

7. (*a*) How do internal combustion engines cause air pollution?

   (*b*) Explain "Greenhouse effect".

8. What is air pollution? What are its major sources? How can air pollution be minimized?

   Or

   Discuss 'Acid rain' and 'Greenhouse effect'.

9. Define noise. Describe various sources of noise pollution.

7

NOISE POLLUTION

7.1 INTRODUCTION

Noise is derived from latin word **nausea,** meaning unwell stomach with an urge to vomit.

Ironically, the subject of noise pollution is one of the most silent of all forms of pollutions. But, this little spoken-about form of pollution is nevertheless causing as many health issues as the other forms of pollution do. Here is a brief synopsis of the various causes, health effects and control measures of noise pollution.

Sound that is undesirable for human hearing is called as noise. When there is a lot of noise in the environment, it constitutes what is known as noise pollution. Noise pollution can be caused due to various sources – there is street noise, traffic noise, noise in public transport places, noise in playgrounds and parks, noise in the shopping malls, noise in workplaces… the list is endless. One of the greatest sources of noise pollution is the airports, and anyone staying close to an airport will attest to that.

Sources of Noise Pollution

Sound is measured in a unit known as decibels. Though there is no fixed particular decibel limit to decide when sound becomes noise, it is understood that a continuously high decibel limit will constitute noise pollution. Some areas do designate their own sound limits, which of course vary from one legislation to another. In the United States, most states have a sound limit of 65 dB in the daytime and 55 dB in the nighttime, applicable to the streets. Anyone crossing this limit would be causing noise pollution.

However, all these designated sound limits are too ambiguous, because most appliances we use in factories as well in the household go much beyond the prescribed limits. The following are some of the sources of noise pollution that we are quite familiar with, but generally ignore:

- Appliances in the home such as mixer grinders, vacuum cleaners, washing machines, etc., together cause a cumulative sound of about 87 dB. This itself is above the sound limits in most areas. On top of that, if loudspeakers, television sets and music systems are used with high volumes, then we can well imagine how much noise pollution is being created.

- Small factories using single unit machines would cause a sound of about 98 dB and above. The sound will definitely go higher as the number of machines increase.
- Airplanes cause the highest sound among all – 150 dB. But road vehicles are also great contributors of noise pollution. These vehicles include the trucks, buses, tractors, SUVs and even motorcycles and most cars.
- Then there are lots of environmental sources of noise pollution that cannot be ignored. Continuous noises are the most distressing. Noise coming from sources such as dripping taps and ticking of clocks can contribute to environmental noise pollution.

The subjective sensation of loudness of a sound varies more nearly in accordance with the sound pressure level. To get a feel for the sound pressure levels is dB, we mention below the rating of a few familiar sounds (Table 7.1)

Table 7.1 Sound pressure levels of some familiar sounds

Nature of Sound	*SPL (Sound Pressure Level) in dB*
A well-designed anechoic chamber	< 0
Rustle of leaves in quiet countryside	10–15
Soft whisper	15–20
Low-volume radio in a country residence	35–45
Normal conversation	55–65
Noisy restaurant (no acoustic treatment)	60–70
Noisy bus	70–80
Lion's roar (at 3 m)	90–100
Propeller-driven aircraft (at 30 m)	110–120
Commercial aircraft (at 30 m)	120–140

The range of normal human hearing is in the range of 50 Hz to 15000 Hz frequencies. Frequencies below 50 Hz are called infrasound and above 15000 Hz, ultrasound. We can feel infrasonic vibrations because some of our body parts resonate in this range. As a result, sonic vibrations can cause harm to our body.

Ultrasonic devices are widely used in industry for sealing packages, welding, cutting, drilling and cleaning. Our knowledge of the harmful effects of certain frequencies of ultrasonic vibrations is still incomplete.

Noise is produced in modern civilization mainly through the following agencies:

1. Industry and Machinery,
2. Transportation (Surface and air),
3. Certain forms of entertainment and community activities.

It is transmitted to us through the surrounding air and the structures in which we live.

1. **Industrial Noise:** For the most part is local, affecting mostly those who work in the industry. While the effects of industrial noise are often more serious than those from the other categories of noise mentioned above, the noise is usually confined to small regions where, with proper care and design, its effects can be minimized. The maximum sound pressure levels considered safe for exposure for workers in a factory are given in Table 7.2.

Table 7.2 Maximum permissible sound pressure levels considered safe for exposure in a factory

Age in years	*Maximum safe sound pressure levels (dB)*	
	Wide band noise	*Narrow band noise*
Less than 30	99	80
30–40	88	78
40–50	84	74
50–60	80	70

2. **Transportation Noise:** On the other hand, is far more widespread, invading vast areas of large cities and affecting hundreds of thousands of city dwellers. It is the most persistent form of noise, growing at the fate of about 10 dB per decade with increase in the number of personal transport vehicles and faster modes of travel. In most of the larger cities in the world, noise levels have frequently reached 90 dB or more. Maximum permissible sound pressure levels (SPL) in dBA is given in Table 7.3.
3. **Noise from Entertainment and Community Activities:** The third category of noise from entertainment and community activities is in a sense peculiar to human beings (Table 7.3). What is entertainment to one man is agony to another, unlike in the rest of the animal world where what is good for one animal is also good for another animal of the species. Aircraft noise has already assumed serious proportions for those living in the neighbourhood of airports. Modern commercial aircraft produce as much as 10 kW of noise power.

Table 7.3 Maximum permissible sound pressure levels

Type of Room	*Maximum permissible SPL in dBA* (under no occupancy)*
Small private offices	45
Conference Rooms	30
Cinema Theatres	35
Drama Theatres	30
Concert Halls	25
Secretarial Offices	50
Homes (sleeping rooms)	40
School Rooms	40
Workshops	70

*Noise measuring instruments have a standard setting– the 'A' scale, which gives relatively less importance to the lower frequencies. The readings on this scale are denoted by dBA.

A new source of noise that is just around the corner is the sonic boom or sonic bang of the supersonic transport (SST) planes which are likely to replace the present day aircraft for long distance travel. The creation of shock waves by objects travelling faster than the local speed of sound is a well-known phenomenon in aerodynamics. The effects of sonic bang on people are, however, not yet seen clearly, but a startled reaction has been widely observed.

Noise Pollution Effects

Noise pollution effects on human health, such as hearing loss and cardiovascular diseases, can bring day-to-day activities to a standstill. To know more about noise pollution, and its effects, read on...

Undesirable sound is referred as noise. In recent years, noise pollution has been a hot topic of discussion among researchers. Noise that is continuous or lasts for a short period can equally be detrimental to our health. Some music enthusiasts say that intermittent noise is not hazardous, but this isn't true, even intermittent noise can negatively impact human hearing in the long run. Common sources of noise include vehicular traffic, factories involved in the manufacturing process, and home appliances like vacuum cleaners and music systems. Fast becoming a major environmental issue, noise pollution effects on health have increased rapidly in the last few years.

Effects of Noise Pollution on Human Health

Noise pollution can take a severe toll on human health in the long run. These effects will not become apparent immediately, but there could be repercussions later on. The following is a list of the kinds of effects noise pollution will have on human health after continuous exposure for months, and even years:

- The most immediate effect is a deterioration of mental health. As an example, people who are living too close to airports will probably be quite jumpy. Continuous noise can create panic episodes in a person and can even increase frustration levels. Also, noise pollution is a big deterrent in focusing the mind to a particular task. Over time, the mind may just lose its capacity to concentrate on things.
- Another immediate effect of noise pollution is a deterioration of the ability to hear things clearly. Even on a short-term basis, noise pollution can cause temporary deafness. But if the noise pollution continues for a long period of time, there's a danger that the person might go stone deaf.
- Noise pollution also takes a toll on the heart. It is observed that the rate at which heart pumps blood increases when there is a constant stimulus of noise pollution. This could lead to side-effects like elevated heartbeat frequencies, palpitations, breathlessness and the like, which may even culminate into seizures.
- Noise pollution can cause dilation in the pupils of the eye, which could interfere in ocular health in the later stages of life.
- Noise pollution is known to increase digestive spasms. This could be the precursor of chronic gastrointestinal problems.

Noise pollution effects on human health is a matter of great concern. The reason being the adverse consequences that high noise levels have on human health. Noise pollution can affect us in several ways, some of which are listed below:

Hearing Problems: Exposure to noise can damage one of the most vital organs of the body, the ear. Hearing impairment due to noise pollution can either be temporary or permanent. When the sound level crosses the 70 dB mark, it becomes noise for the ear. Noise levels above 80 decibels produce damaging effects to the ear. When ear is exposed to extreme loud noise (above 100 decibels) for a considerable period of time, it can cause irreparable damage and lead to permanent hearing loss.

Cardiovascular Issues: A noisy environment can be a source of heart related problems. Studies have shown that high intensity sound cause a dramatic rise in blood pressure as noise levels constrict the arteries, disrupting the blood flow. The heart rate (the number of heartbeats per minute) also increase. These sudden abnormal changes in the blood increase the likelihood of cardiovascular diseases in the long run. Read more on cardiovascular diseases.

Sleep Disturbances: This is one of the noise pollution effects that can deter your overall well being. Noise can interrupt a good night's sleep, and when this occurs, the person feels extremely annoyed and uncomfortable. People deprived of uninterrupted sleep show a sharp dip in their energy levels which often results into extreme fatigue. This can considerably decrease a person's ability to work efficiently.

Interference in Verbal Communication: A noisy environment that produces more than 50–60 decibels simply does not allow 2 people to communicate properly. Interpreting the speech of a second person becomes quite difficult and may lead to misunderstandings.

Mental Health Problems: Exposure to loud sound can lead to elevated stress levels as well as stimulate violent behavior. A constant noise in the vicinity can also trigger headaches, make people tense and anxious, and disturb emotional balance.

Noise Pollution Effects on Animals

Man made noise pollution has made the Earth an uncomfortable place to stay for animals as well. Hearing loss and rapid increase in heart rate are some of the ill-effects of noise pollution on animals. High intensity sound induces fear, forcing them to abandon their habitat. Other effects of noise on animals is given below:

- Anxious behavior in animals is commonly observed in the form of trembling when they are exposed to high decibel levels. Intolerable noise levels can decrease a cow's capacity of milk production. These animals require a calm and relaxed environment to provide a better milk yield.
- Research has proven that aircraft induced noise is responsible for a significant decrease in reproduction activity in a wide range of animals. Noise pollution effects can also be seen on chicken in the form of a sharp drop in egg production. Stunted growth in chickens due to intense noise has also been observed.
- Underwater animals such as whales and dolphins too experience discomfort due to noise pollution caused by submarines, shipping companies, and sonars.
- Birds in urban areas, that use their sharp hearing abilities to hunt prey, are at loss due to intervention of man made noise. High noise levels can also negatively affect the breeding and feeding patterns of some animals. This is one of the reasons why certain species have become extinct.
- Sea animals produce varying sound levels among each other. However, the noise of commercial ships does not allow these animals to communicate properly, which can wreak havoc on their reproductive function. Sonar navy tests can hurt the whale's ear and reduce its ability to breed and feed.

Controlling Noise Pollution

Governments are making their efforts for controlling noise pollution, but we must appreciate the difficulty of the task. Unless and until we take care of ourselves, the problems of noise pollution will always loom large. Here are some ways in which we can make individual efforts at reducing noise pollution for ourselves and for others:

- We must constantly check up on the appliances we use at home. Most of them have rubber insulations that act for soundproofing. But over time, this insulation may wear out, and that is when the noise pollution will begin. Keep track of which appliances need maintenance, and replace insulations if needed.

- Growing trees is a very significant way in which roadside noise can be curtailed. Trees act as buffers for absorbing the sound that is produced on the streets and hence reduce noise pollution. That is the reason why roads with trees on both sides seem to be more silent and peaceful. Grow trees around your house if you can. It will protect you from the noise on the streets. This will also help if you stay close to an airport.
- Do not honk horns in your vehicles unless it is absolutely necessary. We all know how easily traffic sound limits are trespassed when there is a traffic jam. We might be desperate to get through, but honking horns will not solve any issues. It will only add to the noise pollution.

If you are working in a factory that has a lot of noise issues, make it a point to wear earplugs and muffs. If you are the owner of the factory, provide these things to your workers.

Like water and air pollution, noise pollution too needs to be controlled. To bring down noise levels, one has to first identify the sources of noise. Considering the ill effects of noise, some preventive steps need to be taken to reduce noise levels at the source. Here are some tips to limit the noise:

- People living in the heart of city or near the airport, often have to bear the brunt of high noise levels. To decrease noise, one can install dual-paned windows. Offices, too can use windows to curb noise levels. Soundproofing is another alternative that can be looked at to reduce excessive sound levels.
- Use of music systems and television sets with high volumes can cause noise pollution at home. Instead, using these appliances with the volume kept at a moderate level is a better option.
- An effective way to manage noise would be to wear ear protection while working in noisy conditions. Animals should be moved to some other place, away from the noise source, thereby decreasing their noise exposure time.
- Vehicles and factory machines need to be maintained properly and checked from time to time. Lack of maintenance will not only increase noise levels, but also decrease the efficiency of these machines.

Noise pollution effects on the environment have also been studied. It is observed that keeping plants in a noise producing environment can have an adverse impact on their growth and nutrient intake. Read more on environmental pollution.

Remember, noise being unwanted sound, is naturally a disturbance to the ears. Noise pollution effects on mental and physiological health on humans and animals alike, cannot be overlooked. Hence, taking the aforementioned precautions to curtail noise pollution would definitely show reduction in this new urban menace (noise).

Active Noise Control or Active Noise Cancellation or Active Noise Reduction

Introducing a nullifying "antinoise" which has the same amplitude but the opposite polarity to the noise that is being controlled, which combines to form a new wave, does active noise control. This process of the combining of two waves is called Interference and this cancels both the sound waves. The process of cancellation is called Phase Cancellation. A proper cancellation scheme has to be prepared. The result of Active Noise Control is a sound wave that is barely audible. Two waves of equal amplitude and identical phase add together to increase whereas two waves of equal amplitude and opposite phase result in cancellation of the sounds.

After the unwanted noise is identified, active noise control can work to generate correct signals that is send to the loudspeakers which start producing anti sound to nullify the original sound. The zone of silence is dependant on the wavelength of both the sounds and the wavelength would depend on the size of source of unwanted sound and the size, number and position of the loudspeakers, which would produce the anti-sound. It is advisable to keep the audio power level of both the sources, *i.e.*, the source of unwanted sound and the source of anti sound, the same. The source of anti-sound or the "transducer" should be kept exactly where the sound needs to be weakened which would be the ear of the listener or listeners. It works better if done in case of a single listener since it requires lower power level.

Contemporary Active Noise Reduction can be achieved by the use of computers. The computer examines and identifies the waveform and the wavelength of the noise and then generates the polarization and reverse waveform so that interference can be achieved. As stated above the wavelength and amplitude of the sound is the same but the polarity is reversed. This whole process creates the desired result of destroying the noise. Active Noise Reduction is different from any other process of noise control method especially soundproofing since in all the other methods power is not in use. Insulation and sound absorbing tiles are mechanical ways of controlling noise.

There are reasons behind using Active Noise Control method when compared to all the other methods because it requires lesser equipments and lesser power. It works very well on lesser frequencies and in case of a single user who is static, it can be handled well but when it is a dynamic user, noise control becomes a challenge. There are different environments where noise needs to be controlled and depending on that application also differs. Repetitive sounds are easier to handle than continuous sounds. In a noise area where the source of noise is singular, active noise control is an easy process and other processes are also effective like the ear mufflers or insulators. In this case the source and the obstructer maintain a physical contact whereas when the sources of noise are multiple and the listeners are multiple or the single listener is kinetic, low frequency sound has to be used which can be used to counter the high frequency noise. Places, where Active Noise Control is used, are aircrafts, car interiors and factories.

Industrial Noise

Industrial Noise refers to noise that is created in the factories which is jarring and unbearable. Sound becomes noise only it becomes unwanted and when it becomes more than that it is referred to as "noise pollution". Heavy industries like shipbuilding and iron and steel have long been associated with Noise Induced Hearing Loss (NIHL).

Industrial Noise Pollution

Industrial Noise Pollution is becoming a nuisance and measures have to be taken to reduce it or at least control it. It is hampering the physical and mental health of the workers and is alienating them from society...No I am not exaggerating...this is truth that factory workers are suffering from withdrawal symptoms and it is high time that we start taking the issue seriously.

This is posing to be a big challenge with very passing day and is a threat to safety and health of the people who are working in the industry and common people as well. It has been scientifically proved that noise more than 85 decibels can cause hearing impairment and does not meet the standards set for healthy working environment. Moreover it can also cause accidents. The problem has been viewed and analyzed from all the perspectives but the solution probably is not so easy to achieve since there is a lot of contradiction between legislation, guidance and documents.Industrial Noise resulting to noise pollution has many reasons such as industries being close to human habitats which prevents the noise from decaying before it reaches human ear.

Effects of Industrial Noise Pollution

It has already been stated that continuous exposure to noise pollution leads to hearing impairment but it has various other effects as well which are as follows:

- It can result into increase in blood pressure
- Increased stress
- Fatigue
- Stomach ulcers
- vertigo
- Headaches.
- Sleep disturbance
- Annoyance
- Speech Problems
- Dysgraphia, which means writing learning impairment
- Aggression
- Anxiety
- Withdrawal

Industrial Noise adversely affects the workers and they suffer from various health problems as I have listed above. High volume leads to increased adrenaline levels, which leads to the constriction of blood vessels, which normally happens when the individual is tensed, afraid, anxious or extremely happy and excited. The work pressure and the noise pollution both lead to a fairly typical situation, which leads to clinical as well as psychological stress. The blood pressure, due to the constriction of blood vessels (vasoconstriction), remains high for the major part of the day. The human dislike towards anything that is not pleasant, yet compelling leads to emotional stress and depression which is a term for a quite a severe situation. Statistically there is a rise in the number of workers dying of cardiac arrests and cerebral attacks is on an increase considerably due to industrial noise pollution in railway yards, factories etc.

Some people also suffer from headaches, which decrease their efficiency levels and hamper the quality of their work. That leads to crises in workplace as well as home. Workers are always agitated and excited that result into carelessness. Workers also become fatigued and in some cases over fatigued which should ring the alarm for organizations that want to grow. The worst part of Industrial Noise pollution is it affects the unborn baby in a womb and that too in the early days after conception since the fetus is sensitive to sounds and high decibels affects the growth of its organs.

The problems that the Industrial workers face are sleep disorders and behavioral changes. They experience increased levels of stress. They fail to achieve harmony thus leading to a lot of minor psychological problems, which are too common to be noticed. They are irritated and annoyed; therefore fail to interact with a person around them and this gradually leads them to become "loners". They withdraw from the society and some in extreme circumstances might have "tremors", speech problems and many other behavioral problems. It is high time that the rules and regulations that are already made should be applied and followed so that we can cope with the ever increasing problem of noise pollution that is concentrated in the industry environment.

How Loud is "Too Loud" for your Hearing ?

It is essential to recognize a sound loud enough to damage your hearing, and then proceed to protect yourself from injury. Sounds are explained, and different sound sources are rated for potential to damage hearing. Protection and prevention are stressed.

To measure the loudness of sound, a unit of measurement called a decibel (dB) is used. A decibel is one tenth of a bel (B), or the unit of measurement devised by engineers of the Bell Telephone Laboratory to quantify the level of sound reduction when a tone is transmitted over one mile (1.6 km) of standard telephone cable. Since the bel proved to be too large for most common sound measurements, the use of the smaller decibel became more popular.

The decibel is used in a wide variety of scientific measurements related to acoustics and electronics. This unit of measurement places a physical value — usually of noise or light intensity — in which a doubling of actual intensity causes perceived intensity to always increase by roughly the same amount. Specifically, a 3 dB increase is about equal to an approximate doubling of sound. This is important to understand because when you look at the dB values given in the following charts and text, you will understand a relatively small numerical increase of 10 dB between the sound of rainfall and a sewing machine accounts for the significantly greater noise level. It also shows that the 110 dB sound of a screaming child is only several times less intense than a thunderclap or a rock concert.

Sound reference chart

- 0 Rustling leaves
- 20 Ticking watch
- 30 Quiet whisper
- 40 Refrigerator hum
- 50 Rainfall
- 60 Sewing machine, typical conversation level speech
- 70 Washing machine
- 80 Alarm clock (two feet away)
- 85 Average traffic
- 90 Gas lawnmower
- 95 MRI testing
- 100 Tractor, hair dryer, subway train
- 105 Power mower, chainsaw
- 110 Screaming child
- 120 Rock concert, ambulance, thunderclap
- 130 Jackhammer, jet engine plane (100 feet away)
- 140 Fireworks
- 165 12 gauge shotgun blast

Noise in the Workplace

The National Institute for Occupational Safety and Health (NIOSH), a branch of the U. S. Centers for Disease Control and Prevention, reports that ear injury can develop in workers if exposed to 85 dB or more over eight hours. As a result, this 85 dB exposure limit for an eight-hour workday has become a standard in the mining, construction, oil-gas well drilling, and agriculture industries, as well as the U.S. Air Force and Army. Loss of hearing caused by excessively loud or prolonged noises is called NIHL, or noise induced hearing loss.

Injury to the Cilia Cells

Sounds of less than 80 dB do not often result in hearing loss, even after long or repetitive exposure. It is not possible to predict how an individual will respond to loud noises, since each person possesses a different hearing sensitivity. Yet, exposure to a single loud noise or continuous noise exposure can result in temporary hearing loss. This temporary loss is known as temporary threshold shift, and it will typically resolve 16 to 48 hours after the end of exposure.

Hearing loss can be permanent if the offending sound is loud enough to damage or destroy the delicate inner ear cells, called cilia, or stereo cilia, that are found in the cochlea of the inner ear. These cilia cells respond to mechanical sound vibrations by sending an electrical signal to the auditory nerve. The mechanism of injury to cilia is purely physical; excessive vibration results in microscopic tearing or breaking of the cilia. There are many cilia located in the cochlea of the ear; the more that are damaged, the more profound the deafness. Once cilia cells are chemically or physically damaged or destroyed, they will not regenerate or repair themselves.

Protect your Hearing

A commonsense recommendation to avoid hearing loss due to excessively loud noise is to either wear hearing protection or remove yourself from the sound source.

Use either earplugs, earmuffs or other protection devices when exposed to sounds above 85 dB. Since you cannot always be certain when your sound environment places you at risk, use this rule of thumb: If you are talking to someone three away from you and you cannot clearly hear what is being said, the noise level around you could be damaging your hearing.

Do the obvious, and simply walk away when sounds are too loud or turn down the volume of the sound source if possible. It is also possible to limit the sound intensity by not standing directly near its source. Time spent in a loud environment is important to the health of your hearing, so limit the time you expose your ears to loud noises. Cover your ears with your hands or cross the street when you hear someone operating a leaf blower or when you are near a jackhammer.

Tinnitus Treatment Institute

Since NIHL, or noise induced hearing loss, can result in tinnitus due to injury of the cochlear ciliary cells, the Tinnitus Treatment Institute has great interest in prevention of injury by sound exposure. Prevention is always preferable to treatment. For more information about tinnitus and NIHL, visit the website of the Tinnitus Treatment Institute.

What can we do?

Some towns and cities actually have laws and limits as to the noise pollution that its citizens can make. Other places, like golf courses or nature reserves, are also considered to be "quieter" places with less noise pollution. There are some things that all people can do to help lower their own levels of noise pollution that they — and hopefully others around them — produce.

- Rake leaves by hand, don't use a noisy leaf blower.
- Trim bushes or shrubs by hand, don't use a noisy bush trimmer.
- Sound proof rooms that might have music conducted in them, like a room with a piano or if someone in the house plays drums or guitar or whatever. This can be done simply with curtains, window inserts, carpeting, and closing windows and doors.
- Don't blast music on the radio or computer or speakers. Be considerate of your own ears and those of other around you.
- Don't slam doors/car doors, close them easily and with only as much force as needed. (People don't usually think of this, but imagine — how loud is it when you slam your car door? Pretty tolerable. But imagine thousands of people doing so. Now that can start to get loud.
- Turn off the TV or radio when you aren't actually fully listening to it.
- Train your dog not to bark so much.
- Don't yell. Have civil conversations. Call someone or go find them instead of yelling across the street for them, for example.
- Plant trees and bushes around you house. They help keep the air clean, absorb sound, give privacy, and add nice design and looks to a house.
- Do noisy things (dishes, hammering, ect.) over or on a rubber mat to reduce noise.
- Put carpets, rugs, mats, throw rugs, ect. in your house/mats outside.
- Put fabric window coverings instead of plastic or wooden shades/blinds.
- Don't rev up a motorcycle or car unless it is actually needed for the drive.
- Don't beep your carn horn "just because", make sure it is a legitamate reason.

Noise Pollution in India and Other Countries

In India we have yet to think in terms of noise-free surroundings. In Mumbai, most people prefer to live within walking distance of the suburban railway line, regardless of the noise. Jerry builders, out to make a fast buck, make the walls of upper floors thinner. Upper floors are more subject to outdoor noise. In the US, flat-hunters have been known to take transistor radios with them while looking over a flat to test whether sound from the adjoining flats can be heard.

We haven't yet reached the stage of **Rio de Janeiro,** said to be the noisiest city in the world, where noise levels in the streets can hit the 120 dB level. But we are getting near to it. In our larger metropolitan cities like **Mumbai** and **Delhi,** the **Accoustical Division of the National Physical Laboratory, New Delhi,** found an average noise level of over 90 dB, seldom falling below 60 dB.

Our climatic conditions are not helpful; we have to keep door's and windows open most of the year, which means that noise can hardly be shut out.

Noise research in India has barely begun, but studies and trends abroad are good pointers. According to one study reported in **'Psychology Today',** three American workers, **David C. Glass, Sheldon Cohen and Jerome E. Euger,** found that noise did produce stress, but after exposure to the noise. Subjects exposed to noise fared worse than the "controls"; they showed less persistency in working at insoluble problems, which the researchers regard as an index of tolerance for frustration. Soft unpredictable noise produced lower tolerance for frustration than loud predictable noise, suggesting that predictability may be even more important than the loudness of the noise. Ability to control the noise, as by throwing a switch (even if the switch is not actually used), appeared to limit the subsequent adverse effects of the noise.

Anti-noise lobbies and activists are now throwing in their weight on the side of the citizen. Already there is growing public opinion against loudspeakers and ear-splitting Diwali-crackers. In the future it may be possible as a matter of course to claim damages for hearing loss sustained through exposure to intolerable noise. In the UK, legislation is coming which will enable a householder exposed to a noise level exceeding 68 dBA on the L_{10} index to claim from highway authorities the cost of insulation against traffic noise. In the USA, a number of states have made noise-induced hearing loss a compensable disability, even though it might not have meant the loss of a job. The US Federal Government in 1971 set maximum safe noise levels (90 dB for eight-hour exposure) for all workers.

The white-collar worker might appear to be relatively better off. But he has his own irritants telephones, typewriters, buzzers, loud chatter. The telephone bell has an SPL of about 76 dB; typewriters have a more uniform sound level, lacking the 'peaks' of the telephone. But conversation appears to be the most distracting source of noise; 46 per cent of people questioned by two Swiss researchers in a study of 15 "landscaped" offices, found it so. Curiously many reported that it was the content of the conversation, rather than the loudness, that was disturbing. What, then, of a telephone conversation (the hearer being able to hear only half of the dialogue)? —No studies, so far!

The same Swiss researchers report that an average mean noise level of 48 to 53 dBA might be all to the good. It would damp the great variety of single "acoustical events" –which should not exceed 58 to 65 dBA —without affecting speech intelligibility in offices. There was less disturbance due to noise in offices with a higher occupancy, larger total floor space, larger individual floor space and lower room height.

EXERCISE

1. What is noise? Explain the sources of noise.
2. Which agencies of modern civilization are mainly responsible for noise pollution?
3. Write an Essay an "Effects of Noise Pollution on human heath."
4. What are remedial steps to control Noise pollution?
5. Write short notes on —
   (*i*) Effects of industrial Noise pollution
   (*ii*) Noise pollution in India.

8

SOLID WASTE MANAGEMENT

8.1 INTRODUCTION

*Waste management is the collection, transport, processing, recycling or disposal, and monitoring of waste materials.*The term usually relates to materials produced by human activity, and is generally undertaken to reduce their effect on health, the environment or aesthetics. Waste management is also carried out to recover resources from it. Waste management can involve solid, liquid, gaseous or radioactive substances, with different methods and fields of expertise for each.

Waste management practices differ for developed and developing nations, for urban and rural areas, and for residential and industrial producers. Management for non-hazardous residential and institutional waste in metropolitan areas is usually the responsibility of local government authorities, while management for non-hazardous commercial and industrial waste is usually the responsibility of the generator.

Waste management concepts

There are a number of concepts about waste management which vary in their usage between countries or regions. Some of the most general, widely-used concepts include:

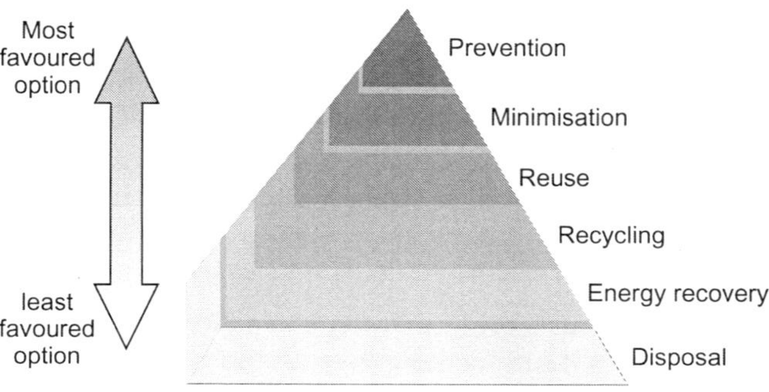

Fig. 8.1

- **Waste hierarchy:** The waste hierarchy refers to the "3 Rs" reduce, reuse and recycle, which classify waste management strategies according to their desirability in terms of waste minimization. The waste hierarchy remains the cornerstone of most waste minimization strategies. The aim of the waste hierarchy is to extract the maximum practical benefits from products and to generate the minimum amount of waste.
- **Extended producer responsibility:** Extended Producer Responsibility (EPR) is a strategy designed to promote the integration of all costs associated with products throughout their life cycle (including end of life disposal costs) into the market price of the product. Extended producer responsibility is meant to impose accountability over the entire lifecycle of products and packaging introduced to the market. This means that firms which manufacture, import and/or sell products are required to be responsible for the products after their useful life as well as during manufacture.
- **Polluter pays principle:** The Polluter Pays Principle is a principle where the polluting party pays for the impact caused to the environment. With respect to waste management, this generally refers to the requirement for a waste generator to pay for appropriate disposal of the waste.

8.2 SOLID WASTE MANAGEMENT (COLLECTION, CONVEYANCE AND DISPOSAL)

Solid waste can be disposed to land or oceans. Solid wastes can also be recovered and reprocessed, a procedure popularly known as recycling. Before disposal or recovery, however, the waste must be collected. All these *i.e.,* **collection**, **disposal** and/or **recovery** form a part of the solid waste management system.

Collection

- **Access to waste collection points.** Many sources of waste might only be reached by roads or alleys which may be inaccessible to certain methods of transport because of their width, slope, congestion of surface.
- **Public awareness and attitudes to waste.** This can affect the readiness to carry waste to a shared container, the willingness to accept the proximity of a shared container, the willingness to segregate waste to assist recycling, the frequency at which wastes should be collected, the amount of litter and animal excreta that are left on the street the willingness to pay for waste management services and the social groups from which waste management staff can be drawn.
- **Collection includes all the activities.** Associated with gathering of solid wastes and hauling of the waste collected to the location from where collection vehicles will transport it to disposal site. There are three basic methods of collection.
    1. *Community storage point.* The municipal refuge is taken to fixed storage bins and stored till the waste collection agency collect it daily for disposal in vehicle.
    2. *Kerb side collection.* In advance of collection time the refuse is brought in containers and placed on the foot way, from it is collected by waste collection agency.
    3. *Block collection.* Individuals bring the waste in containers and hand it over to the collection staff who empties it into waiting vehicle and return the containers to individuals.

The collection truck and crew are most important members of the collection system. Collection truck commonly used are of the enclosed, compacting type. Compaction in a collection vehicle temporarily reduces the refuse volume by about 80%. Mechanical collection systems should be used. These systems consist of standardized containers and truck mounted lifting mechanism. The crew parks its truck near the container placed in road side and lift it at the backside of the truck. By mechanical handling the container is emptied into the truck.

Modern collection systems involve pneumatic pipe line transport. In pneumatic systems, refuse is pulled by suction or vacuum through underground pipes to central underground collection site or it can go directly to the processing site.

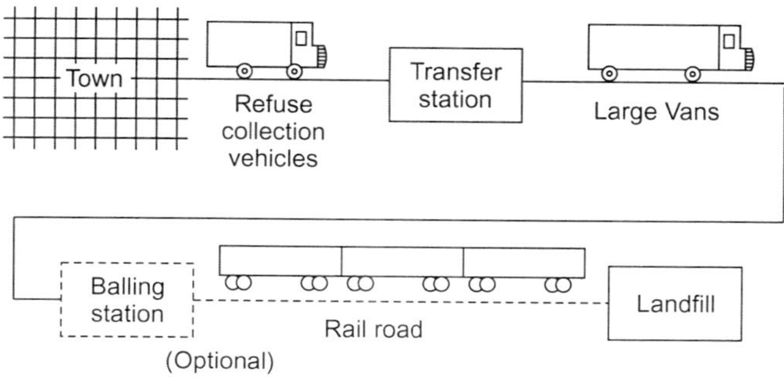

Fig. 8.2 Block Diagram of Transfer Station Method

Transportation

The method of transport involves two steps- (1) The transfer of waste from the smaller collection vehicle to the large transport equipment and (2) the subsequent transport of the wastes, to a long distance disposal site (Fig. 8.2).

Transfer stations- It is not always feasible for individual collection trucks to haul refuse to a waste processing plant or final disposal site when the destination is not immediate vicinity of the community in which waste is collected. Therefore, to solve this problem, there may be several transfer stations areawise where the refuse will be collected and then it will be transported to the disposal or processing sites by bigger transport vehicles.

Individual transfer station capacities may vary from less than 100 tons to more than 500 tons of wastes per day, depending upon the size of community. There are two basic modes of operation: **direct discharge or storage discharge.** In a storage discharge transfer station, the refuse is first emptied from the collection trucks into a storage pit or on a large platform. But in a direct discharge station, each refuse truck empties directly into the larger transport vehicle. The trailers have a capacity of 75 m^3 and hold the solid waste from four collection vehicles if it is not compacted and from eight collection vehicles if it is compacted.

As an alternative to large (often international) companies that can provide most or all of the solid waste services in a city, Microenterprises or Small Enterprises (MSEs) can be involved. They often use simple equipment and labour-intensive methods, and therefore can collect waste in places where the conventional trucks of large companies cannot enter. These MSEs may be stated as a business, to create income and employment, or they may be initiated by community members who wish to improve the immediate environment of their homes.

A recurring problem with collection schemes that operate at the community level is that the community scheme generally takes the waste a relatively short distance to a transfer point, from where the waste is supposed to be collected by another organisation *i.e.,* often municipality.

Problems of coordination and payment often result in the waste being left at transfer points for a long time. Another solution is to recycle as much of the waste locally so that there is very little need for ongoing transport to collected waste.

Waste collection methods vary widely between differnt countries and regions, and it would be impossible to describe them all. Many areas, especially those in less developed countries, do not have a formal waste-collection system in place.

For example, in Australia most urban domestic households have a 240-litre (63.4 U.S gallon) bin that is emptied weekly from the curb using side-or rear-loading compactor trucks. In Europe and a few other places around the world, a few communities use a proprietary collection system known as Envac, which conveys refuse via underground conduits using a vacuum system. Roosevelt Island has had this system since 1975. In Canadian urban centres curbside collection is the most common method of disposal, whereby the city collects waste and/or recyclables and/or organics on a scheduled basis. In rural areas people usually dispose of their waste by hauling it to a transfer station. Waste collected is then transported to a regional landfill.

Waste Disposal Methods

Waste management methods for vary widely between areas for many reasons, including type of waste material, nearby land uses, and the area available.

1. **Processing of Solid Wastes (For volume reduction).** Processing is the second fundamental function of solid waste management. Processing improves the efficiency of solid waste disposal and prepares solid waste for subsequent recovery of materials and energy. In the not too distant past, disposal of solid wastes included open dumping, sanitary land filling and disposal at sea. Because of environmental problems associated with open dumps and sea disposal, the only acceptable method of solid waste disposal at present is sanitary land filling.

   (*a*) **Processing for recovery of materials for recycling:** Processing to segregate solid waste components may be done at the point of generation (on site processing) or at central processing facility. Onsite processing needs cooperation of waste producer; homes, commercial establishments, industries and the like. In onsite processing, wastes are segregated into types at the point of generation. For example, paper into one container, cans into another, glass, plastic and so on. In absence of onsite processing segregation into components may be done at central facility. Unit operations, in a central facility involve screening, air classifying and magnetic separations size reduction using shredders is also used to produce a more uniformily sized products.

   Screening. Screening is a unit operation of separating a feed into oversize and undersize products. Oversize products are those that do not pass the openings of the screen, undersized products are those that do pass the opening of the screen. Solids like plastic, glass, rubber, iron scraps, tin cans, paper etc., are picked up and sent for recycling to the factories.

   (*b*) **Shredding and pulverising:** First, in the processing is the size reduction, so that total volume and weight of wastes and reduced. Volume reduction helps in utilizing less land for disposal. It also reduces cost of transportation. Alongwith volume and weight reduction, waste processing changes its form and improves its handling character. Size reduction is effected by the physical processes of shredding or pulverizing. Shredding means cutting and tearing whereas pulverisation means crushing and grinding. Shredding and pulverising reduce the overall volume of the original or raw waste materials about 40%. This is needed, because the production of refuse derived fuel on (RDF), requires processing of the raw solid waste.

(*c*) **Hammer mills:** One of the most common types equipment used for processing municipal solid waste (MSW) into a uniform and homogeneous mass is the *hammer mill. A* hammer mill is a mechanical impact device in which the waste material is cramped with a force to crush or tear individual pie of the waste. Impact is provided by several hammers that rotate at high speeds around a central horizontal or vertical shaft. It is possible to reduce the size of waste material components to uniform fragments between 25 and 50 mm with proper operation.

(*d*) **Baling:** Compacting the solid waste into the form of rectangular blocks or bales is called *baling,* MSW bales are typically about 1.5 m^3 in size and weight about 1 kN. Solid waste can be compacted high pressure (about-700 k Pa) in either vertical or horizontal presses, the bales are frequently wrapped with steel wire to help retain their rectangular shape during handling. They may also be enclosed in hot a sphalt, plastic or port-land cement bags on tied with metal bands, depending on the intended use or disposal method. In this process, the volume reduction is about 90%. The basic advantage of an MSW baling process in the significant decrease in waste volume, the ease of handling the compacted refuse and the reduction of litter and nuisance potential.

(*e*) **Incineration:** Incineration is a waste disposal method that involves the combustion of waste at high temperature. Incineration and other high temperature waste treatment systems are described as "thermal treatment". In effect, incineration of waste materials converts the waste into heat, gaseous emissions, and residual solid ash. Other types of thermal treatment include pyrolysis and gasification.

Fig. 8.3 A Waste-to-energy Plant (Incinerator plant)

A waste-to-energy plant (WtE) is a modern term for an incinerator that burns wastes in high-efficiency furnace/boilers to produce steam and/or electricity and incorporates modern air pollution control systems and continuous emissions monitors. This type of incinerator is sometimes called an energy-from-waste (EfW) facility. Incineration is popular is countries such as Japan where land is a scarce resource, as incinerators do not consume as much area as a landfill. Sweden has been a leader in using the energy generated from incineration over the past 20 years. Denmark also extensively used waste-to-energy incineration in localized combined heat and power facilities supporting district heating schemes.

Incineration is carried out both on a small scale by individuals, and on a large scale by industry. It is recognized as a practical method of disposing of certain hazardous waste materials (such as biological medical waste), though it remains a controversial method of waste disposal in many places due to issues such as emission of gaseous pollutants. Breaking down complex chemical chains such a dioxin through the application of heat usually cannot be done by simply burning the material at the temperature seen in an open-air fire. It is often necessary to supplement the combustion process with gas or oil burners and air blowers to raise the temperature high enough to result in molecular breakdown. Alternately, the exhaust gases from a natural air fire may pass through tubes heated to sufficiently high temperatures to trigger thermal breakdown. Thermal breakdown of pollutant molecules can indirectly create other pollution problems. Dioxin breakdown begins at 1000 °C, but at the same time poisonous nitrogen oxides and ozone begin to form when atmospheric nitrogen and oxygen break down at 1600 °C. This undesired oxide formation may require further catalytic treatment of the exhaust gases.

(*f*) **Composting and anaerobic digestion:** Waste materials that are organic in nature, such as plant material, food scraps and paper products, are increasingly being recycled using biological composting and/or digestion processes to decompose the organic matter and kill pathogens. The resulting organic material is then recycled as mulch or compost for agricultural or landscaping purposes.

(*g*) **Mechanical biological treatment:** Mechanical Biological Treatment (MBT) is a technology category for combinations of mechanical sorting and biological treatment of the organic fraction of municipal waste. MBT is also sometimes called BMT (Biological Mechanical Treatment), however this simply refers to the order of processing. The "mechanical" element is usually a bulk handling mechanical sorting state. This removes recyclable elements from the waste (such as metals, plastics and glass), and/or processes it to produce Refuse Derived Fuel (RDF) that is burnt in power plants, boilers or kilns. The "biological" element refers to a biological digestion process, which breaks down the biodegradable component of the waste to produce biogas and/or organic matter. The biogas can be used to generate energy; and organic matter recycled as compost.

An example of large-scale biological treatment facility is the composting facility in Edmonton, Canada, where 200,000 tonnes of residential solid waste and 22,500 tonnes of biosolids are composted each year to produce 80,000 tonnes of compost. The co-composter itself is 38,690 square metres in size, equivalent to 8 football fields.

(*h*) **Pyrolysis and gasification:** Pyrolysis and gasification are two related forms of thermal treatment where waste materials are heated to high temperatures with limited oxygen availability. The process typically occurs in a sealed vessel under high pressure. Converting material to energy in a sealed environment is potentially more efficient than direct incineration, with more energy able to be recovered and used.

Pyrolysis of solid waste converts the material into solid, liquid and gas products. The liquid oil and gas can be burnt to produce energy or refined into other products. The solid residue (char) can be further refined into products such as activated carbon. Gasification is used to convert organic materials directly into a synthetic gas (syngas) composed of carbon monoxide and hydrogen. The gas is then burnt to produce electricity and steam. Gasification is used in biomass power stations to produce renewable energy and heat.

8.3 RECOVERY OF RESOURCES

Resource recovery means the obtaining of some economic benefit from material that someone has regarded as waste. It includes:

- Reuse- being used for the same purpose again (such as refilling a soft drinks bottle)
- Recovery-processing material so that it can be used again as the same material, such as the processing of waste paper to make pulp and then paper
- Conversion- processing the material to make something different (such as producing padding for clothing and sleeping bags from plastic bottles, or producing compost from food waste)
- Energy recovery-usually referring to the burning of waste so that the heat can be used. Another method of energy recovery is to collect the gas is produced in very large sanitary landfills and use it as a fuel or to generated electricity.

Some key factors that affect the potential for resource recovery are the cost of the separated material, its purity, its quantity and its location. The costs of storage and transport are major factors that decide the economic potential for resource recovery. In many low-income countries, the fraction of material that is won for resource recovery is very high, because this work is done in a very labour-intensive way, and for very low incomes. In such situations the creation of employment is the main economic benefit of resource recovery.

A relatively recent idea is waste management has been to treat the waste material as a resource to be exploited, instead of simply a challenge to be managed and disposed of. There are a number of different methods by which resources may be extracted from waste: the materials may be extracted and recycled, or the calorific content of the waste may be converted to electricity.

The process of extracting resources or values from waste is variously referred to as secondary resource recovery, recycling, and other terms. The practice of treating waste materials as a resource is becoming more common, especially in metropolitan areas where space for new landfills is becoming scarcer. There is also a growing acknowledgment that simply disposing of waste materials is unsustainable in the long term, as there is a finite supply of most raw materials.

There are a number of methods of recovering resources from waste materials, with new technologies and methods being developed continuously.

Recycling

Recycling means to recover for other use a material that would otherwise be considered waste. The popular meaning of 'recycling' in most developed countries has come to refer to the widespread collection and reuse of various everyday waste materials, such as newspapers and drink bottles. They are collected and sorted into common types so that the raw materials from these items can be used again to create new products. In many areas, material for recycling is collected separately from general waste using dedicated bins and collection vehicles. Other waste management processes can recover materials from mixed waste streams.

In developed countries, the most common consumer items recycled include aluminium beverage cans, steel, food and aerosol cans, HDPE and PET bottles, glass bottles and jars, paperboard cartons, newspapers, magazines, and cardboard. Other types of plastic (PVC, LDPE, PP, and PS) are also recyclable, although these are not as commonly collected. These items are usually composed of a single type of material, making them relatively easy to recycle into new products. The recycling of complex products (such as computers and electronic equipment) is more difficult and costly, due to the separation and reprocessing required.

Landfilling (Disposal Methods)

For hundreds of years, people have used garbage dumps to get rid of their trash. Yesterday's garbage dump was nothing more than a pit of field just outside of town where people left their garbage.

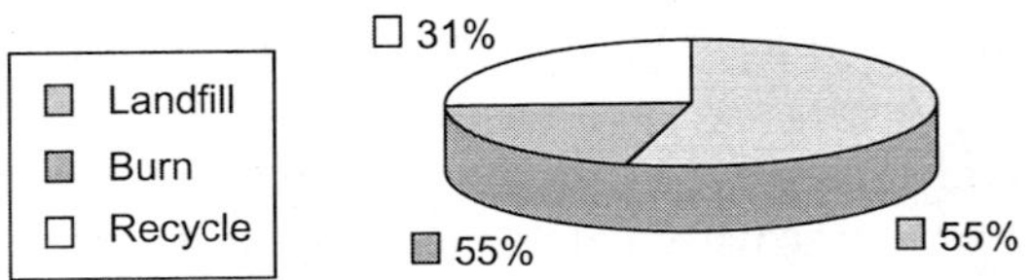

Fig. 8.4 What we do our trash

People tossed all sorts of waste into these dumps. The dumps were breeding grounds for disease carrying pests such as flies, mosquitoes, and rats. Rainwater flushed filthy, and sometimes poisonous, liquids from the dump into nearby streams and groundwater supplies that people used for drinking, bathing, and clothes washing.

Later, some towns spread dirt to contain the dumped waste and to discourage vermin. This helped, but it was little more than a cover-up for unsanitary dumping.

Today, we still bury our garbage, although not in the open dumps of yesterday. About 55 percent of our garbage is hauled off in garbage trucks and packed into sanitary landfills making landfilling America's number one way of getting rid of its trash. (The other 45 percent is either recycled or burned.)

Although the nation as a whole has plenty of space to build landfills, some areas in the North-east may be running out of room for new landfills. Obtaining permits to build new landfills has become increasingly difficult because of public opposition people don't want landfills built in their backyards, and besides, a new landfill costs up to $10 million to build.

That's why some communities are looking for new ways to deal with solid waste-recycling and burning, for instance. But there will always be a need for landfills. Why ? because not all waste can be recycled or burned. How do you recycle a broken light bulb, and why burn it if it doesn't provide any heat energy ?

Landfill burial is the only feasible way to dispose of some types of waste, and sometimes it's the safest way, too. Generally, the best disposal method for hazardous wastes-batteries, paints, pesticides, and the like are state-of-the-art landfills. These landfills are designed to prevent hazardous wastes from seeping into underground water supplies.

Now that open dumping is illegal ,deciding where to put a landfill requires careful planning. Skilled engineers inspect potential landfill sites. They look at a number of thing including:

- The geology of the area
- The nature of the local environment
- how easy the site is to reach
- how far the site is from the area that generates the waste.

Work on a landfill site begins only after the site passes strict legal, environmental, and engineering tests. It is not a quick procedure; landfills can take five years to complete.

A landfill, also known as a dump or tip (and historically as a midden), is a site for the disposal of waste materials by burial and is the oldest form of waste treatment. Historically, landfills have been the most common method of organized waste disposal and remain so in many places around the world.

Landfills may include internal waste disposal sites (where a producer of waste carries out their own waste disposal at the place of production) as well as sites used by many producers. Many Landfills are also used for other waste management purposes, such as the temporary storage, consolidation and transfer, or processing of waste material (sorting, treatment or recycling).

A landfill also may refer to ground that has been filled in with soil and rocks instead of waste materials, so that it can for a specific purpose, such as for building houses. Unless they are stabilized, these areas may experience severe shaking or liquefaction of the ground in a large earthquake.

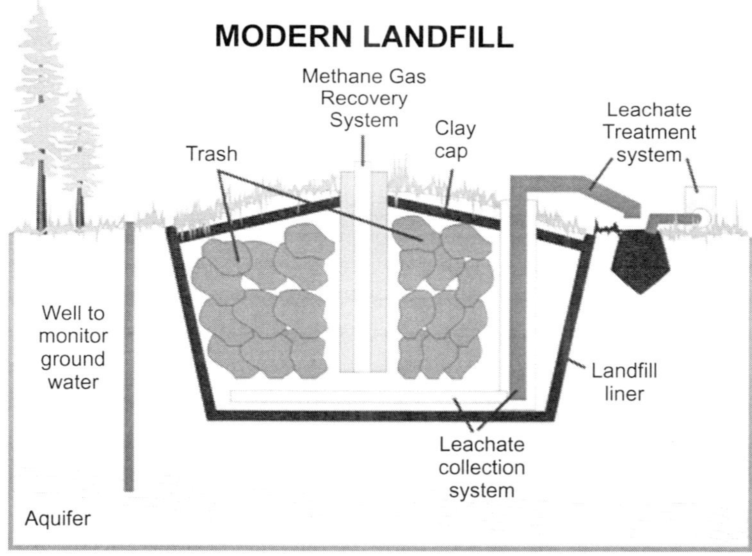

Fig. 8.5 Landfill

Site Construction Requirements

The construction of a landfill requires a staged approach. Landfill designers are primarily concerned with the viability of a site. To be commercially and environmentally viable a landfill must be constructed in accord with specific requirements, which are related to:

Location:

- Easy access to transport by road
- Transfer stations if rail network is preferred
- Land value (price)
- Cost of meeting government requirements
- Location of community served
- Type of construction (more than one may be used at single site)

- Pit-filling existing holes in the ground, typically left behind by mining
- Canyon-filling in naturally occurring valleys or canyons
- Mound-piling the waste up above the ground

Stability:

- Underlying geology
- Nearby earthquake faults
- Water table
- Location of nearby rivers, streams, and flood plains

Capacity. The available void space must be calculated by comparison of the landform with aproposed restoration profile.

- This calculation of capacity is based on:
- Density of the wastes
- Amount of intermediate and daily cover
- Amount of settlement that the waste will undergo following tipping
- Thickness of capping
- Construction of lining and drainage layers.

Protection of soil and water through:

- Installation of liner and collection systems.
- Storm water control
- Leachate management.
- Landfill gas migration.

Nuisances and hazards management.

Costs:

- Feasibility studies
- Site after care
- Site investigations (costs involved may make small sites uneconomical).
- Site respect

During landfill operations the waste collection vehicles are weighed at a weighbridge on arrival and their load is inspected for wastes that do not accord with the landfill's waste acceptance criteria. Afterwards, the waste collection vehicles use the existing road network on their way to the tipping face or working front where they unload their load. After loads are deposited, compactors or dozers are used to spread and compact the waste on the working face. Before leaving the landfill boundaries, the waste collection vehicles pass through the wheel cleaning facility. If necessary, they return to the weighbridge in order to be weighed without their load. Through the weighing process, the daily incoming waste tonnage can be calculated and listed in databases. In addition to rucks, some landfills may be equipped to handle railroad containers. The use of 'rail-haul' permits landfills to be located at more remote sites, without the problems associated with many truck trips.

Typically, in the working face, the compacted waste is covered with soil daily. Alternative waste-cover materials are several sprayed on foam products and temporary blankets. Blankets can be lifted into place with tracked excavators and then removed the following day prior to waste placement. Chipped wood and chemically 'fixed' biosolids, may also be used as an alternate daily cover. The space that is occupied daily by the compacted waste and the cover material is called a daily cell. Waste compaction is critical to extending the life of the landfill. Factors such as waste compressibility, waste layer thickness and the number of passes of the compactor over the waste affect the waste densities.

Land Reclamation

As human overcrowding of developed areas intensified during the 20th century, it has become important to develop land reuse strategies for completed landfills. Some of the most common usages are for parks, golf courses and other sports fields. Increasingly, however, office buildings and industrial uses are made on a completed landfill. In these latter uses, methane capture is customarily carried out to minimize explosive hazard within the building.

An example of a Class A office building constructed over a landfill is the Dakin Building at Sierra Point, Brisbane, California. The underlying fill was deposited from 1965 to 1985, mostly consisting of construction debris from San Francisco and some municipal wastes. Aerial photographs prior to 1965 show this area to be tidelands of the San Francisco Bay. A clay cap was constructed over the debris prior to building approval.

A notable example is Sydney Olympic Park, the primary venue for the 2000 Summer Olympic Games, which was built atop an industrial wasteland that included landfills.

Another strategy for landfill reclamation is the incineration of landfill trash at high temperature via the plasma arc gasification process, which is currently used at two facilities in Japan, and will be used at a planned facility in St. Lucie County, Florida.

Impacts

A large number of adverse impacts occur from landfill operations. These impacts can vary fatal accidents (*e.g.*, scavengers buried under waste piles); infrastructure damage (*e.g.*, damage to access roads by heavy vehicles); pollution of the local environment (such as contamination of groundwater and/or aquifers by leakage and residual soil contamination during landfill usage, as well as after landfill closure); offgassing of methane generated by decaying organic wastes (methane is a greenhouse gas many times more potent than carbon dioxide, and can itself be a danger to inhabitants of an area); harbouring of disease vectors such as rats and flies, particularly from improperly operated landfills, which are common in Third-world countries; injuries to wildlife; and simple nuisance problems (*e.g.*, dust, odour, vermin, or noise pollution).

Environmental noise and dust are generated from vehicles accessing a landfill as well as from working face operations. These impacts are best to intercept at the planning stage where access routes and landfill geometrics can be used to mitigate such issues. Vector control is also important, but can be managed reasonably well with the daily cover protocols.

Most modern landfills in industrialized countries are operated with controls to attempt manage problems such as these. Analyses of common landfill operational problems are available.

Some local authorities have found it difficult to locate new landfills. Communities may charge a fee or levy in order to discourage waste and/or recover the costs of site operations. Some landfills are operated for profit as commercial businesses. Many landfills, however, are publicly operated and funded.

Trash Dump Communities

In many developing countries around the world, communities exist in and around landfills. Residents of these communities, such as La Chureca in Nicaragua, often live in conditions of extreme poverty and use the landfills as a source of food and income. Scavengers work in the garbage in search of recyclables and other valuables.

Landfill Gas

Gasses are produced in landfills due to the anaerobic digestion by microbes on any organic matter. This gas can be collected and flared off or used to generate electricity in a gas fired power plant. Landfill gas monitoring can be carried out to alert for the the presence of a build-up of gasses to a harmful level.

What is an Incinerator?

An incinerator is a furnace for burning waste. Modern incinerators include pollution mitigation equipment such as flue gas cleaning. There are various types of incinerator plant design: moving grate, fixed grate, rotary kiln, fluidised bed.

Private Burn Pile/Barrel

The burn pile is one of the simplest and earliest forms of waste disposal, essentially consisting of a mound of combustible materials piled on bare ground and set on fire. Indiscriminate piles of household waste are strongly discouraged and may be illegal in urban areas, but are permitted in certain rural situations such as clearing forested land for farming, where the stumps are uprooted and burned. Rural burn piles of organic yard waste are also sometimes permitted, though not asphalt shingles, plastics, or other petroleum products.

Burn piles have the potential to spread uncontrolled fires, if for example wind blows burning material off the pile into surrounding combustible grasses or onto buildings. As interior structures of the pile are consumed, the pile can shift and collapse, spreading the burn area. Even in a situation of no wind, small lightweight ignited embers can lift off the pile via convection, and waft through the air into grasses or onto buildings, igniting them.

The burn barrel is a somewhat more controlled form of private waste incineration, containing the burning material inside a metal barrel, with a metal grating over the exhaust. The barrel prevents the spread of burning material in windy conditions, and as the combustibles are reduced they can only settle down into the barrel. The exhaust grating helps to prevent the spread of burning embers. Typically steel 55-gallon drums are used as burn barrels, with air vent holes cut or drilled around the base for air intake. Over time the very high heat of incineration causes the metal to oxidize and rust, and eventually the barrel itself is consumed by the heat and must be replaced.

Private burning of dry cellulosic/paper products is generally clean-burning, producing no visible smoke, but the large amount of plastics in household waste can cause private burning to create a public nuisance and health hazard, generating acrid odors and fumes that make eyes burn and water. The temperatures in a burn barrel are not regulated, and usually do not reach high enough or for enough time to completely break down chemicals such as dioxin in plastics and other waste chemicals. Plastics and other petroleum products must therefore be separated and sent to commercial waste disposal facilities.

Private rural incineration is typically only permitted so long as it is not a nuisance to others, does not pose a risk of fire such as in dry conditions, and the fire is clean-burning, producing no visible smoke. People intending to burn waste may be required to contact a state agency in advance to check current fire risk and conditions, and to alert officials of the controlled fire that will occur.

Moving Grate

The typical incineration plant for municipal solid waste is a moving grate incinerator. The moving grate enables the movement of waste through the combustion chamber to be optimised to allow a more efficient and complete combustion. A single moving grate boiler can handle up to 35 metric tons (39 short tons) of waste per hour, and can operate 8,000 hours per year with only one scheduled stop for inspection and maintenance of about one month's duration. Moving grate incinerators are sometimes referred to as Municipal Solid Waste Incinerators (MSWIs).

The waste is introduced by a waste crane through the "throat" at one end of the grate, from where it moves down over the descending grate to the ash pit in the other end. Here the ash is removed through a water lock.

Municipal solid waste in the furnace of a moving grate incinerator capable of handling 15 metric tons (17 short tons) of waste per hour. The holes in the grate elements supplying the primary combustion air are visible.

Part of the combustion air (primary combustion air) is supplied through the grate from below. This air flow also has the purpose of cooling the grate itself. Cooling is important for the mechanical strength of the grate, and many moving grates are also water cooled internally.

Secondary combustion air is supplied into the boiler at high speed through nozzles over the grate. It facilitates complete combustion of the flue gases by introducing turbulence for better mixing and by ensuring a surplus of oxygen. In multiple/stepped hearth incinerators, the secondary combustion air is introduced in a separate chamber downstream the primary combustion chamber.

According to the European Waste Incineration Directive, incineration plants must be designed to ensure that the flue gases reach a temperature of at least 850 °C (1,560 °F) for 2 seconds in order to ensure proper breakdown of toxic organic substances. In order to comply with this at all times, it is required to install backup auxiliary burners (often fueled by oil), which are fired into the boiler in case the heating value of the waste becomes too low to reach this temperature alone.

The flue gases are then cooled in the superheaters, where the heat is transferred to steam, heating the steam to typically 400 °C (752 °F) at a pressure of 40 bars (580 psi) for the electricity generation in the turbine. At this point, the flue gas has a temperature of around 200 °C (392 °F), and is passed to the flue gas cleaning system.

At least in Scandinavia scheduled maintenance is always performed during summer, where the demand for district heating is low. Often incineration plants consist of several separate 'boiler lines' (boilers and flue gas treatment plants), so that waste can continue to be received at one boiler line while the others are subject to revision.

Fixed Grate

The older and simpler kind of incinerator was a brick-lined cell with a fixed metal grate over a lower ash pit, with one opening in the top or side for loading and another opening in the side for removing incombustible solids called clinkers. Many small incinerators formerly found in apartment houses have now been replaced by waste compactors.

Rotary-Kiln

The rotary-kiln incinerator is used by municipalities and by large industrial plants. This design of incinerator has 2 chambers: a primary chamber and secondary chamber. The primary chamber in a rotary kiln incinerator consist of an inclined refractory lined cylindrical tube. Movement of the cylinder on its axis facilitates movement of waste. In the primary chamber, there is conversion of solid fraction to gases, through volatilization, destructive distillation and partial combustion reactions. The secondary chamber is necessary to complete gas phase combustion reactions.

The clinkers spill out at the end of the cylinder. A tall flue gas stack, fan, or steam jet supplies the needed draft. Ash drops through the grate, but many particles are carried along with the hot gases. The particles and any combustible gases may be combusted in an "afterburner". A diagram of a rotary-kiln incinerator can be found here.

Fluidized Bed

A strong airflow is forced through a sandbed. The air seeps through the sand until a point is reached where the sand particles separate to let the air through and mixing and churning occurs, thus a fluidised bed is created and fuel and waste can now be introduced.

The sand with the pre-treated waste and/or fuel is kept suspended on pumped air currents and takes on a fluid-like character. The bed is thereby violently mixed and agitated keeping small inert particles and air in a fluid-like state. This allows all of the mass of waste, fuel and sand to be fully circulated through the furnace.

Specialized Incineration

Furniture factory sawdust incinerators need much attention as these have to handle resin powder and many flammable substances. Controlled combustion, burn back prevention systems are essential as dust when suspended resembles the fire catch phenomenon of any liquid petroleum gas.

Use of Heat

The heat produced by an incinerator can be used to generate steam which may then be used to drive a turbine in order to produce electricity. The typical amount of net energy that can be produced per tonne municipal waste is about 2/3 MWh of electricity and 2 MWh of district heating.

Thus, incinerating about 600 metric tons (660 short tons) per day of waste will produce about 17 MW of electrical power and 1200 MWh district heating each day.

Pollution

Incineration has a number of outputs such as the ash and the emission to the atmosphere of flue gas. Before the flue gas cleaning system, the flue gases may contain significant amounts of particulate matter, heavy metals, dioxins, furans, sulfur dioxide, and hydrochloric acid.

In a study from 1994, Delaware Solid Waste Authority found that, for same amount of produced energy, incineration plants emitted fewer particles, hydrocarbons and less SO_2, HCl, CO and NO_x than coal fired power plants, but more than natural gas fired power plants. According to

Germany's Ministry of the Environment, waste incinerators reduce the amount of some atmospheric pollutants by substituting power produced by coal fired plants with power from waste fired plants.

Gaseous Emissions

Dioxin and Furans

The most publicized concerns from environmentalists about the incineration of Municipal Solid Wastes (MSW) involve the fear that it produces significant amounts of dioxin and furan emissions. Dioxins and furans are considered by many to be serious health hazards. Older generation incinerators that were not equipped with adequate gas cleaning technologies were indeed significant sources of dioxin emissions. Today, however, due to advances in emission control designs and stringent new governmental regulations, incinerators emit virtually no dioxins. In 2005, The Ministry of the Environment of Germany, where there were 66 incinerators at that time, estimated that"...whereas in 1990 one third of all dioxin emissions in Germany came from incineration plants, for the year 2000 the figure was less than 1%. Chimneys and tiled stoves in private households alone discharge approximately twenty times more dioxin into the environment than incineration plants." According to the United States Environmental Protection Agency, incineration plants are no longer significant sources of dioxins and furans. In 1987, before the governmental regulations required the use of emission controls, there was a total of 10,000 grams (350 oz) of dioxin emissions from U.S., incinerators. Today, the total emissions from the 87 plants are only 10 grams (0.35 oz) yearly, a reduction of 99.9 %. Backyard barrel burning of household and garden wastes, still allowed in some rural areas, generates 580 grams (20 oz) of dioxins yearly. Studies conducted by EPA demonstrate that the emissions from just one family using a burn barrel produces more emissions than an incineration plant disposing of 200 metric tons (220 short tons) of waste per day.

Generally the breakdown of dioxin requires exposure of the molecule to a sufficiently high temperature so as to trigger thermal breakdown of the molecular bonds holding it together. When burning of plastics outdoors in a burn barrel or garbage pit such temperatures are not reached, causing high dioxin emissions as mentioned above. While the plastic does burn in an open air fire, the dioxins remain after combustion and float off into the atmosphere.

Modern municipal incinerator designs include a high temperature zone, where the flue gas is ensured to sustain a temperature above 850 °C (1,560 °F) for at least 2 seconds before it is cooled down. They are equipped with auxiliary heaters to ensure this at all times. These are often fueled by oil, and normally only active for a very small fraction of the time. A side effect controlling dioxin is the potential for generation of reactive oxides (NO_x) in the flue gas, which must be removed with SCR or SNCR (see below).

For very small municipal incinerators, the required temperature for thermal breakdown of dioxin may be reached using a high-temperature electrical heating element, plus an SCR stage.

CO_2

As for other complete combustion processes, nearly all of the carbon content in the waste is emitted as CO_2 to the atmosphere. MSW contain approximately the same mass fraction of carbon as CO_2 itself (27%), so incineration of 1 tonn of MSW produce approximately 1 ton of CO_2.

In the event that the waste was landfilled, 1 ton of MSW would produce approximately 62 cubic metres (2,200 cu ft) methane via the anaerobic decomposition of the biodegradable part of the waste. This amount of methane has more than twice the global warming potential than the 1 ton of CO_2, which would have been produced by incineration. In some countries, large amounts of landfill gas are collected, but still the global warming potential of the landfill gas emitted to atmosphere in the US in 1999 was approximately 32 % higher than the amount of CO_2 that would have been emitted by incineration.

In addition, nearly all biodegradable waste has biological origin. This material has been formed by plants using atmospheric CO_2 typically within the last growing season. If these plants are regrown the CO_2 emitted from their combustion will be taken out from the atmosphere once more.

Such considerations are the main reason why several countries administrate incineration of the biodegradable part of waste as renewable energy. The rest - mainly plastics and other oil and gas derived products — is generally treated as non-renewables.

Different results for the CO_2 footprint of incineration can be reached with different assumptions. Local conditions (such as limited local district heating demand, no fossil fuel generated electricity to replace or high levels of aluminum in the waste stream) can decrease the CO_2 benefits of incineration. The methodology and other assumptions may also influence the results significantly. For example, the methane emissions from landfills occurring at a later date may be neglected or given less weight, or biodegradable waste may not be considered CO_2 neutral. A recent study by Eunomia Research and Consulting on potential waste treatment technologies in London demonstrated that by applying several of these (according to the authors) unusual assumptions the average existing incineration plants performed poorly for CO_2 balance compared to the theoretical potential of other emerging waste treatment technologies.

Other Emissions

Other gaseous toxins in the flue gas from incinerator furnaces include sulfur dioxide, hydrochloric acid, heavy metals and fine particles.

The steam content in the flue may produce visible fume from the stack, which can be perceived as a visual pollution. It may be avoided by decreasing the steam content by flue gas condensation and reheating, or by increasing the flue gas exit temperature well above its dew point. Flue gas condensation allows the latent heat of vaporization of the water to be recovered, subsequently increasing the thermal efficiency of the plant.

Flue Gas Cleaning

The quantity of pollutants in the flue gas from incineration plants is reduced by several processes.

Particulate is collected by particle filtration, most often electrostatic precipitators (ESP) and/or baghouse filters. The latter are generally very efficient for collecting fine particles. In an investigation by the Ministry of the Environment of Denmark in 2006, the average particulate emissions per energy content of incinerated waste from 16 Danish incinerators were below 2.02 g/GJ (grams per energy content of the incinerated waste). Detailed measurements of fine particles with sizes below 2.5 micrometres ($PM_{2.5}$) were performed on three of the incinerators: One incinerator equipped with an ESP for particle filtration emitted 5.3 g/GJ fine particles, while two incinerators equipped with baghouse filters emitted 0.002 and 0.013 g/GJ $PM_{2.5}$. For ultra fine particles ($PM_{1.0}$), the numbers were 4.889 g/GJ $PM_{1.0}$ from the ESP plant, while emissions of 0.000 and 0.008 g/GJ $PM_{1.0}$ were measured from the plants equipped with baghouse filters.

Acid gas scrubbers are used to remove hydrochloric acid, nitric acid, hydrofluoric acid, mercury, lead and other heavy metals. Basic scrubbers remove sulfur dioxide, forming gypsum by reaction with lime.

Wastewater from scrubbers must subsequently pass through a wastewater treatment plant.

Sulfur dioxide may also be removed by dry desulfurisation by injection limestone slurry into the flue gas before the particle filtration.

NO_x is either reduced by catalytic reduction with ammonia in a catalytic converter (selective catalytic reduction, SCR) or by a high temperature reaction with ammonia in the furnace (selective non-catalytic reduction, SNCR).

Heavy metals are often adsorbed on injected active carbon powder, which is collected by the particle filtration.

Other Pollution Issues

Odor pollution can be a problem with oldstyle incinerators, but odors and dust are extremely well controlled in newer incineration plants. They receive and store the waste in an enclosed area with a negative pressure with the airflow being routed through the boiler which prevents unpleasant odors from escaping into the atmosphere. However, not all plants are implemented this way, resulting in inconveniences in the locality.

An issue that affects community relationships is the increased road traffic of waste collection vehicles to transport municipal waste to the incinerator. Due to this reason, most incinerators are located in industrial areas. This problem can be can avoided to an extent through the transport of waste by rail from transfer stations.

The Debate over Incineration

Use of incinerators for waste management is controversial. The debate over incinerators typically involves business interests (representing both waste generators and incinerator firms), government regulators, environmental activists and local citizens who must weigh the economic appeal of local industrial activity with their concerns over health and environmental risk.

People and organizations professionally involved in this issue include the U.S. Environmental Protection Agency and a great many local and national air quality regulatory agencies worldwide.

The Argument for Incineration

- The concerns over the health effects of dioxin and furan emissions have been significantly lessened by advances in emission control designs and very stringent new governmental regulations that have resulted in large reductions in the amount of dioxins and furans emissions.
- The U.K. Health Protection Agency concluded in 2009 that "Modern, well managed incinerators make only a small contribution to local concentrations of air pollutants. It is possible that such small additions could have an impact on health but such effects, if they exist, are likely to be very small and not detectable."
- Incineration plants can generate electricity and heat that can substitute power plants powered by other fuels at the regional electric and district heating grid, and steam supply for industrial customers. Incinerators and other waste-to-energy plants generate at least partially

biomass based renewable energy that offsets greenhouse gas pollution from coal-, oil and gas fired power plants. The E.U. considers energy generated from biogenic waste (waste with biological origin) by incinerators as non-fossil renewable energy under its emissions caps. These greenhouse gas reductions are in addition to those generated by the avoidance of landfill methane.

- The bottom ash residue remaining after combustion has been shown to be a non-hazardous solid waste that can be safely put into landfills or recycled as construction aggregate. Samples are tested for ecotoxic metals.
- In densely populated areas, finding space for additional landfills is becoming increasingly difficult.
- Fine particles can be efficiently removed from the flue gases with baghouse filters. Even though approximately 40% of the incinerated waste in Denmark was incinerated at plants with no baghouse filters, estimates based on measurements by the Danish Environmental Research Institute showed that incinerators were only responsible for approximately 0.3% of the total domestic emissions of particulate smaller than 2.5 micrometres ($PM_{2.5}$) to the atmosphere in 2006.
- Incineration of municipal solid waste avoids the release of methane. Every ton of MSW incinerated, prevents about one ton of carbon dioxide equivalents from being released to the atmosphere.
- Incineration of medical waste and sewage sludge produces an end product ash that is sterile and non-hazardous.
- Most municipalities that operate incineration facilities have higher recycling rates than neighboring cities and counties that do not send their waste to incinerators. This is in part due to enhanced recovery of ceramic materials reused in construction, as well as ferrous and in some cases non-ferrous metals that can be recovered from combustion residue.
- Volume of combusted waste is reduced by approximately 90%, increasing the life of landfills. Ash from modern incinerators is vitrified at temperatures of 1,000 °C (1,830 °F) to 1,100 °C (2,010 °F), reducing the leachability and toxicity of residue. As a result, special landfills are generally no longer required for incinerator ash from municipal waste streams, and existing landfills can see their life dramatically increased by combusting waste, reducing the need for municipalities to site and construct new landfills.

The Argument Against Incineration

- The highly toxic fly ash must be safely disposed of. This usually involves additional waste miles and the need for specialist toxic waste landfill elsewhere, sometimes with concerns for local residents. This has been the case in Bishops Cleeve, Gloucestershire, UK.
- There are still concerns by many about the health effects of dioxin and furan emissions into the atmosphere from old incinerators; especially during start up and shut down events, or where filter bypass events are required.
- Incinerators emit varying levels of heavy metals such as vanadium, manganese, chromium, nickel, arsenic, mercury, lead, and cadmium, which can be toxic at very minute levels.
- Incinerator Bottom Ash (IBA) has high levels of heavy metals with ecotoxicity concerns if not reused properly. Some people have the opinion that IBA reuse is still in its infancy and is still not considered to be a mature or desirable product, despite additional engineering treatments.
- Alternative technologies are available or in development such as Mechanical Biological Treatment, Anaerobic Digestion (MBT/AD), Autoclaving or Mechanical Heat Treatment (MHT)

using steam or plasma arc gasification PGP, or combinations of these treatments. Erection of incinerators block out the development and introduction of other emerging technologies. A UK government WRAP report, August 2008 found that in the UK median incinerator costs per ton were generally higher than those for MBT treatments by £ 18 per metric ton; and £ 27 per metric ton most for modern (post 2000) incinerators.

- Building and operating an incinerator requires long contract periods to recover initial investment costs, causing a long term lock in. Incinerator lifetimes normally range from 25 – 30 years.
- Incinerators produce fine particles in the furnace. Even with modern particle filtering of the flue gases, a small part of these is emitted to the atmosphere. $PM_{2.5}$ is not separately regulated in the European Waste Incineration Directive, even though they are repeatedly correlated spatially to infant mortality in the UK (M.Ryan's ONS data based maps around the EfW/CHP waste incinerators at Edmonton, Coventry, Chineham, Kirklees and Sheffield). Under WID there is no requirement to monitor stack top or downwind incinerator $PM_{2.5}$ levels. Several European doctors associations (including cross discipline experts such as physicians, environmental chemists and toxicologists) in June 2008 representing over 33,000 doctors wrote a keynote statement directly to the European Parliament citing widespread concerns on incinerator particle emissions and the absence of specific fine and ultrafine particle size monitoring or in depth industry/government epidemiological studies of these minute and invisible incinerator particle size emissions.
- Local communities are often opposed to the idea of locating incinerators in their vicinity. (The Not In My Back Yard phenomenon). Studies in Andover, Massachusetts strongly correlated 10% property devaluations with close incinerator proximity.
- Prevention, waste minimisation, reuse and recycling of waste should all be preferred to incineration according to the waste hierarchy. Supporters of zero waste consider incinerators and other waste treatment technologies as barriers to recycling and separation beyond particular levels, and that waste resources are sacrificed for energy producion.
- A recent Eunomia report found that under some circumstances and assumptions, incinera-tion causes less CO_2 reduction than other emerging EfW and CHP technology combinations for treating residual mixed waste. The authors found that CHP incinerator technology without waste recycling ranked 19 out of 24 combinations (where all alternatives to incineration were combined with advanced waste recycling plants); being 228% less efficient than the ranked 1 Advanced MBT maturation technology; or 211% less efficient than plasma gasification/autoclaving combination ranked 2.
- Some incinerators are visually undesirable. In many countries they require a visually intrusive chimney stack.

Trends in Incinerator Use

The history of municipal solid waste (MSW) incineration is linked intimately to the history of landfills and other waste treatment technology. The merits of incineration are inevitably judged in relation to the alternatives available. Since the 1970s, recycling and other prevention measures have changed the context for such judgements. Since the 1990s alternative waste treatment technologies have been maturing and becoming viable.

Incineration is a key process in the treatment of hazardous wastes and clinical wastes. It is often imperative that medical waste be subjected to the high temperatures of incineration to destroy pathogens and toxic contamination it contains.

Small Incinerator Units

Small scale incinerators exist for special purposes. For example, the small scale incinerators are aimed for hygienically safe destruction of medical waste in developing countries. Simple, mobile incinerators are becoming more widely used in developing countries where the threat of avian influenza is high. Small incinerators can be quickly deployed to remote areas where an outbreak has occurred to dispose of infected animals quickly and without the risk of cross contamination.

Recycling Methods

PVC, LDPE, PP, and PS (see resin identification code) are also recyclable, although these are not commonly collected. These items are usually composed of a single type of material, making them relatively easy to recycle into new products. The recycling of complex products (such as computers and electronic equipment) is more difficult, due to the additional dismantling and separation required.

Composting is the purposeful biodegradation of organic matter, such as yard and food waste. The decomposition is performed by microorganisms, mostly bacteria, but also yeasts and fungi. In low temperature phases a number of macroorganisms, such as springtails, ants, nematodes, isopods and earthworms also contribute to the process, as well as soldier fly, fruit flies and fungus gnats. There are a wide range of organisms in the decomposer community.

- A *biodegradable* material is capable of being completely broken down under the action of microorganisms into carbon dioxide, water and biomass. It may take a very long time for some material to biodegrade depending on its environment *(e.g.,* wood in an arid area versus paper in water), but it ultimately breaks down completely. Many contaminating materials not dealt with in common composting are in fact "biodegradeable", and may be dealt with via bioremediation, or other special composting approaches.
- A *compostable* material biodegrades substantially under specific composting conditions. It is metabolized by the microorganisms, being incorporated into the organisms or converted into humus. The size of the material is a factor in determining compostability, and mechanical particle size reduction can speed the process. Large pieces of hardwood may not be compostable under a specific set of composting conditions, whereas sawdust of the same type of wood may be. Some biodegradable materials are only compostable under very specific conditions, usually with an industrial process.

Importance

Composting upcycles organic kitchen and yard waste and manures into an extremely useful humus like, soil end product, permitting the return of vital organic matter, nutrients, and particularly bacteria, that are vital to plant nutrition to the soil. Managed aerobic composting arranges environmental conditions so they are optimal for the natural processes to take place. There is a popular expression: "compost happens", but it is helpful to engineer the best possible circumstances for large amounts of organic waste to decompose quickly and efficiently, with the greatest conservation of useful nutrients and mass. Uncontrolled composting is when compost "happens", and although that may be functional in some circumstances, as with forest floor detritus, a neglected heap of kitchen and yard wastes will more likely result in "smells happen", or "rodents happen" long before useful compost does.

Long used in subsistence farming and home gardening for creating garden ready soil, composting is becoming increasingly important and better understood as a tool for reducing municipal solid waste, and reducing the amount of green waste going into landfills. The decomposition of organic material sent to landfills is a principal cause of methane, an important greenhouse gas, making reduction of organic waste being landfilled a key element in the fight against climate change. In suburban and rural areas, much of the organic waste could be removed from the waste stream by promoting home composting, where consumers compost their yard waste and kitchen scraps on their own land, regardless of whether the material is ever actively re-used as "soil". In urban areas with dwellings predominantly lacking individual yard space, there are indoor small scale composting alternatives, such as vermicomposting and bokashi composting.

Composting organisms require four equally important things to work effectively:

- Carbon ("C" or carbohydrates), for energy - the microbial oxidation of carbon produces the heat.
  - High carbon materials tend to be brown and dry.
- Nitrogen ("N" or protein), to grow and reproduce more organisms to oxidize the carbon.
  - High nitrogen materials tend to be green (or colorful, like fruits and vegetables) and wet.
- Oxygen, for oxidizing the carbon, the decomposition process.
- Water, in the right amounts to maintain activity without causing anaerobic conditions.

Certain ratios of these elements will provide beneficial bacteria with the nutrients to work at a rate that will heat up the pile. In that process much water will be released as vapor ("steam"), and the oxygen will be quickly depleted, explaining the need to actively manage the pile. The hotter the pile gets, the more often added air and water is necessary; the air/water balance is critical to maintaining high temperatures until the materials are broken down. At the same time, too much air or water also slows the process, as does too much C (or too little N).

The most efficient composting occurs with a C:N mix of about 30 to 1. All organics have both carbon and nitrogen, but amounts vary widely, with characteristics noted above (dry/wet, brown/green). Fresh grass clippings have an average ratio of about 15 to 1 and dry autumn leaves about 50 to 1 depending on species. Mixing equal parts by volume approximates the ideal C:N range. Few individual situations will provide the ideal mix of materials at any point in time in this respect, home composting is like horseshoes, perfect is great, but close still works. Observation of amounts, and consideration of different materials as a pile is built over time, can quickly achieve a workable technique for the individual situation.

Ingredients that are primarily carbon include:

- Dry, straw type material, such as cereal straws and corn stalks
- Dry leaves (best shredded, as with a rotary mower, to prevent matting)
- Wood, as coarse or fine (may compact) sawdust, or ground wood waste.

Paper and card board, both unprinted and printed are not recommended as both the inks and paper contain materials such as pigments, clays, binders, etc, that are not biodegradable. While these insoluble ingredients are not toxic, they will not readily break down as other biodegradable materials. In addition, paper will decompose very slowly interfering with the composting process.

Ingredients with relatively high nitrogen content include:

- Green plant material, like crop residues, new shoots, hay (especially alfalfa), grass clippings and weeds.
- Manure from poultry, humans, pets, and herbivorous animals such as horses, cows and llamas.
- Kitchen waste, fruit and vegetable cooked waste and trimmings, juicing-pulp residue, tea and coffee grounds, meat, bones, eggs.

For "back yard" composting, mixing the materials as they are added increases the rate of decomposition, as does reduced particle size *(i.e.,* chopped, shredded), or materials can be added in alternating layers, about 15 centimetres (6 in) thick. Keeping a carbon "cache" handy to the pile for covering and mixing with fresh wet additions (lawn clippings, kitchen scraps) is simplest. Special additions or activators are not necessary, although some sprinklings of good garden loam as a first pile is built will aid more rapid working by *inoculating* it with beneficial soil bacteria, and some of the material from the first finished batch can be used in the subsequent mixes. Adding soil also provides grit to help earthworms digest, as well as providing particles for finished compost to aggregate with to create humus. Agricultural lime is not necessary the bacteria prefer a slightly acidic pH, and their processing invariably results in a nearneutral product. Seaweed meal, rock dust or rock flour, and other trace element amendments are best added to the finished compost, or directly to the garden.

Approaches

Active (aerobic):

An active compost heap, steaming on a cool morning. The heap is kept warm by the exothermic action of the bacteria as they decompose the organic matter.

Hot thermophilic composting is essential with some materials, such as meat and other animal products, dairy products, eggs, grease, cooking oil, manure of non-herbivores, and residuals from the treatment of wastewater, in order to kill pathogens; but these materials are not generally recommended in home composting because of the likelihood of creating odors and attracting rodents. Human waste can be composted by industrial methods as well as composting toilets. When high temperatures are reached, the resulting compost can be safely used for agricultural or horticultural purposes, providing local health regulations are met. Humanure fertilizer (as opposed to night soil) is used throughout the developing world and is becoming more accepted as a garden amendment in the developed world.

Hot, aerobic composting is conducted at close to the ideal conditions noted above, allowing thermophilic bacteria to thrive. These aerobic bacteria break down material faster, producing less odor, fewer pathogens, and less greenhouse gas than cool, uncontrolled, or accidental anaerobic methods. Commercial scale composting operations actively control the composting conditions (C:N ratio, moisture level and air), usually in a closed environment (in-vessel composting, tunnel composting or aerated static pile composting), where air is fan forced through the mass, and moisture added with sprayers, or conserved via the enclosure, with computer monitored probes detecting conditions.

In Thailand an aerated static pile system is in use by farmer groups at over 400 sites. The process needs only 30 days to finish without manual turning, with 10 metric tons of compost produced per month. A 38 centimetres (15 in) squirrelcage blower with 2.2 kilowatts (3.0 hp) motor is used to force air through 10 covered static piles of compost twice a day. The raw materials consist of agricultural wastes and animal manure in the ratio of 3:1 by volume.

High temperatures destroy insects, larvae, and weed seeds, but no compost will be totally sterilized by high temperatures alone. In a hot compost where the temperature exceeds 55 °C (131 °F) for several days, the ability of most organisms to survive is compromised, and there are temperature standards set by various regulating authorities for commercial products Nevertheless, many organisms in nature can survive extreme temperatures, including extremophiles such as *Thermus thermophilus* which play an important role in thermogenic composting, as well as pathogens such as Clostridium. The necessary second stage of hot composting is maturation, a period allowing the dissipation of any phyto-toxins remaining from the process or contaminating ingredients (*e.g.*, chemical residues), and achieving a state of nutrient stability (low C:N ratio) that will not have an impact on Nitrogen availability in the receiving soil.

For backyard composters, carbon and nitrogen ratios in various ingredients and the calculations required to get the ideal mixture can be intimidating, so rules of thumb exist for approximating it by ingredient types and condition, as mentioned in Materials above. If the pile is built in a short period, and has a good mix of materials (C:N) and a coarse structure, with about 50% moisture ("like a squeezed out sponge"), the temperature should rise within days to as high as 60 °C (140 °F). When the temperature begins to fall, more air is needed, usually added by turning the pile or using an agitating tool, and moisture may be needed at the same time. Turning or other aeration is usually needed about every 6–10 days to maintain the highest heat levels until the material is fairly uniformly broken down to unrecognizability, and temperatures no longer rebound. A pile that has been maintained at peak temperatures may be ready for maturing in as little as 30 days. Another 30–60 days maturing should suffice to allow passing the "germination test".

To achieve thermophilic decomposition, a compost bin is best about 1 cubic metre (1.3 cu yd), or 1 metre (3 ft) wide, 1 metre (3 ft) tall, and as long as desired for windrow composting. This provides enough insulating mass to build up heat but also allows oxygen infiltration. The center of the pile heats up the most, so regular turning/mixing is needed for insuring all material spends some time in the hottest area. When turning the pile results in no further temperature rise, the active aerobic phase is complete, and the mass may be turned out to a maturing pile. When the matured material has a dark brown crumbly appearance and the smell of rich damp earth, it is ready to use.

The natural sequence of the decomposition community involved will be:

- 0–15 °C (32–59 °F): Psychrophiles predominate, beginning the heating process as they multiply.
- 15–40 °C (60–104 °F): Mesophiles take over, psychrophiles die off or are relegated to the borders.
- 40–70 °C (105–160 °F): Thermophiles work at their peak, including consuming many other bacteria.

At the lower temperatures and around the borders, there will also be various fungal activity, as well as larger organisms, getting their share — a very dry, cooler pile may be attractive to ants, and gastropods may visit very wet piles. As the temperature returns to ambient at the end of the process, the sequence reverses, including new organisms that prefer the more degraded materials. Added heat and pile insulation may be useful in the coldest weather, but is not ordinarily necessary, and is not desirable if it interferes with aeration or natural convective evaporation. Keeping the top dry and burying fresh additions in the center of a pile will be effective during winter conditions until heating resumes in spring.

Passive (anaerobic)

Cool or *ambient* temperature composting, when the level of physical intervention is minimal, usually results in temperatures never reaching above 30 °C (86 °F). It is slower but effective, and is the more common type of composting in domestic gardening. Such composting systems may be in open or closed containers of wood or plastic, or in open exposed piles. Kitchen scraps are put in the garden compost bin and left untended. This scrap bin can have a very high water content which reduces aeration, and may become odorous. To improve drainage and airflow, and reduce odor, carbon-rich materials, or 'browns', such as wood chips, shredded bark, leaves, or twigs may be added to mix and cover each wet addition, or holes made occasionally in the pile. The amount of attention may vary from none through occasional to "regular".

In Nature

An unusual form of composting in nature is seen in the case of the mound-builders (megapodes) of the Australasian region. These Megapodes are fowl-sized birds famous for building nests in the form of compost heaps containing organic litter, in which they incubate their eggs. The male birds work assiduously to maintain the correct incubation temperatures, by adding and removing litter from the compost pile.

Forest floor detritus is the natural fertilizing material in temperate forests. Soil dwelling organisms slowly decompose the continuously deposited litter from below, returning the nutrients to the soil for bacterial conversion to forms useful to the native vegation. This is a natural form of mulching or sheet composting.

Vermicomposting

Worm composting or vermicomposting is a method using Red Wiggler worms in a container to process kitchen waste. Moistened high carbon bedding such as shredded paper is used as a base to which the food waste is added, and the worms and microorganisms convert the materials to rich compost called worm castings, a nutrient and microbially rich material. Worm composting can be done indoors, allowing year round composting, and providing apartment dwellers with a means of recycling food waste.

Industrial Systems

Industrial composting systems are increasingly being installed as a waste management alternative to landfills, along with other advanced waste processing systems. Mechanical sorting of mixed waste streams combined with anaerobic digestion or in-vessel composting, is called mechanical biological treatment, increasingly used in developed countries due to regulations controlling the amount of organic matter allowed in landfills. Treating biodegradable waste before it enters a landfill reduces global warming from fugitive methane; untreated waste breaks down anaerobically in a landfill, producing landfill gas that contains methane, a potent greenhouse gas.

Large scale composting systems are used by many urban centers around the world. Co-composting is a technique which combines solid waste with de-watered biosolids, although difficulties controlling inert and plastic contamination from Municipal solid waste makes this approach less attractive. The world's largest MSW cocomposter is the Edmonton Composting Facility in Edmonton, Alberta, Canada, which turns 220,000 tonnes of residential solid waste and 22,500 dry tonnes of biosolids per year into 80,000 tonnes of compost. The facility is 38,690 meters^2 (416,500 ft^2), equivalent to $4^{1/2}$ Canadian football fields, and the operating structure is the largest stainless steel building in North America, the size of 14 NHL rinks.

Agriculture

Compost is an important source of nutrients commonly used in modern agriculture. Through steaming, compost can be sanitized and prepared for further use.

Home composting is the small scale domestic application of the principles of sustainable, biodegradable waste management, *i.e.*, composting. The general principles involved in composting apply to any scale, from "backyard" to industrial, but the techniques will vary for each with the size of the waste stream, the cost, amount of effort, and the organization required. Industrial scale systems are invariably capital and/or labor intensive, but on the home or small farm scale, composting can be managed to require varying outlay of capital and labor. For the small urban household, an indoor Bokashi or worm bin may suffice, for a suburban property with a larger yard and a food garden, a bin system would be preferable, while in small farm settings, a seasonal windrow system might be called for.

Purpose

A home garden can be primarily fertilized with compost

Past practices of biodegradable waste *disposal* not only remove valuable nutrients from the local biocycle, but when buried in a landfill, the anaerobic decomposition that results contributes to ground water contamination and uncontrolled methane generation, increasing the occurrence of greenhouse gas emissions. Organic waste contains valuable nutrients that can contribute to the soil health in the immediate area of their production if recycled. (see Compost.) *Brown waste* such as paper, cardboard, and dry plant material like leaves, are "captured carbon", and *green waste* such as fresh plant residues and fruit and vegetable scrap contain much of the nitrogen used to generate their growth, as well as other essential macro and micronutrients, such as boron, copper, or iodine, lacking in some locales, so otherwise replaced with purchased commercial amendments. It also factors for minimum ammonium losses during composting of the residues of trimming. From the compost at home, it can help improve your soil, prevent plant disease, and reduce green house gases. The compost can also help soften plant material, increase the water holding capacity of the plant, and convert the ammonia in the plant into protein.

Container composting is a common style of small scale composting, using any of a wide variety of plastic, wood, masonry, or wire screen containers. Vented or closed sided compost bins each have proponents for the effect on air circulation and heat loss or retention in the compost. The *Indore* method developed by Sir Albert Howard, as well as the Shewell Cooper method favour spaced slats, while the *New Zealand Box* method advocates the use of closed sides. Combinations of the two are also used.

A **compost bin** is the container used to make compost. Commercially these bins may be made of hard plastic, and is commonly cylindrical in shape, although a variety of shapes and sizes are available. The manufactured bins are commonly found in urban and suburban areas, and in some districts of North America and the U.K., local authorities subsidise the cost to encourage public waste stream reduction. They also may be user made out of wood, or whatever salvaged, recycled materials are at hand, particularly in rural areas or where larger amounts of yard and garden waste are generated. In some areas, community bin sites are provided. Compost bins can be as simple as a square slatted wood enclosure or as sophisticated as a tumbler, which allows for the "pile" to be turned for aeration purposes.

Sheet composting is the process of placing the organic matter for the compost directly onto the soil as a mulch and letting it decay there, rather than in a heap or container. One or more layers of organic material are spread over the growing area, watered thoroughly, and left to decompose until planting time. The next season's garden is usually planted without tilling the compost into the soil.

More layers of organic material are placed as the bottom layers decompose thoroughly. This method is occasionally called "lasagna gardening" because of the layered structure.

Proponents of this system argue that sheet composting causes fewer nutrients to be lost through leaching than heap methods, also that fresh organic matter rather than decayed, provides a slower release of minerals when applied. It is also said that, in the long term, sheet composting leads to higher nitrogen levels in the soil, as some may be lost by leaching or vaporisation in a traditional heap.

Field composting is a large-scale composting method used in no till farming. Commonly, this is achieved by growing a 'green manure' cover crop such as mustard, alfalfa, or buckwheat, which is cut, preferably before seed setting, and left on the field to decompose. The next crop is sown without tilling the previous crop into the field. To avoid temporary nitrogen depletion, leguminous green manure crops such as lupin, winter tares, field beans, or clover, which are able to fix their own nitrogen supply in root nodules may be used. The nitrogen is released as the plants decay.

Anaerobic digestion is a series of processes in which microorganisms break down biodegradable material in the absence of oxygen and is widely used to treat wastewater. As part of an integrated waste management system, anaerobic digestion reduces the emission of landfill gas into the atmosphere. Anaerobic digestion is widely used as a renewable energy source because the process produces a methane and carbon dioxide rich biogas suitable for energy production helping replace fossil fuels. Also, the nutrient rich digestate can be used as fertiliser.

The digestion process begins with bacterial hydrolysis of the input materials in order to break down insoluble organic polymers such as carbohydrates and make them available for other bacteria. Acidogenic bacteria then convert the sugars and amino acids into carbon dioxide, hydrogen, ammonia, and organic acids. Acetogenic bacteria then convert these resulting organic acids into acetic acid, along with additional ammonia, hydrogen, and carbon dioxide. Methanogens, finally are able to convert these products to methane and carbon dioxide.

Previously, the technical expertise required to maintain anaerobic digesters coupled with high capital costs and low process efficiencies had limited the level of its industrial application as a waste treatment technology. Anaerobic digestion facilities have, however, been recognised by the United Nations Development Programme as one of the most useful decentralised sources of energy supply, as they are less capital intensive than large power plants.

Applications

Anaerobic digestion is particularly suited to wet organic material and is commonly used for effluent and sewage treatment. Anaerobic digestion is a simple process that can greatly reduce the amount of organic matter which might otherwise be destined to be landfilled or burnt in an incinerator.

Almost any organic material can be processed with anaerobic digestion. This includes biodegradable waste materials such as waste paper, grass clippings, leftover food, sewage and animal waste. The exception to this is woody wastes that are largely unaffected by digestion as most anaerobes are unable to degrade lignin. The exception being xylophalgeous anaerobes (lignin consumers), as used in the process for organic breakdown of cellulosic material by a cellulosic ethanol start up company in the U.S. Anaerobic digesters can also be fed with specially grown energy crops such as silage for dedicated biogas production. In Germany and continental Europe these facilities are referred to as *biogas plants. A co-digestion* or *co-fermentation* plant is typically an agricultural anaerobic digester that accepts two or more input materials for simultaneous digestion.

In developing countries simple home and farm based anaerobic digestion systems offer the potential for cheap, low-cost energy for cooking and lighting. Anaerobic digestion facilities have been recognized by the United Nations Development Programme as one of the most useful decentralized sources of energy supply. From 1975, China (See Bioenergy in China) and India have both had large government backed schemes for adaptation of small biogas plants for use in the house hold for cooking and lighting. Presently, projects for anaerobic digestion in the developing world can gain financial support through the United Nations Clean Development Mechanism if they are able to show they provide reduced carbon emissions.

Pressure from environmentally related legislation on solid waste disposal methods in developed countries has increased the application of anaerobic digestion as a process for reducing waste volumes and generating useful by products. Anaerobic digestion may either be used to process the source separated fraction of municipal waste, or alternatively combined with mechanical sorting systems, to process residual mixed municipal waste. These facilities are called mechanical biological treatment plants.

Utilising anaerobic digestion technologies can help to reduce the emission of greenhouse gases in a number of key ways:

- Replacement of fossil fuels
- Reducing methane emission from landfills
- Displacing industrially-produced chemical fertilizers
- Reducing vehicle movements
- Reducing electrical grid transportation losses.

Methane and power produced in anaerobic digestion facilities can be utilized to replace energy derived from fossil fuels, and hence reduce emissions of greenhouse gases. This is due to the fact that the carbon in biodegradable material is part of a carbon cycle. The carbon released into the atmosphere from the combustion of biogas has been removed by plants in order for them to grow in the recent past. This can have occurred within the last decade, but more typically within the last growing season. If the plants are regrown, taking the carbon out of the atmosphere once more, the system will be carbon neutral. This contrasts to carbon in fossil fuels that has been sequestered in the earth for many millions of years, the combustion of which increases the overall levels of carbon dioxide in the atmosphere.

If the putrescible waste processed in anaerobic digesters was disposed of in a landfill, it would break down naturally and often anaerobically. In this case the gas will eventually escape into the atmosphere. As methane is about twenty times more potent as a greenhouse gas as carbon dioxide this has significant negative environmental effects.

Digestate liquor can be used as a fertilizer supplying vital nutrients to soils. The solid, fibrous component of digestate can be used as a soil conditioner. The liquor can be used as a substitute for chemical fertilizers which require large amounts of energy to produce and transport. The use of manufactured fertilizers is therefore more carbon intensive than the use of anaerobic digestate fertiliser. This solid digestate can be used to boost the organic content of soils. There are some countries, such as Spain where there are many organically depleted soils, and here the markets for the digestate can be just as important as the biogas.

In countries that collect household waste, the utilization of local anaerobic digestion facilities can help to reduce the amount of waste that requires transportation to centralized landfill sites or incineration facilities. This reduced burden on transportation has and will reduce carbon emissions from the collection vehicles. If localized anaerobic digestion facilities are embedded within an electrical distribution network, they can help reduce the electrical losses that are associated with transporting electricity over a national grid.

In Oakland, California at the East Bay Municipal Utility District's (EBMUD) Main Wastewater Treatment Plant(MWWTP), food waste is currently codigested with primary and secondary municipal wastewater solids and other high strength wastes. Compared to municipal wastewater solid digestion, food waste digestion has many benefits. Anaerobic digestion of food waste pulp from the EBMUD food waste process provides a higher normalized energy benefit, compared to municipal wastewater solids:

- 730 to 1,300 kWh per dry ton of food waste applied.
- 560 to 940 kWh per dry ton of municipal wastewater solids applied.

Power Generation

Biogas from sewage works is sometimes used to run a gas engine to produce electrical power; some or all of which can be used to run the sewage works. Some waste heat from the engine is then used to heat the digester. It turns out that the waste heat is generally enough to heat the digester to the required temperatures. The power potential from sewage works is limited — in the UK there are about 80 MW total of such generation, with potential to increase to 150 MW, which is insignificant compared to the average power demand in the UK of about 35,000 MW. The scope for biogas generation from non-sewage waste biological matter — energy crops, food waste, abattoir waste etc., is much higher, estimated to be capable of about 3,000 MW (needs ref). Farm biogas plants using animal waste and energy crops are expected to contribute to reducing CO_2 emissions and strengthen the grid while providing UK farmers with additional revenues.

Gas engine manufacturers include Schmitt Enertec, GE Jenbacher or Caterpillar Inc, and more.

The process

There are a number of microorganisms that are involved in the process of anaerobic digestion including acetic acid-forming bacteria (acetogens) and methane-forming archaea (methanogens). These organisms feed upon the initial feedstock, which undergoes a number of different processes converting it to intermediate molecules including sugars, hydrogen and acetic acid before finally being converted to biogas.

Different species of bacteria are able to survive at different temperature ranges. Ones living optimally at temperatures between 35–40 °C are called mesophiles or mesophilic bacteria. Some of the bacteria can survive at the hotter and more hostile conditions of 55–60 °C, these are called thermophiles or thermophilic bacteria. Methanogens come from the primitive group of archaea. This family includes species that can grow in the hostile conditions of hydrothermal vents. These species are more resistant to heat and can therefore operate at thermophilic temperatures, a property that is unique to bacterial families.

As with aerobic systems the bacteria in anaerobic systems the growing and reproducing microorganisms within them require a source of elemental oxygen to survive.

In an anaerobic system there is an absence of gaseous oxygen. Gaseous oxygen is prevented from entering the system through physical containment in sealed tanks. Anaerobes access oxygen from sources other than the surrounding air. The oxygen source for these microorganisms can be the organic material itself or alternatively may be supplied by inorganic oxides from within the input material. When the oxygen source in an anaerobic system is derived from the organic material itself, then the 'intermediate' end products are primarily alcohols, aldehydes, and organic acids plus carbon dioxide. In the presence of specialised methanogens, the intermediates are converted to the 'final' end products of methane, carbon dioxide with trace levels of hydrogen sulfide. In an anaerobic system the majority of the chemical energy contained within the starting material is released by methanogenic bacteria as methane.

Populations of anaerobic microorganisms typically take a significant period of time to establish themselves to be fully effective. It is therefore common practice to introduce anaerobic microorganisms from materials with existing populations. This process is called 'seeding' the digesters and typically takes place with the addition of sewage sludge or cattle slurry.

Stages

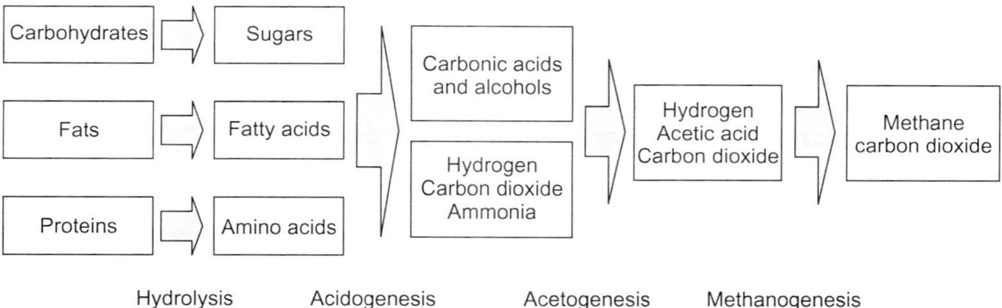

The key process stages of anaerobic digestion

There are four key biological and chemical stages of anaerobic digestion:

1. Hydrolysis
2. Acidogenesis
3. Acetogenesis
4. Methanogenesis

In most cases biomass is made up of large organic polymers. In order for the bacteria in anaerobic digesters to access the energy potential of the material, these chains must first be broken down into their smaller constituent parts. These constituent parts or monomers such as sugars are readily available by other bacteria. The process of breaking these chains and dissolving the smaller molecules into solution is called hydrolysis. Therefore hydrolysis of these high molecular weight polymeric components is the necessary first step in anaerobic digestion. Through hydrolysis the complex organic molecules are broken down into simple sugars, amino acids, and fatty acids.

Acetate and hydrogen produced in the first stages can be used directly by methanogens. Other molecules such as Volatile Fatty Acids (VFA's) with a chain length that is greater than acetate must first be catabolised into compounds that can be directly utilised by methanogens.

The biological process of acidogenesis is where there is further breakdown of the remaining components by acidogenic (fermentative) bacteria. Here VFAs are created along with ammonia, carbon dioxide and hydrogen sulfide as well as other by products. The process of acidogenesis is similar to the way that milk sours.

The third stage anaerobic digestion is acetogenesis. Here simple molecules created through the acidogenesis phase are further digested by acetogens to produce largely acetic acid as well as carbon dioxide and hydrogen.

The terminal stage of anaerobic digestion is the biological process of methanogenesis. Here methanogens utilise the intermediate products of the preceding stages and convert them into methane, carbon dioxide and water. It is these components that makes up the majority of the biogas emitted from the system. Methanogenesis is sensitive to both high and low pHs and occurs between pH 6.5 and pH 8. The remaining, non-digestable material which the microbes cannot feed upon, along with any dead bacterial remains constitutes the digestate.

A simplified generic chemical equation for the overall processes outlined above is as follows:

$$C_6H_{12}O_6 \rightarrow 3CO_2 + 3CH_4$$

Feedstock

The most important initial issue when considering the application of anaerobic digestion systems is the feedstock to the process. Digesters typically can accept any biodegradable material, however if biogas production is the aim, the level of putrescibility is the key factor in its successful application. The more putrescible the material the higher the gas yields possible from the system.

Substrate composition is a major factor in determining the methane yield and methane production rates from the digestion of biomass. Techniques are available to determine the compositional characteristics of the feedstock, whilst parameters such as solids, elemental and organic analyses are important for digester design and operation.

Anaerobes can breakdown material to varying degrees of success from readily in the case of short chain hydrocarbons such as sugars, to over longer periods of time in the case of cellulose and hemicellulose. Anaerobic microorganisms are unable to break down long chain woody molecules such as lignin. Anaerobic digesters were originally designed for operation using sewage sludge and manures. Sewage and manure are not, however, the material with the most potential for anaerobic digestion as the biodegradable material has already had the energy content taken out by the animal that produced it.

A second consideration related to the feedstock will be moisture content. The wetter the material the more suitable it will be to handling with standard pumps instead of energy intensive concrete pumps and physical means of movement. Also the wetter the material, the more volume and area it takes up relative to the levels of gas that are produced. The moisture content of the target feedstock will also affect what type of system is applied to its treatment. In order to use a high solids anaerobic digester for dilute feedstocks, bulking agents such as compost should be applied to increase the solid content of the input material. Another key consideration is the carbon:nitrogen ratio of the input material. This ratio is the balance of food a microbe requires in order to grow. The optimal C:N ratio for the 'food' a microbe is 20–30:1. Excess N can lead to ammonia inhibition of digestion.

The level of contamination of the feedstock material is a key consideration. If the feedstock to the digesters has significant levels of physical contaminants such as plastic, glass or metals then preprocessing will be required in order for the material to be used. If it is not removed then the digesters can be blocked and will not function efficiently. It is with this logic in mind that mechanical biological treatment plants are designed. The higher the level of pretreatment a feedstock requires, the more processing machinery will be required and hence the project will have higher capital costs.

After sorting or screening to remove any physical contaminants, such as metals and plastics, from the feedstock the material is often shredded, minced and mechanically or hydraulically pulped to increase the surface area available to microbes in the digesters and hence increase the speed of digestion. The feedstock material is then fed into the airtight digester where anaerabic treatment takes place.

Configurations

Anaerobic digesters can be designed and engineered to operate using a number of different process configurations:

- Batch or continuous
- Temperature: Mesophilic or thermophilic
- Solids content: High solids or low solids
- Complexity: Single stage or multistage.

Batch or continuous

A batch system is the simplest form of digestion. Biomass is added to the reactor at the start of the process in a batch and is sealed for the duration of the process. Batch reactors suffer from odour issues that can be a severe problem when they are emptied. Typically biogas production will be formed with a normal distribution pattern over time. The operator can use this fact to determine when they believe the process of digestion of the organic matter has completed. As the batch digestion is simple and requires less equipment and lower levels of design work it is typically a cheaper form of digestion.

In continuous digestion processes organic matter is constantly or added in stages to the reactor. Here the end products are constantly or periodically removed, resulting in constant production of biogas. Examples of this form of anaerobic digestion include, Continuous Stirred-Tank Reactors (CSTRs), Upflow Anaerobic Sludge Blanket (UASB), Expanded Granular Sludge Bed (EGSB) and Internal Circulation reactors (IC).

Temperature

There are two conventional operational temperature levels for anaerobic digesters, which are determined by the species of methanogens in the digesters:

- *Mesophilic* which takes place optimally around 37°–41 °C or at ambient temperatures between 20°–45 °C where mesophiles are the primary microorganism present.
- *Thermophilic* which takes place optimally around 50°–52° at elevated temperatures up to 70 °C where thermophiles are the primary microorganisms present.

A limit case has been reach in Bolivia, with anaerobic digestion in temperature working conditions less than 10 °C. The anaerobic process is very slow, taking more than three times the mesophilic normal time process.

There are a greater number of species of mesophiles than thermophiles. These bacteria are also more tolerant to changes in environmental conditions than thermophiles. Mesophilic systems are therefore considered to be more stable than thermophilic digestion systems.

As mentioned above, thermophilic digestion systems are considered to be less stable, however the increased temperatures facilitate faster reaction rates and hence faster gas yields. Operation at higher temperatures facilitates greater sterilisation of the end digestate. In countries where legislation, such as the Animal By Products Regulations in the European Union, requires end products to meet certain levels of reduction in the amount of bacteria in the output material, this may be a benefit.

A drawback of operating at thermophilic temperatures is that more heat energy input is required to achieve the correct operational temperatures. This increase in energy may not be out-weighed by the increase in the outputs of biogas from the systems. It is therefore important to consider an energy balance for these systems.

Solids

Typically there are two different operational parameters associated with the solids content of the feedstock to the digesters:

- High-solids
- Low-solids

Digesters can either be designed to operate in a high solids content, with a Total Suspended Solids (TSS) concentration greater than ~20%, or a low solids concentration less than ~15%.

High-solids digesters process a thick slurry that requires more energy input to move and process the feedstock. The thickness of the material may also lead to associated problems with abrasion. High-solids digesters will typically have a lower land requirement due to the lower volumes associated with the moisture.

Low-solids digesters can transport material through the system using standard pumps that require significantly lower energy input. Low-solids digesters require a larger amount of land than high-solids due to the increase volumes associated with the increased liquid-to-feedstock ratio of the digesters. There are benefits associated with operation in a liquid environment as it enables more thorough circulation of materials and contact between the bacteria and their food. This enables the bacteria to more readily access the substances they are feeding off and increases the speed of gas yields.

Number of stages

Digestion systems can be configured with different levels of complexity:

- One-stage or single-stage
- Two-stage or multi–stage

A single-stage digestion system is one in which all of the biological reactions occur within a single sealed reactor or holding tank. Utilising a single–stage reduces construction costs, however facilitates less control of the reactions occurring within the system. Acidogenic bacteria, through the production of acids, reduce the pH of the tank. Methanogenic bacteria, as outlined earlier, operate in a strictly defined pH range. Therefore the biological reactions of the different species in a single stage reactor can be in direct competition with each other. Another one-stage reaction system is an anaerobic lagoon. These lagoons are pond like earthen basins used for the treatment and long-term storage of manures. Here the anaerobic reactions are contained within the natural anaerobic sludge contained in the pool.

In a two-stage or multi-stage digestion system different digestion vessels are optimised to bring maximum control over the bacterial communities living within the digesters. Acidogenic bacteria produce organic acids and more quickly grow and reproduce than methanogenic bacteria. Methanogenic bacteria require stable pH and temperature in order to optimise their performance.

Typically hydrolysis, acetogenesis and acidogenesis occur within the first reaction vessel. The organic material is then heated to the required operational temperature (either mesophilic or thermophilic) prior to being pumped into a methanogenic reactor. The initial hydrolysis or acidogenesis tanks prior to the methanogenic reactor can provide a buffer to the rate at which feedstock is added. Some European countries require a degree of elevated heat treatment in order to kill harmful bacteria in the input waste. In this instance there may be a pasteurisation or sterilisation stage prior to digestion or between the two digestion tanks. It should be noted that it is not possible to completely isolate the different reaction phases and often there is some biogas that is produced in the hydrolysis or acidogenesis tanks.

Residence

The residence time in a digester varies with the amount and type of feed material, the configuration of the digestion system and whether it be one-stage or two-stage.

In the case of single-stage thermophilic digestion residence times may be in the region of 14 days, which comparatively to mesophilic digestion is relatively fast. The plug-flow nature of some of these systems will mean that the full degradation of the material may not have been realised in this timescale. In this event digestate exiting the system will be darker in colour and will typically have more odour.

In two-stage mesophilic digestion, residence time may vary between 15 and 40 days.

In the case of mesophilic UASB digestion hydraulic residence times can be (1hour-1day) and solid retention times can be up to 90 days. In this manner the UASB system is able to separate solid an hydraulic retention times with the utilisation of a sludge blanket.

Continuous digesters have mechanical or hydraulic devices, depending on the level of solids in the material, to mix the contents enabling the bacteria and the food to be in contact. They also allow excess material to be continuously extracted to maintain a reasonably constant volume within the digestion tanks.

Products

There are three principal products of anaerobic digestion: Biogas, Digestate and Water.

Biogas

Biogas holder with lightning protection rods and back-up gas flare

Typical composition of biogas:

Matter	%
Methane, CH_4	50–75
Carbon dioxide, CO_2	25–50
Nitrogen, N_2	0–10
Hydrogen, H_2	0–1
Hydrogen sulfide, H_2S	0–3
Oxygen, O_2	0–2

Biogas is the ultimate waste product of the bacteria feeding off the input biodegradable feedstock, and is mostly methane and carbon dioxide, with a small amount hydrogen and trace hydrogen sulfide. (As produced, biogas also contains water vapor, with the fractional water vapor volume a function of biogas temperature.) Most of the biogas is produced during the middle of the digestion, after the bacterial population has grown, and tapers off as the putrescible material is exhausted. The gas is normally stored on top of the digester in an inflatable gas bubble or extracted and stored next to the facility in a gas holder.

The methane in biogas can be burned to produce both heat and electricity, usually with a reciprocating engine or microturbine often in a cogeneration arrangement where the electricity and waste heat generated are used to warm the digesters or to heat buildings. Excess electricity can be sold to suppliers or put into the local grid. Electricity produced by anaerobic digesters is considered to be renewable energy and may attract subsidies. Biogas does not contribute to increasing atmospheric carbon dioxide concentrations because the gas is not released directly into the atmosphere and the carbon dioxide comes from an organic source with a short carbon cycle.

Biogas may require treatment or 'scrubbing' to refine it for use as a fuel. Hydrogen sulfide is a toxic product formed from sulfates in the feedstock and is released as a trace component of the biogas. National environmental enforcement agencies such as the U.S. Environmental Protection Agency or the English and Welsh Environment Agency put strict limits on the levels of gasses containing hydrogen sulfide, and if the levels of hydrogen sulfide in the gas are high, gas scrubbing and cleaning equipment (such as amine gas treating) will be needed to process the biogas to within regionally accepted levels. An alternative method to this is by the addition of ferrous chloride $FeCl_2$ to the digestion tanks in order to inhibit hydrogen sulfide production.

Volatile siloxanes can also contaminate the biogas; such compounds are frequently found in household waste and wastewater. In digestion facilities accepting these materials as a component of the feedstock, low molecular weight siloxanes volatilise into biogas. When this gas is combusted in a gas engine, turbine or boiler, siloxanes are converted into silicon dioxide (SiO_2) which deposits internally in the machine, increasing wear and tear. Practical and cost-effective technologies to remove siloxanes and other biogas contaminants are available at the present time. In certain applications, *in situ* treatment can be used to increase the methane purity by reducing the carbon dioxide content.

In countries such as Switzerland, Germany and Sweden the methane in the biogas may be concentrated in order for it to be used as a vehicle transportation fuel or alternatively input directly into the gas mains. In countries where the driver for the utilisation of anaerobic digestion are renewable electricity subsidies, this route of treatment is less likely as energy is required in this processing stage and reduces the over all levels available to sell.

Digestate

Digestate is the solid remnants of the original input material to the digesters that the microbes cannot use. It also consists of the mineralised remains of the dead bacteria from within the digesters. Digestate can come in three forms; fibrous, liquor or a sludge-based combination of the two fractions. In two-stage systems the different forms of digestate come from different digestion tanks. In single stage digestion systems the two fractions will be combined and if desired separated by further processing.

The second by product (acidogenic digestate) is a stable organic material comprised largely of lignin and cellulose, but also of a variety of mineral components in a matrix of dead bacterial cells; some plastic may be present. The material resembles domestic compost and can be used as compost or to make low grade building products such as fibreboard.

The third by product is a liquid (methanogenic digestate) that is rich in nutrients and can be used as a fertiliser dependent on the quality of the material being digested. Levels of Potentially Toxic Elements (PTEs) should be chemically assessed. This will be dependent upon the quality of the original feedstock. In the case of most clean and source-separated biodegradable waste streams the levels of PTEs will be low. In the case of wastes originating from industry the levels of PTEs may be higher and will need to be taken into consideration when determining a suitable end use for the material.

Digestate typically contains elements such as lignin that cannot be broken down by the anaerobic microorganisms. Also the digestate may contain ammonia that is phytotoxic and will hamper the growth of plants if it is used as a soil improving material. For these two reasons a maturation or composting stage may be employed after digestion. Lignin and other materials are available for degradation by aerobic microorganisms such as fungi helping reduce the overall volume of the material for transport. During this maturation the ammonia will be broken down into nitrates, improving the fertility of the material and making it more suitable as a soil improver. Large composting stages are typically used by dry anaerobic digestion technologies.

Wastewater

The final output from anaerobic digestion systems is water. This water originates both from the moisture content of the original waste that was treated but also includes water produced during the microbial reactions in the digestion systems. This water may be released from the dewatering of the digestate or may be implicitly separate from the digestate.

The wastewater exiting the anaerobic digestion facility will typically have elevated levels of Biochemical Oxygen Demand (BOD) and Chemical Oxygen Demand (COD), these are measures of the reactivity of the effluent and show an ability to pollute. Some of this material is termed 'hard COD' meaning it cannot be accessed by the anaerobic bacteria for conversion into biogas. If this effluent was put directly into watercourses it would negatively affect them by causing eutrophication. As such further treatment of the wastewater is often required. This treatment will typically be an oxidation stage where air is passed through the water in a sequencing batch reactors or reverse osmosis unit.

Energy recovery

The energy content of waste products can be harnessed directly by using them as a direct combustion fuel, or indirectly by processing them into another type of fuel. Recycling through thermal treatment ranges from using waste as a fuel source for cooking or heating, to fuel for boilers to generate steam and electricity in a turbine. Pyrolysis and gasification are two related forms of thermal treatment where waste materials are heated to high temperatures with limited oxygen availability. The process typically occurs in a sealed vessel under high pressure. Pyrolysis of solid waste converts the material into solid, liquid and gas products. The liquid and gas can be burnt to produce energy or refined into other products. The solid residue (char) can be further refined into products such as activated carbon. Gasification and advanced Plasma arc gasification are used to convert organic materials directly into a synthetic gas (syngas) composed of carbon monoxide and hydrogen. The gas is then burnt to produce electricity and steam.

8.4 HAZARDOUS MATERIALS/HAZARDOUS WASTES AND ITS MANAGEMENT

What is Hazardous Material?

Hazardous materials are part of our daily lives. We all make daily contacts with a considerable number of chemicals, many of which are known carcinogens. Others are suspected carcinogens, some occur naturally in the foods we eat; some are man-made products, which have benefits desired by members of our society because of our lifestyles.

What is Hazardous Wastes?

Hazardous wastes are also around us. Some hazardous wastes are the inevitable by-products of an industrialized society. The staggering variety and amounts of chemicals in production is a key of the types and amounts of hazardous wastes generated by our society. Automobiles, televisions, office chairs, and ball point pens all get their shine from manufacturing processes that produce hazardous wastes. These wastes have been called our **Gross National By-products.**

*Hazardous wastes are legally defined the **Resource Conservation and Recovery Act** (RCRA) as a solid, or combination of solid wastes, which because of its quantity or concentration ; or physical, chemical or infectious characteristics may cause, or significantly contribute to an increase in mortality or an increase in serious irreversible or incapacitating reversible illness; or pose a substantial present or potential hazard to human health or the environment when improperly treated, stored, transported, or disposed of, or otherwise managed.*

For a material to be defined by RCRA as a hazardous waste, it must be a solid waste. The RCRA definition of solid waste distinguishes between waste and all other materials. For example, a hazardous material used in a manufacturing process is not covered under RCRA because it is not a waste. A solid waste has been defined by the RCRA;

1. A solid, liquid, semi-solid, or contained gas
2. Any garbage, refuse, sludge, or other waste material not specifically excluded under 40 CFR part 261, s. 261.4 [a]
3. Any other waste material resulting from industrial, commercial, mining, or agricultural operations or community activities, which:

(*a*) Is discarded or accumulated, stored, or physically, chemically, or biologically treated prior to being discarded; or

(*b*) Has served its originally intended use and sometimes is discarded; or

(*c*) Is a manufacturing or mining by-product and sometimes is discarded.

Specifically, a solid waste is determined to be a hazardous waste if; it is not exempted from the regulation; it has certain specific characteristics; it is listed in the regulations as hazardous; or it is broadly classified as a statutory waste because it can cause harm to humans or the environment, regardless of whether it is presently regulated.

The term hazardous waste has a specific legal definition, which is defined in the RCRA law. When hazardous waste is used in a regulatory sense (as is used here) it refers to a specific waste that is regulated by RCRA. Thus, the regulatory term hazardous waste is not interchangeable with any other phrase. For example, the phrases chemical waste and wastes that have chemical hazards carry no regulatory meaning.

Exempted waste

Certain categories or wastes are not subject to RCRA hazardous waste controls. The following examples are waste that are excluded.

1. Waste discharge regulated under the Clean Water Act, *e.g.*, industrial effluents (Note: This exclusion applies only to the actual point source discharge. It does not exclude industrial wastewaters while they are being collected, stored or treated before discharge, not does it exclude sludges that are generated by industrial wastewater treatment.
2. Nuclear wastes regulated under the Atomic Energy Act.
3. Household wastes (only if they do not contain hazardous waste)
4. Coal combustion wastes
5. Soil fertilizers used in agriculture
6. Drilling muds and brines used in oil, gas and geothermal explorations
7. Mining wastes, as from extraction, treatment, and processing of ores and minerals
8. Cement kiln dust
9. Wastes produced by very small quantity generators.

Characteristic waste

A waste is classified as hazardous, and is subject to federal regulations, if it exhibits one or more of the following characteristics: Ignitability, Corrosivity, Reactivity, or Extraction Potential (EP) toxicity

(*a*) **Ignitable.** A solid waste is ignitable if it has a flashpoint of less than 140 °F. (A flashpoint is the lowest temperature to which a material can be heated to give off sufficient vapour to form an ignitable mixture with the air near the surface of the material).

(*b*) **Corrosive.** A solid waste is corrosive if it has a pH of less than or equal to 2.0 (acidic), or equal to or greater than 12.5 (basic); or it corrodes standard steel containers at a rate greater than 6.35 mm (0.250 inch) per year.

(*c*) **Reactive.** A solid waste is reactive if it is normally unstable and readily undergoes violent with water, forms potentially explosive mixture with water; or forms fumes, toxic gases or vapors in sufficient concentrations to endanger human health or the environment.

Extraction Procedure (EP) Toxic Wastes

EP toxic wastes contain certain materials (toxic metals and/or certain organic compounds) that can be released when mixed with slightly acidic water. The presence of toxic materials is determined by using an Extraction Procedure (EP) specified by the EPA, which is designed to simulate the leaching action that occur in landfills. The EPA proposed to amend its hazardous waste identification regulations by modifying the exiting **Extraction Procedure Toxicity Characteristics** (EPTC) an introducing a new leach procedure, the Toxicity Characteristic Leaching Procedure (TCLP).

Listed Wastes

A waste is subject to the RCRA regulations if it is included on lists of wastes considered hazardous by the EPA. There are three following lists:

1. The F lists of hazardous wastes from nonspecific sources (*e.g.*, wastewater treatment sludges from electroplating operations).
2. The K list of hazardous waste from specific sources (*e.g.*, bottom sediment sludge from treatment of wastewaters from wood preserving).
3. Discarded commercial chemical products (the U and P lists) containers, and spill residues. The products on the U list are called toxic wastes (*e.g.*, vinyl chloride). Those on the P list are called acute hazardous wastes (*e.g.*, cyanides), and are subject to more rigorous controls that the other listed hazardous wastes.

Statutory Waste

A statutory waste may be legally hazardous as defined by Congress; however, since the EPA has not yet formulated all the specific rules to control hazardous waste, a statutory waste may not yet be regulated.

The statutory definition of a hazardous waste is particularly important because the legal liability of the generator is long-term, extending through the life of the waste. Some hazardous wastes are presistent, slow to decompose chemically and/or biochemically. As a consequence, for its own protection, an institution may classify and manage a waste as hazardous charcteristics specified in the federal regulations and is not a listed waste. The ration able is that, if the institution did not declare the waste as hazardous, the institution may run the risk of future liability if this originally non-hazardous material were found to cause harm.

Large Quantity Generators of Hazardous Wastes

Manufacturing industries, small businesses, households, hospitals, research and testing laboratories and the agriculture industry are among the generators of hazardous wastes. A survey was conducted in 1981; the data are from larger quantity generators sources generating more than 1000 kilogram per month of hazardous waste.

Table 8.1 Types of Hazardous Waste Generated

Waste Group Generated	*Percent of Generators*
Spent solvents (halogenated/non-halogenated)	51.0
lgnitable wastes	43.4
Corrosive wastes	33.4
Spilled, discarded, or off-specification commercial chemical products or manufacturing chemical intermediates	28.8
EP toxic wastes	27.8
Electro planting and coating wastewater treatment sludges and cyanidebearing solutions and sludges	16.4
Statutory hazardous wastes (*i.e.*, not listed or regulated by EPA or state)	12.2
Listed industry waste from specific sources	10.2
Acutely hazardous wastes	10.2
Reactive wastes	7.1

Where Do the Hazardous Wastes Go ?

About 96 percent of the total hazardous wastes produced (about 255 million metric tons in 1983) are managed on-site. In many cases, industry has chosen to install specialized treatment facilities to handle the hazardous waste. The other four percent (about 10 million metric tons in 1983) goes off-site either for storage, treatment of disposal.

Who are the Small Quantity Generators (SQGs) ?

An office of solid waste survey estimated that there were more than 650,000 businesses that generate hazardous waste in the U.S. Of this number, more than 100,000 produce over 2.20 lacs lbs., of hazardous waste per month.

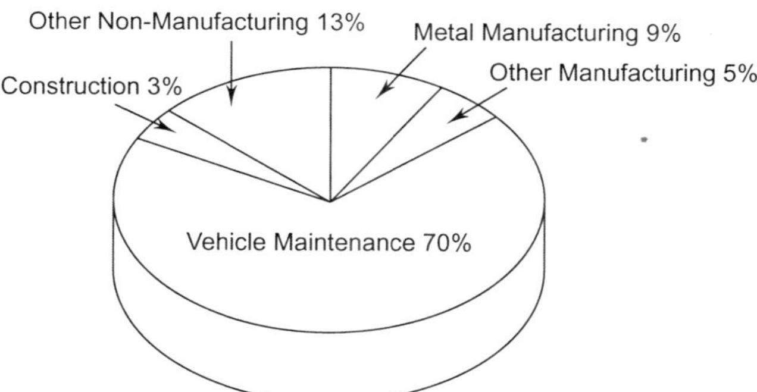

Fig. 8.6 Small Quantity Generators of Hazardous Wastes

Another report, the EPA estimates that the new RCRA will increase the number of federally regulated waste generators from about 15,000 to about 175,000. An EPA survey in March, 1985 suggested that 85% of the SQGs are in the vehicle maintenance, equipment repair, construction, printing, photography, laboratories, schools, laundries, dry cleaning, and pesticides application industries, Most of the remainders are in manufacturing of finishing of metals.

What Types of Hazardous Wastes do SQGs Generate?

Small quantity generators are responsible for less than 0.5% of all domestic hazardous wastes, according to John Skinner, Director, U.S. EPA office of solid waste. They produce only 800,000 metric tons per year of hazardous waste compared to 250 million metric tons from large companies. The types of hazardous waste generated by SQGs are essentially similar to those generated by larger quantity generators. To the extent that this is true, the move to regulate small quantity generators is appropriate. More than 10 million tons per year of hazardous wastes. Most of this waste is disposed of in municipal landfills not licensed to accept hazardous waste, or are burned in mass burn and energy recovery in cinerators, which may not be licensed to burn hazardous wastes.

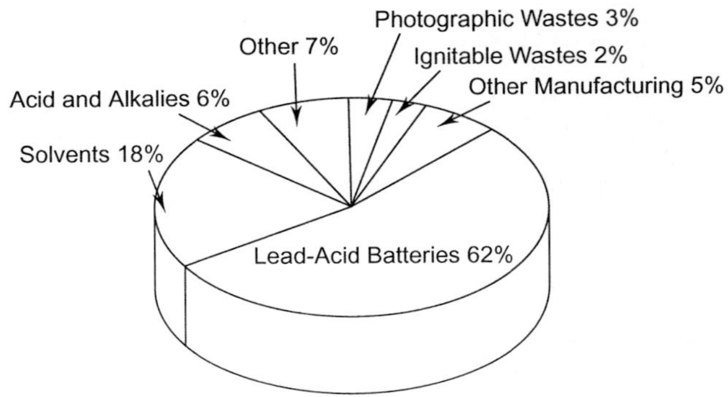

Fig. 8.7 Types of Hazardous Wastes from Small Quantity Generators

Household solid waste is not legally classified as hazardous. Present federal low exempts house hold solid wastes from RCRA regulations. However, Congress was apparently ambivalent when it incorporated the household waste exclusion into the law. For example, it provided that a resource recovery facility recovering energy from burning municipal waste if the facility receives and burns only household waste and non-hazardous waste from other sources, and does not accept hazardous wastes.

There appears to be a trend mass burn solid waste incineration and energy recovery. This trend is spurred by the increasing costs of land filling and by the difficulty of sitting new landfills. However, the concerns over household hazardous wastes and the increasing use of plastics in products and packaging have led to a lively debate over the human health and environmental impacts of household waste incineration, especially in non-RCRA, permitted *mass-burn* and energy recovery facilities. Many areas have already taken measures to properly manage household hazardous wastes. Table 4 shows examples of household and small business hazardous waste disposal programs in many areas.

Table. 8.2 Household and small business hazardous waste disposal

Industry	*Waste Streams*	*Industry*	*Waste Streams*
Laboratories	Acids/Bases Heavy metals/Inorganic Ignitable Wastes Relatives solvents	Vehicle Maintenance	Acids/Bases Heavy Metals/Inorganic Ignitable Wasfles Lead Acid Batteries Solvents

(Table Continue)

Printing and Allied Industries	Acids/Bases Heavy Metals/Inorganic Ink Sludges Spent Planting Wastes Solvents	Wood Preserving	Preserving Agents
Pesticide End Users and Application Services	Heavy Metals/Inorganic Pesticides Solvents	Motor Freight Terminals and Railroad, Transportation	Acids/Bases Heavy Metals/Inorganic Ignitable Wastes Lead Acid Batteries Solvents
Construction	Acids/Bases Ignitable Wastes	Metal Manufacturing	Acid/Bases Cyanide Wastes Heavy Metals/Inorganic Ignitable wastes Relatives Solvents
Furniture/Wood Manufacturing and Refinishing	Ignitable Wastes Solvents	Cleaning Agents and Cosmetics	Acids/Bases Heavy Metals/Inorganic Initable Wastes Pesticides Solvents
Other Manufacturing Textiles Plastics Leather	Heavy metals/Inorganics Solvents	Formulators	Acids/Bases Cyanide Wastes Heavy Metals/Inorganic Ignitable Wastes Pesticides Relatives Solvents
Laundries and Dry Cleaners	Dry cleaning filtration Residues Solvents		
Educational and Vocational Shops	Acids/Bases Ignitable Wastes Pesticides Reactives Solvents		
Building Cleaning and Maintenance	Acids/Bases Solvents		

Risks of Hazardous Wastes

Many ecologic incidents that called attention to environmental concerns more than 30 years ago are behind the hazardous waste concerns in Agenda 21. These events played fundamental roles in the enactment of legislation to control hazardous waste, because of evidence of both negative health impacts and elevated costs associated with their inadequate disposal.

Such is the case of the first-reported major incident of mercury poisoning, which arose from discharge of industrial wastes into Minamata Bay in Japan. The resulting effects on individuals disease of the central nervous system were described as early as 1956. Other well-publicized releases of toxic chemical from chemical waste disposal sites include Love Canal, New York, and Lekkekerk in the Netherlands. In both incidents, evacuation of residents from the affected communities and clean-up of the sites occasioned great social and financial expenses.

Objectives for Prevention of Hazardous Wastes

Four program areas are proposed in Agenda 21 to attain the overall objective: Within the framework of integrated life cycle management, to prevent to the extent possible, and minimize, the generation of hazardous wastes, as well as to manage those wastes in such a way that they do not cause harm to health and the environment. These program areas include.

1. Avoid or reduce generation of hazardous wastes.
2. Dispose of wastes as close as possible to the place where they are generated.
3. Manage wastes in an environmentally sound and efficient way.
4. Optimize environmentally sound recovery of wastes.
5. Preventing illegal international traffic in hazardous wastes.
6. Promoting and strengthening institutional capacities in hazardous waste management.
7. Promoting and strengthening international cooperation in the management of transboundary movements of hazardous wastes.
8. Promoting the prevention and minimization of hazardous wastes.
9. Reduce to a minimum or eliminate transboundary movements of hazardous wastes.

Prevention and Minimization of Hazardous Wastes

The specific economic and environmental benefits of industrial source reduction preventing the generation of wastes in the first place rather than cleaning them up after they have been created. Source reduction activities conducted by this group of chemical plants resulted in a significant decreases in releases of toxic and hazardous wastes, saved money and were associated with increases in product yields; and the capital cost of implementing them was non existent, reduced, or was paid back in less than 2.5 years.

Transportation

Transportation of hazardous substances are regulated by law. EPA has given instructions regarding packaging, marking and labelling. The material container should be inert to hazardous waste. Colour labels are being used to indicate hazardous substances while transporting.

Table. 8.3

Colour code	*Hazardous Substance (Number index)*
Orange	Explosives, Compressed gas (1)
Green	Non flammable (2)
Red	Flammable
White	Poisonous
White and Red strips	Flammable solid
Yellow	Peroxides
White	Peroxides
Yellow and magenta	Infections
White and Black	Corrosive

Hazardous Wastes from Hospitals

Healthcare wastes are generated as a result of activities related to the practice of medicine (including veterinary medicine and dentistry). Often this term is used to refer only to solid wastes (*i.e.*, not to wastewater). Some of the healthcare wastes coming from any particular hospital or institution are similar in nature to domestic solid wastes, and may be called "general healthcare wastes". The remaining wastes pose serious health hazards because of their physical, chemical or biological nature, and so are known as "hazardous healthcare wastes" or "healthcare risk wastes". Wastes which are particularly offensive because of their appearance or smell may also be classed with the hazardous wastes.

Healthcare wastes have attracted considerable attention because of the emotional impact of seeing body parts amidst solid waste, and because of the increasing concern about AIDS and hepatitis. In many cases the most dangerous items in healthcare wastes are needles from syrings and drips, because the needles shield the viruses from chemical disinfectants and a harsh external environment, and the sharp point allows easy access for the viruses into the blood stream of anyone who is pricked by the needle.

Many attempts to upgrade healthcare waste management rely solely of the provision of incinerators or other treatment technologies. Such a strategy has several weaknesses in that:

- Often the hospitals and healthcare facilities are not able to afford the operating costs of the plant, and so the plants are left unused or not repaired when the break down
- Many of the risks occur before the waste gets to this final stage, and so they are not reduced by the provision of treatment equipment
- The real need is often provide better methods of storage to train the staff to adopt safer working practices.

Hazards Associated with Biomedical Wastes

Proper disposal of bio medical wastes is necessary as these are infectious and hazardous. Improper disposal can lead to the following dangerous situations:

1. Organic portion ferments and attracts fly breeding.
2. Injuries from sharp broken instrument, syringe etc., to the healthcare and waste handling persons.
3. Increased risk of infections to medical nursing and other staff.
4. Poor infection control may lead to infections of patients with HIV, hepatitis B and C.
5. Increases risk of drugs being handled by persons handling wastes.
6. Poor waste management encourages unscruprous persons to recycle the materials from waste.
7. Development of resistant strains by microorganisms.

Mainly the medical persons, nurses, patients, health workers, laboratory personnel are at risk of infections. If the waste is not disposed properly, pathogens can pollute, soil, water and air and cause health hazard for general public.

Table 8.4. Types of waste generated by hospitals

Types of waste	*Treatment and disposal advice*
1. Human anatomical wastes (Tissues, organs, body parts)	Incineration/deep burial
2. Animal waste (animal, tissues, organs, body parts, carcasses, bleeding parts, blood and experimental animals).	Incineration/deep burial
3. Microbiology and biotechnology wastes (Lab culture, specimen of microorganisms vaccines, cell culture, toxins, devices)	Local autoclaving/micro waving incineration
4. Wastes sharp (needle, syringes, scalpels blades, glass	Chemical disinfection, autoclaving microwaving multilation and shredding
5. Discarded medicines and cytotoxic drugs (outdated contaminated, discarded drugs)	Incineration destruction and disposal in landfills
6. Soiled waste contaminated with blood, body, fluid, including cotton, dressings soiled plaster etc.	Autoclave/microwaving incineration
7. Solid wastes (tubes, catheters)	Chemical disinfection autoclaving, microwaving, mutilation and shredding
8. Liquid waste (from laboratory, hospital drains, washing, cleaning disinfection then discharge into drains.	Disinfection by chemical treatment and then discharge into drain
9. Incineration ash	Landfills
10. Chemical waste	Chemical disinfection discharge into drains

Other hazardous and special wastes

Some waste materials need special care because their properties make them more hazardous or problematic than general wastes. Used oil can be refined for reuse of burned in properly equipped furnaces. Slaughterhouse wastes should be buried in special trenches as suitable sites. Car tyres should be reused as much as possible, and carefully protected from open burning. Chemical wastes from some industries (including tanning, dry cleaning, photographic processing and many chemical production industries) and unwanted pesticides and other agricultural chemicals, should be collected under close supervision and treated in appropriate ways. The management of hazardous chemicals is not only a matter of technology and legislation, but also of enforcement, funding and financial instruments. Some wastes are so hazardous and expensive to treat that priority attention should be focussed on changing to processes that use substitutes that are less hazardous; and to minimising the quantities that are discarded. Indeed, minimisation and substitution should be seen as the preferred options in dealing with any difficult waste.

The Basel Convention seeks to control the movement of hazardous wastes across international boundaries. This instrument is necessary because the high cost of treating hazardous wastes in industrialized countries makes it financially attractive to ship the wastes to another country where no special requirements for their disposal will be applied.

EXERCISE

1. What is concept of solid waste management? Explain different stages of the system.
2. Write short notes on –
   (*i*) Recycling
   (*ii*) Landfilling
   (*iii*) Land Reclamation
   (*iv*) Incineration
3. Explain Aerobic and Anaerobic digestion.
4. What is Hazardous waste?
5. How can we prevent ourselves from Hazardous wastes?

9

Disaster and Disaster Management

9.1 INTRODUCTION

Disaster is a sudden, calamitous event bringing great damage, loss, and destruction and devastation to life and property. The damage caused by disasters is immeasurable and varies with the geographical location, climate and the type of the earth surface/degree of vulnerability. This influences the mental, socio-economic, political and cultural state of the affected area. Generally, disaster has the following effects in the concerned areas,

1. It completely disrupts the normal day to day life
2. It negatively influences the emergency systems
3. Normal needs and processes like food, shelter, health, etc., are affected and deteriorate depending on the intensity and severity of the disaster.

It may also be termed as "a serious disruption of the functioning of society, causing widespread human, material or environmental losses which exceed the ability of the affected society to cope using its own resources."

Thus, a disaster may have the following main features:

- Unpredictability
- Unfamiliarity
- Speed
- Urgency
- Uncertainty
- Threat

Thus, in simple terms we can define disaster as a hazard causing heavy loss to life, property and livelihood.e.g., a cyclone killing 10,000 lives and a crop loss of one crore can be termed as disaster.

9.2 TYPES OF DISASTER

Generally, disasters are of two types – Natural and Man-made. Based on the devastation, these are further classified into major/minor natural disaster and major/minor man-made disasters. Some of the disasters are listed below,

Table. 9.1

Major Natural Disasters:	*Minor Natural Disasters:*
• Flood	• Cold wave
• Cyclone	• Thunderstorms
• Drought	• Heat waves
• Earthquake	• Mud sides
	• Storm
Major Man-made Disaster:	*Minor Man-made Disaster:*
• Setting of fires	• Road/train accidents, riots
• Epidemic	• Food poisoning
• Deforestation	• Industrial disaster/crisis
• Pollution due to prawn cultivation	• Environmental pollution
• Chemical pollution	
• Wars	

The following types of disasters with the respective safety tips are explained in detail:

1. Earthquake
2. Tsunami
3. Cyclone
4. Flood
5. Drought
6. Landslides
7. Nuclear Disaster
8. Chemical Disaster
9. Fire
10. Environmental Hazards

1. Earthquakes

"An earthquake is a sudden motion or trembling of the ground produced by the abrupt displacement of rock masses".

Most earthquakes result from the movement of one rock mass past another in response to tectonic forces, viz some terms related to description of earthquakes are:

- **The focus** is the point where the earthquake's motion starts,
- **The epicenter** is the point on the earth's surface that is directly above the focus.
- **Earthquake Magnitude** is a measure of the strength of an earthquake as calculated from records of the event made on a calibrated seismograph.In 1935, Charles Richter first defined local magnitude, and the Richter scale is commonly used today to describe an earthquake's magnitude.

- **Earthquake Intensity** is a measure of the effects of an earthquake at a particular place. It is determined from observations of the earthquake's effects on people, structures and the earth's surface.

Among the many existing scales, the Modified Mercalli Intensity Scale of 12 degrees, symbolized as MM, is frequently used.

Earthquakes usually give no warning at all.

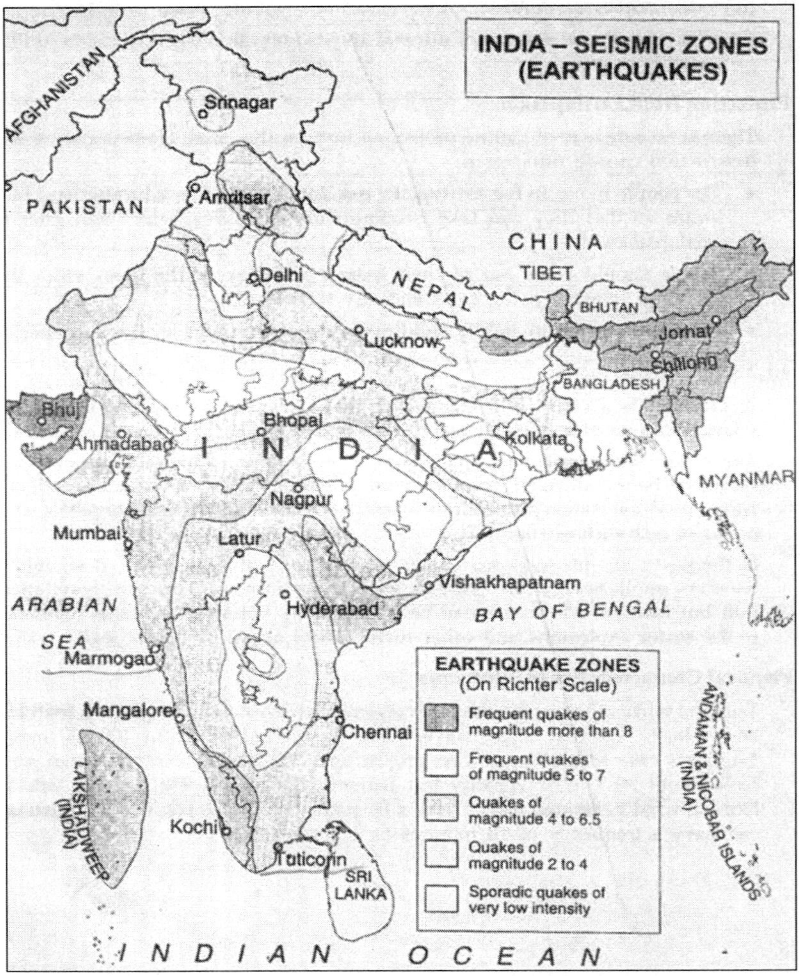

Fig. 9.1 Areas in India Prone to Earthquake

Earthquake Hazards

Earthquake hazards can be categorized as either direct hazards or indirect hazards.

Direct Hazards:

- Ground shaking
- Differential ground settlement

- Soil liquefaction
- Immediate landslides or mud slides, ground lurching and avalanches
- Permanent ground displacement along faults
- Floods from tidal waves, Sea Surges & Tsunamis.

Indirect Hazards:

- Dam failures
- Pollution from damage to industrial plants
- Delayed landslides.

Most of the damage due to earthquakes is the result of strong ground shaking. For large magnitude events, trembling has been felt over more than 5 million sq. km.

Safty Tips

Before the Earthquake

Now is the time to formulate a safety plan for you and your family. If you wait until the earth starts to shake, it may be too late. Consider the following safety measures:

- Always keep the following in a designated place: bottled drinking water, non-perishable food (chura, gur, etc), first-aid kit, torchlight and battery-operated radio with extra batteries.
- Teach family members how to turn off electricity, gas, etc.
- Identify places in the house that can provide cover during an earthquake.
- It may be easier to make long distance calls during an earthquake. Identify an out of town relative or friend as your family's emergency contact. If the family members get separated after the earthquake and are not able to contact each other, they should contact the designated relative/friend. The address and phone number of the contact person/relative should be with all the family members.

Safeguard your house

- Consider retrofitting your house with earthquake-safety measures. Reinforcing the foundation and frame could make your house quake resistant. You may consult a reputable contractor and follow building codes.
- Kutcha buildings can also be retrofitted and strengthened.

During quake

Earthquakes give no warning at all. Sometimes, a loud rumbling sound might signal its arrival a few seconds ahead of time. Those few seconds could give you a chance to move to a safer location. Here are some tips for keeping safe during a quake.

- Take cover. Go under a table or other sturdy furniture; kneel, sit, or stay close to the floor. Hold on to furniture legs for balance. Be prepared to move if your cover moves.
- If no sturdy cover is nearby, kneel or sit close to the floor next to a structurally sound interior wall. Place your hands on the floor for balance.
- Do not stand in doorways. Violent motion could cause doors to slam and cause serious injuries. You may also be hit be flying objects.
- Move away from windows, mirrors, bookcases and other unsecured heavy objects.

- If you are in bed, stay there and cover yourself with pillows and blankets
- Do not run outside if you are inside. Never use the lift.
- If you are living in a kutcha house, the best thing to do is to move to an open area where there are no trees, electric or telephone wires.

If outdoors

- Move into the open, away from buildings, streetlights, and utility wires. Once in the open, stay there until the shaking stops.
- If your home is badly damaged, you will have to leave. Collect water, food, medicine, other essential items and important documents before leaving.
- Avoid places where there are loose electrical wires and do not touch metal objects that are in touch with the loose wires.
- Do not re-enter damaged buildings and stay away from badly damaged structures.

If in a moving vehicle

Move to a clear area away from buildings, trees, overpasses, or utility wires, stop, and stay in the vehicle. Once the shaking has stopped, proceed with caution. Avoid bridges or ramps that might have been damaged by the quake.

After the quake

Here are a few things to keep in mind after an earthquake. The caution you display in the aftermath can be essential for your personal safety.

- Wear shoes/chappals to protect your feet from debris
- After the first tremor, be prepared for aftershocks. Though less intense, aftershocks cause additional damages and may bring down weakened structures. Aftershocks can occur in the first hours, days, weeks, or even months after the quake.
- Check for fire hazards and use torchlights instead of candles or lanterns.
- If the building you live in is in a good shape after the earthquake, stay inside and listen for radio advises. If you are not certain about the damage to your building, evacuate carefully. Do not touch downed power line.
- Help injured or trapped persons. Give first aid where appropriate. Do not move seriously injured persons unless they are in immediate danger of further injury. In such cases, call for help.
- Remember to help your neighbours who may require special assistance-infants, the elderly, and people with disabilities.
- Listen to a battery operated radio for the latest emergency information.
- Stay out of damaged buildings.
- Return home only when authorities say it is safe. Clean up spilled medicines, bleaches or gasoline or other flammable liquids immediately. Leave the area if you smell gas or fumes from other chemicals. Open closet and cupboard doors cautiously.
- If you smell gas or hear hissing noise, open windows and quickly leave the building. Turn off the switch on the top of the gas cylinder.
- Look for electrical system damages, if you see sparks, broken wires, or if you smell burning of amber, turn off electricity at the main fuse box. If you have to step in water to get to the fuse box, call an electrician first for advice.

- Check for sewage and water lines damage. If you suspect sewage lines are damaged, avoid using the toilets. If water pipes are damaged, avoid using water from the tap.
- Use the telephone only for emergency calls.
- In case family members are separated from one another during an earthquake (a real possibility during the day when adults are at work and children are at school), develop a plan for reuniting after the disaster. Ask an out of state/district relative or friend to serve as the "family contact". Make sure everyone in the family knows the name, address, and phone numbers of the contact persons.

2. Tsunamis

Tsunamis are Ocean Waves produced by Earth Quakes or Underwater land slides. This word has been derived from Japanese and means "Harbor Waves"

Tsunami is actually a series of waves that can travel at speeds from 400–600 mph in the open ocean. As the waves approach the coast, their speed decreases, but their amplitude increases. Unusual wave heights of 10–20 ft high can be very destructive and cause many deaths and injuries. Most deaths caused by Tsunamis are because of Drowning.

Associated risks include:

- Flooding
- Contamination of Drinking Water
- Fires from ruptured gas lines and tanks
- Loss of vital Community Infrastructure [police, fire, medical]
- Areas of greatest risks are
  - Less than 25 feet above sea level
  - Within 1 mile of the shore line.

Environmental Conditions left by the Tsunamis may contribute to the transmission of the following diseases From Food or Water

- Diarrhea illnesses; Cholera, Acute Diarrhea, Dysentery
- Hepatitis-A, Hepatitis-E
- Typhoid Fever
- Food borne illnesses; Bacterial, Viral, Parasitic, Non-infections
- From Animals or Mosquitoes
- Leptospirosis, Plague, Malaria, J.E, Dengue, Rabies
- Respiratory Diseases; Avian flu, Influenza, Measles

Safety Tips

Tsunami-specific planning should include the following:

Learn about tsunami risk in your community. Contact your local emergency management office or American Red Cross chapter. Find out if your home, school, workplace or other frequently visited locations are in tsunami hazard areas. Know the height of your street above sea level and the distance of your street from the coast or other high-risk waters. Evacuation orders may be based on these numbers.

If you are visiting an area at risk from tsunamis, check with the hotel, motel, or campground operators for tsunami evacuation information and how you would be warned. It is important to know designated escape routes before a warning is issued.

If you are at risk from tsunamis, do the following:

Plan an evacuation route from your home, school, workplace, or any other place you'll be where tsunamis present a risk. If possible, pick an area 100 feet above sea level or go up to two miles inland, away from the coastline. If you can't get this high or far, go as high as you can. Every foot inland or upwards may make a difference. You should be able to reach your safe location on foot within 15 minutes. After a disaster, roads may become impossible or blocked. Be prepared to evacuate by foot if necessary. Footpaths normally lead uphill and inland, while many roads parallel coastlines. Follow posted tsunami evacuation routes; these will lead to safety. Local emergency management officials can help advise you as to the best route to safety and likely shelter locations.

Practice your evacuation route. Familiarity may save your life. Be able to follow your escape route at night and during inclement weather. Practicing your plan makes the appropriate response more of a reaction, requiring less thinking during an actual emergency situation.

Use a NOAA Weather Radio with a tone-alert feature to keep you informed of local watches and warnings. The tone alert feature will warn you of potential danger even if you are not currently listening to local radio or television stations.

Talk to your insurance agent. Homeowners' policies do not cover flooding from a tsunami. Ask about the National Flood Insurance Program.

Discuss tsunami with your family. Everyone should know what to do in case all family members are not together. Discussing tsunamis ahead of time will help reduce fear and anxiety, and let everyone know how to respond. Review flood safety and preparedness measures with your family.

How to Protect Your Property?

Avoid building or living in buildings within several hundred feet of the coastline. These areas are more likely to experience damage from tsunamis, strong winds, or coastal storms.

Make a list of items to bring inside in the event of a tsunami. A list will help you remember anything that can be swept away by tsunami waters.

Elevate coastal homes. Most tsunami waves are less than 10 feet. Elevating your house will help reduce damage to your property from most tsunamis.

Follow flood preparedness precautions. Tsunamis are large amounts of water that crash onto the coastline, creating floods.

Have an engineer check your home and advise about ways to make it more resistant to tsunami water. There may be ways to divert waves away from your property. Improperly built walls could make your situation worse. Consult with a professional for advice.

3. Cyclone

The term "cyclone" refers to all classes of storms with low atmospheric pressure at the centre, are formed when an organized system of revolving winds, clockwise in the Southern Hemisphere, anti-clockwise in the Northern Hemisphere, develops over tropical waters.

Cyclones are classified on the basis of the average speed of the wind near the centre of the system as follows:

Wind Speed	Classification
• Up to 61 km/hr	Tropical Depression
• 61 km/hr – 115 km/hr	Tropical Storm
• Greater than 115 km/hr	Hurricane

Hurricane

A hurricane is a low pressure, large scale weather system which derives its energy from the latent heat of condensation of water vapor over warm tropical seas. A mature hurricane may have a diameter ranging from 150 to 1000 km with sustained wind speeds often exceeding 180 km/hr near the centre with still higher gusts.

A unique feature of a hurricane is the Eye. The eye provides a convenient frame of reference for the system, and can be tracked with radar, aircraft or satellite.

Safety Tips

Before the Cyclone Season:

- Keep watch on weather and listen to radio or TV. Keep alert about the community warning systems – loudspeakers, bells, conches, drums or any traditional warning system.
- Get to know the nearest cyclone shelter/safe houses and the safest route to reach these shelters.
- Do not listen to rumours.
- Prepare an emergency kit containing:
  - A portable radio, torch and spare batteries
  - Stocks dry food – Chura, Chhatua, Mudhi, gur, etc.
  - Matches, fuel lamp, portable stove, cooking utensils, waterproof bags
  - A first aid kit, manual, etc.
  - Katuri, pliers, small saw, axe and plastic rope.
- Check the roof and cover it with net or bamboo. Check the walls, pillars, doors and windows to see if they are secure. If not, repair those at the earliest. In case of tin roofs, check the condition of the tin and repair the loose points. Cover the mud walls with polythene or coconut leaves mats or straw mats on a bamboo frame. Bind each corner of the roof with a plastic rope in case of thatched roof.
- Trim dry tree branches, cut off the dead trees and clear the place/courtyard of all debris, including coconuts and tree branches.
- Clear your property of loose materials that could blow about and cause injury or damage during extreme winds.
- If your area is prone to storm surge, locate safe high ground or shelter.
- Keep important documents, passbook, etc., in a tight plastic bag and take it along with your emergency kits if you are evacuating.
- Identify the spot where you can dig holes to store food grains, seeds, etc., in polythene bags.
- Keep a list of emergency addresses and phone numbers on display. Know the contact telephone number of the government offices/agencies, which are responsible for search, rescue and relief operations in your area.

Upon a cyclone warning:

- Store loose items inside. Put extra agricultural products/ stock like paddy in plastic bags and store it by digging up a hole in the ground, preferably at a higher elevation and then cover it properly. Fill bins and plastic jars with drinking water.
- Keep clothing for protection, handy
- Prepare a list of assets and belongings of your house and give information to volunteers and other authorities about your near and dear ones.
- Fill fuel in your car/motorcycle and park it under a solid cover. Tie bullock carts, boats securely to strong posts in an area, which has a strong cover and away from trees. Fallen trees can smash boats and other assets.
- Close shutters or nail all windows. Secure doors. Stay indoors, with pets.
- Pack warm clothing, essential medications, valuables, papers, water, dry food and other valuables in waterproof bags, to be taken along with your emergency kit.
- Listen to your local radio/TV, local community warning system for further information.
- In case of warning of serious storm, move with your family to a strong pucca building. In case of warning of cyclones of severe intensity, evacuate the area with your family, precious items and documents and emergency kit. Take special care for children, elders, sick, pregnant women and lactating mothers in your family. Do not forget your emergency food stock, water and other emergency items. **GO TO THE NEAREST CYCLONE SHELTER.**
- Do not venture into the sea for fishing.

On warning of local evacuation

Based on predicted wind speeds and storm surge heights, evacuation may be necessary. Official advice may be given on local radio/ TV or other means of communication regarding safe routes and when to move.

- Wear strong shoes or chappals and clothing for protection.
- Lock your home, switch off power, gas, water, and take your emergency kit.
- If evacuating to a distant place take valuable belonging, domestic animals, and leave early to avoid heavy traffic, flooding and wind hazards.
- If evacuating to a local shelter or higher grounds carry the emergency kit and minimum essential materials.

When the cyclone strikes:

- Disconnect all electrical appliances and turn off gas.
- If the building starts crumbling, protect yourself with mattresses, rugs or blankets under a strong table or bench or hold on to a solid fixture (*e.g.*, a water pipe)
- Listen to your transistor radio for updates and advice.
- Beware of the calm 'eye'. If the wind suddenly drops, don't assume the cyclone is over, violent winds will soon resume from the opposite direction. Wait for the official **"all clear"**.
- If driving, stop – but well away from the sea and clear of trees, power lines and watercourses. Stay in the vehicle.

After the cyclone:

- Do not go outside until officially advised it is safe.
- Check for gas leaks. Do not use electric appliances, if wet.
- Listen to local radio for official warnings and advice.
- If you have to evacuate, or did so earlier, do not return until advised. Use a recommended route for returning and do not rush.
- Be careful of snake bites and carry a stick or bamboo.
- Beware of fallen power lines, damaged bridges, buildings and trees, and do not enter the floodwaters.
- Heed all warnings and do not go sightseeing.

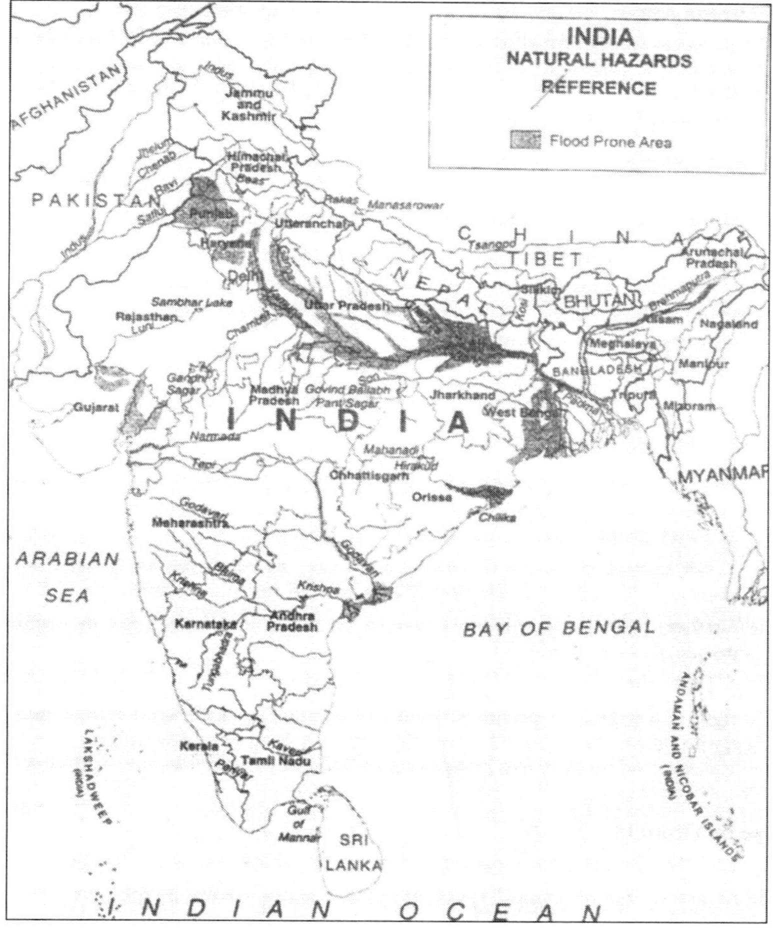

Fig. 9.2 Flood Prone in India

4. Floods

Floods occur when the ground can't hold any more rain and the rivers and lakes crest over, their banks, this normally happens during the very heavy rain seasons that occur in most places where there is no place for the water to go like after severe thunderstorms, monsoons, hurricanes and typhones the waters receed once the rains stop and sun comes out to help with evaporation however the recession of the waters takes days and weeks to do on it's own so people normall end up pumping the water out of their homes also occurs when poorly built dams fail to hold the water in.

Safety Tips

The simple things you and your family can do to stay safe and protect your property from floods.

Before flooding occurs:

- All your family members should know the safe route to nearest shelter/raised pucca house.
- If your area is flood-prone, consider alternative building materials. Mud walls are more likely to be damaged during floods. You may consider making houses where the walls are made of local bricks up to the highest known flood level with cement pointing.
- Have an emergency kit on hand which includes a:
    - A portable radio, torch and spare batteries
    - Stocks of fresh water, dry food (chura, mudi, gur, biscuits), kerosene, candle and match-boxes
    - Waterproof or polythene bags for clothing and valuables, an umbrella and bamboo stick (to protect from snake), salt and sugar.
    - A first aid kit, manual and strong ropes for tying things.

When you hear a flood warning or if flooding appears likely:

- Tune to your local radio/TV for warnings and advice.
    - Keep vigil on flood warning given by local authorities
    - Don't give any importance to rumours and don't panic
    - Keep dry food, drinking water and clothes ready.
- Prepare to take bullock carts, other agricultural equipments, and domestic animals to safer places or to higher locations.
- Plan which indoor items you will raise or empty if water threatens to enter your house.
- Check your emergency kit.

During floods:

- Drink boiled water.
- Keep your food covered, don't take heavy meals.
- Use raw tea, rice-water, tender coconut-water, etc., during diarrhoea; contact your ANM/AWW for ORS and treatment.
- Do not let children remain on empty stomach.
- Use bleaching powder and lime to disinfect the surrounding.
- Help the officials/volunteers distributing relief materials.

If you need to evacuate:

- Firstly pack warm clothing, essential medication, valuables, personal papers, etc., in water-proof bags, to be taken with your emergency kit.

- Take the emergency kit
- Inform the local volunteers (if available), the address of the place you are evacuating to.
- Raise furniture, clothing and valuables onto beds, tables and to the top of the roof (electrical items highest).
- Turn off power.
- Whether you leave or stay, put sandbags in the toilet bowl and over all laundry/bathroom drain-holes to prevent sewage back-flow.
- Lock your home and take recommended/known evacuation routes for your area.
- Do not get into water of unknown depth and current.

  If you stay or on your return.
- Stay tuned to local radio for updated advice.
- Do not allow children to play in, or near, flood waters.
- Avoid entering floodwaters. If you must, wear proper protection for your feet and check depth and current with a stick. Stay away from drains, culverts and water over knee-deep.
- Do not use electrical appliances, which have been in floodwater until checked for safety.
- Do not eat food, which has been in floodwaters.
- Boil tap water (in cities) until supplies have been declared safe. In case of rural areas, store tube well water in plastic jars or use halogen tablets before drinking.
- Be careful of snakes, snakebites are common during floods.

5. Droughts

Drought is a period or condition of unusually dry weather within a geographic area where rainfall is normally present. During a drought there is a lack of precipitation. Droughts occur in all climatic zones. However, its characteristics vary significantly from one region to another.

Drought usually results in a water shortage that seriously interferes with human activity. Water supply reservoirs empty, wells dry up, and crop damage ensues. Its seriousness depends on the degree of the water shortage, size of area affected, and the duration and warmth of the dry period. In many developing countries, such as India, people place a great demand on water supply. During a drought period there is a lack of water, and thus many of the poor die.

Most precipitation depends on water vapor carried by winds from an ocean or other source of moisture. If these moisture carrying winds are replaced by winds from a dry region, or if they are modified by downward motion, as in the center of an anticyclone, the weather is abnormally dry and often persistently cloudless. If the drought period is short, it is known as a dry spell. A dry spell is usually more than 14 days without precipitation, whereas a severe drought may last for years.

Statistics indicate that every 22 years a major drought occurs in the United States, most seriously affecting the midwestern states. The drought of 1933-35, during which large areas of the Great Plains became known as the Dust Bowl, is an example of a disastrous drought that took place in the United States. The effect of the drought was brought about by overcropping, overpopulation, and lack of relief measures.

Although drought cannot be reliably predicted, certain precautions can be taken in drought-risk areas. These include construction of reservoirs to hold emergency water supplies, education to avoid overcropping and overgrazing, and programs to limit settlement in drought-prone areas. The Southern Africa Development Community monitors the crop and food situation in the region and alerts the people during periods of crisis.

Types of Drought

Meteorological Drought

Meteorological drought is the amount of dryness and the duration of the dry period. Atmospheric conditions that result in deficiencies of precipitation change from area to area.

Agricultural Drought

Agricultural drought mainly effects food production and farming. Agricultural drought and precipitation shortages bring soil water deficits, reduced ground water or reservoir levels, and so on. Deficient topsoil moisture at planting may stop germination, leading to low plant populations.

Hydrological Drought

Hydrological drought is associated with the effects of periods of precipitation shortages on water supply. Water in hydrologic storage systems such as reservoirs and rivers are often used for multiple purposes such as flood control, irrigation, recreation, navigation, hydropower, and wildlife habitat. Competition for water in these storage systems escalates during drought and conflicts between water users increase significantly.

Socioeconomic Drought

Socioeconomic drought occurs when the demand for an economic good exceeds supply as a result of a weather-related shortfall in water supply. The supply of many economic goods, such as water, forage, food grains, fish, and hydroelectric power, depends on weather. Due to variability of climate, water supply is sufficient in some years but not satisfactory to meet human and environmental needs in other years. The demand for economic goods is increasing as a result of increasing population. Supply may also increase because of improved production efficiency and technology.

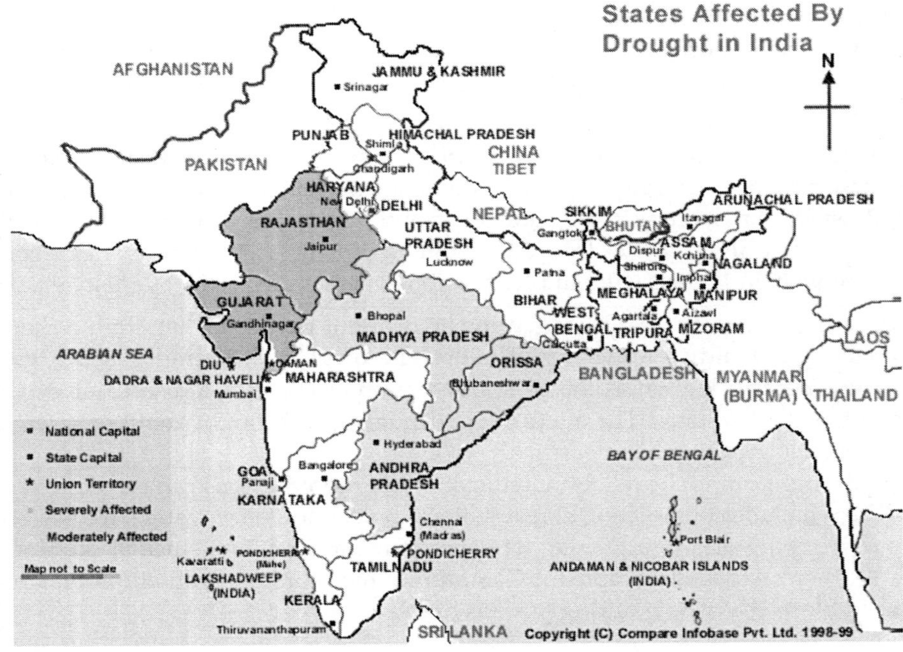

6. Landslide

Landslides are rock, earth, or debris flows on slopes due to gravity. They can occur on any terrain given the right conditions of soil, moisture, and the angle of slope. Integral to the natural process of the earth's surface geology, landslides serve to redistribute soil and sediments in a process that can be in abrupt collapses or in slow gradual slides. Such is the nature of the earth's surface dynamics. Also known as mud flows, debris flows, earth failures, slope failures, etc., they can be triggered by rains, floods, earthquakes, and other natural causes as well as human-made causes, such as grading, terrain cutting and filling, excessive development, etc. Because the factors affecting landslides can be geophysical or human-made, they can occur in developed areas, undeveloped areas, or any area where the terrain was altered for roads, houses, utilities, buildings, and even for lawns in one's backyard. They occur in all fifty states with varying frequency and more than half the states have rates sufficient to be classified as a significant natural hazard.

During a Landslide:

- Stay alert and awake. Many debris flow fatalities occur when people are sleeping. Listen to a weather radio or portable, battery-powered radio or television for warnings of intense rainfall. Be aware that intense, short bursts of rain may be particularly dangerous, especially after longer periods of heavy rainfall and damp weather.
- If you are in areas susceptible to landslides and debris flows, consider leaving if it is safe to do so. Remember that driving during an intense storm can be hazardous. If you remain at home, move to a second story if possible. Staying out of the path of a landslide or debris flow saves lives.
- Listen for any unusual sounds that might indicate moving debris, such as trees cracking or boulders knocking together. A trickle of flowing or falling mud or debris may precede larger landslides. Moving debris can flow quickly and sometimes without warning.
- If you are near a stream or channel, be alert for any sudden increase or decrease in water flow and for a change from clear to muddy water. Such changes may indicate landslide activity upstream, so be prepared to move quickly. Don't delay! Save yourself, not your belongings.
- Be especially alert when driving. Embankments along roadsides are particularly susceptible to landslides. Watch the road for collapsed pavement, mud, fallen rocks, and other indications of possible debris flows.

What to Do if You Suspect Imminent Landslide Danger?

- Contact your local fire, police, or public works department. Local officials are the best persons able to assess potential danger.
- Inform affected neighbors. Your neighbors may not be aware of potential hazards. Advising them of a potential threat may help save lives. Help neighbors who may need assistance to evacuate.
- Evacuate getting out of the path of a landslide or debris flow is your best protection.

After the Landslide:

- Stay away from the slide area. There may be danger of additional slides.
- Check for injured and trapped persons near the slide, without entering the direct slide area. Direct rescuers to their locations.

- Help a neighbor who may require special assistance - infants, elderly people, and people with disabilities. Elderly people and people with disabilities may require additional assistance. People who care for them or who have large families may need additional assistance in emergency situations.
- Listen to local radio or television stations for the latest emergency information.
- Watch for flooding, which may occur after a landslide or debris flow. Floods sometimes follow landslides and debris flows because they may both be started by the same event.
- Look for and report broken utility lines to appropriate authorities. Reporting potential hazards will get the utilities turned off as quickly as possible, preventing further hazard and injury.
- Check the building foundation, chimney, and surrounding land for damage. Damage to foundations, chimneys, or surrounding land may help you assess the safety of the area.
- Replant damaged ground as soon as possible since erosion caused by loss of ground cover can lead to flash flooding.
- Seek the advice of a geotechnical expert for evaluating landslide hazards or designing corrective techniques to reduce landslide risk. A professional will be able to advise you of the best ways to prevent or reduce landslide risk, without creating further hazard.

Media and Community Education Ideas:

- In an area prone to landslides, publish a special newspaper section with emergency information on landslides and debris flows. Localize the information by including the phone numbers of local emergency services offices, the American Red Cross chapter, and hospitals.
- Report on what city and county governments are doing to reduce the possibility of landslides. Interview local officials about local land-use zoning regulations.
- Interview local officials and major insurers regarding the National Flood Insurance Program. Find out if debris flow is covered by flood insurance policies from the National Flood Insurance Program and contact your local emergency management office to learn more about the program.
- Work with local emergency to prepare special reports for people with mobility impairments on what to do if evacuation is ordered.
- Support your local government in efforts to develop and enforce land use and building ordinances that regulate construction in areas susceptible to landslides and debris flows. Buildings should be located away from steep slopes, streams and rivers, intermittent-stream channels, and the mouths of mountain channels.

Before a Landslide: How to Plan?

Develop a Family Disaster Plan. Learn about landslide risk in your area. Contact local officials, state geological surveys or departments of natural resources, and university departments of geology. Landslides occur where they have before, and in identifiable hazard locations. Ask for information on landslides in your area, specific information on areas vulnerable to landslides, and request a professional referral for a very detailed site analysis of your property, and corrective measures you can take, if necessary.

If you are at risk from landslides:

- Talk to your insurance agent.
- Develop an evacuation plan.
- Discuss landslides and debris flow with your family. Everyone should know what to do in case all family members are not together. Discussing disaster ahead of time helps reduce fear and lets everyone know how to respond during a landslide or debris flow.

7. Nuclear Hazards

The effects of nuclear holocaust will result into blasts, heat storms, secondary fires, fire, ionizing radiation and fall outs.

These effects fall into 3 categories;

1. Immediate,
2. Short term and
3. Long term effects.

- **The immediate effects;** include blast effects, heat effects, electromagnetic pulse (EMP) effects and radiation effects.
- **The short term effects;** include problems connected with water supply, sanitation, food, dispersal of excreta, wastes and dead bodies, break down of vector control measures and outbreak of infections. Radioactive contamination of water and food are major concerns. The affected area creates a lot of other problems for the survivors and the rescue teams.

  Major problem among survivors is of bone marrow depression resulting in leucopenia, which increases their susceptibility to infections.
- **Long term effects**; the knowledge about the long-term effects is still incomplete. Some well known effects include radiation injuries due to radiation fallout, suppression of body immunity, chronic infection and other associated illnesses.

Persistent radiation hazards will lead to prolonged contamination of water supply, increased ultraviolet radiation, climatic and ecological disturbances, psychological disturbances and genetic abnormalities.

Current World Concern

In the light of the above facts the current world concern about the use of nuclear weapons is justified.The world already possess an estimated total of 30,000 megatons of nuclear weapons with a total explosive power 50-100 times greater than that of all the explosives used during the Second World War.

Even if 1% of the nuclear weapons now possessed are used on urban populations, they can cause more deaths in a few hours than during the entire period of the Second World War.

8. Chemical Hazards

By their nature, the manufacture, storage, and transport of chemicals are accidents waiting to happen. Chemicals can be corrosive, toxic, and they may react, often explosively. The impacts of chemical accidents can be deadly, for both human beings and the environment.

Many if not most products we use in everyday life are made from chemicals and thousands of chemicals are used by manufacturing industries to make these products. The source of many of these chemicals is petroleum, which is refined into two main fractions: fuels and the chemical feedstocks that are the building blocks of plastics, paints, dyes, inks, polyester, and many of the products we buy and use every day. Fuels and chemical feedstocks made from petroleum are called organic chemicals. The other important class of chemicals is inorganics, which include acids, caustics, cyanide, and metals. Commercial products made from inorganics range from car bodies to computer circuit boards.

Of the more than forty thousand chemicals in commercial use, most are subject to accidental spills or releases. Chemical spills and accidents range from small to large and can occur anywhere chemicals are found, from oil drilling rigs to factories, tanker trucks to fifty-five-gallon drums and all the way to the local dry cleaner or your garden tool shed.

"Severe chemical accident that causes large-scale damage to an environment and/or injuries to or death of many animals, plants, or humans is called chemical disaster."

One of the worst industrial chemical disasters occurred without warning early on the morning of December 3, 1984, at Union Carbide's pesticide plant in Bhopal, India. While most people slept, a leak, caused by a series of mechanical and human failures, released a cloud of lethal methyl isocyanate over the sleeping city. Some two thousand people died immediately and another eight thousand died later. Health officials, not informed about chemicals at the factory, were completely unprepared for the tragedy.

Congressional hearings that followed the Bhopal accident revealed that U.S. companies routinely discharged hazardous chemicals into the air, while emergency planners knew little about the potential for disaster at local industrial facilities. Less than a year later, a Union Carbide plant that produced methyl isocyanate in Institute, West Virginia, leaked a toxic cloud in the Kanawha Valley. While the West Virginia incident was not another tragedy, it was a shocking reminder that an accident such as the one that occurred at Bhopal could happen in the United States.

The hearings and media attention to institute led to enactment of the Emergency Planning and Community Right to Know Act of 1986 (EPCRA), requiring companies to provide information about their potentially toxic chemicals. At the same time, states were required to establish emergency planning districts and local committees to prepare for any emergency—a fire, an explosion, a flood that might result in the release of chemicals into the environment. In 2003, more than 31,000 industrial facilities must report more than 650 individually listed toxic chemicals and chemical categories to the U.S. Environmental Protection Agency (EPA) that is made public in the Toxic Release Inventory.

In 1990, amendments to the clean Air Act required industrial chemical companies to submit a risk management plan that included "worst case" chemical accident scenarios. Industry leaders did not want these potential disasters made public and argued that they could alert terrorists which facilities to target. In July 2002, the Senate's Environment and Public Works Committee approved a bill to identify plants vulnerable to terrorist attacks that produce hazardous chemicals. Congress also voted against a landmark community right to know law that would have required some 6,600 chemical facilities to reveal their "worst case" accident scenarios.

Although the major chemical accidents seem most threatening because they often kill people outright, it is the smaller, more routine accidents and spills that affect most people. Some of the most common spills involve tanker trucks and railroad tankers containing gasoline, chlorine, acid, or other industrial chemicals. Many spills occur during the transportation of hazardous materials; one study found that 18,000 hazardous materials spills occurred during 1976. In 1983, spills from 4,829 highway and 851 railroad accidents resulted in eight deaths, 191 injuries, and damages exceeding more than $110,000,000. The National Environmental Law Center reported that 34,500 accidents involving toxic chemicals were reported to the EPA's Emergency Response and Notification System between 1988 and 1992, meaning that on average, a toxic chemical accident was reported nineteen times a day in the United States, or nearly once every hour.

Emergency response workers are especially at risk. In 1988 six firemen were killed minutes after arriving at the scene of two burning pick-up trucks in Missouri, when more than 30,000 pounds of ammonium nitrate stored in a nearby trailer exploded. This incident led to the formation of the hazardous materials division of the Kansas City, Missouri, Fire Department, specializing in hazardous materials handling.

Safety Tips

The safety from the chemical Hazards remain out of bounds in normal routine. Their handling require professional expertise, knowledge of chemicals, their reactions etc.

1. Emergency response personnel are involved in assessing the risk of hazardous material releases and working to avoid any harmful effects. Teams of workers evaluate the concentrations of the chemicals, where and how people might be exposed, and potential toxic effects on the exposed people.
2. In many cases, emergency response teams are on twenty-four-hour call; if a spill occurs, they use source data (such as the hazmat placards on trucks and tanker cars), databases of chemical properties, and chemical movement models to rapidly predict the movement of contaminants and the toxicity of the spilled chemicals.
3. If rapid spill cleanup is necessary, the emergency response team designs and implements cleanup measures to protect exposed populations and ecosystems from toxic responses. A wide range of cleanup systems has been developed for chemical spills. Small spills on land are cleaned up by simply excavating the contaminated soil and moving it to a secure landfill. Oil spills on water are contained using floating booms and adsorbents, or solid materials that capture the soil, so that it can be disposed of in landfills. Newer, more innovative methods for spill cleanup include bioremediation (using bacteria to metabolize the contaminants) and chemical oxidation (using oxidants, such as hydrogen peroxide and ozone to break the chemicals down).
4. Although chemical spills represent potentially very large environmental problems from a wide range of chemicals, emergency response procedures developed by environmental scientists and engineers are providing solutions to the resulting human health and ecological effects.
5. Chemical accidents and spills can be devastating to humans, wildlife, and the environment. The best way to reduce the harm caused by chemical accidents is to design plants with better safety controls that operate at lower temperatures and pressures, and to use and manufacture less toxic compounds, a field that is being pursued by "green" chemists and engineers. But until toxic chemicals are routinely replaced by less harmful substitutes, the emergency response procedures developed by environmental scientists and engineers help lessen the human health and ecological effects of chemical spills and accidents.

9. Fire Accidents

Fire may be as useful as it is deadly. Although, every precaution may be taken, devastating accidents can happen. Special care should be taken with common household appliances such as heaters, curling irons, electric grills and other items such as candles and outdoor barbecues. Cigarettes and fireworks are also common problems.

(*a*) **High-Rise Fires:**

- Calmly leave the apartment, closing the door behind you. Remember the keys!
- Pull the fire alarm near the closest exit, if available, or raise an alarm by warning others.
- Leave the building by the stairs.
- Never take the elevator during fire!

If the exit is blocked by smoke or fire:

- Leave the door closed but do not lock it.
- To keep the smoke out, put a wet towel in the space at the bottom of the door.
- Call the emergency fire service number and tell them your apartment number and let them know you are trapped by smoke and fire. It is important that you listen and do what they tell you.
- Stay calm and wait for someone to rescue you.

If there is a fire alarm in your building which goes off:

- Before you open the door, feel the door by using the back of our hand. If the door is hot or warm, do not open the door.
- If the door is cool, open it just a little to check the hallway. If you see smoke in the hallway, do not leave.
- If there is no smoke in the hallway, leave and close the door. Go directly to the stairs to leave. Never use the elevator.

If smoke is in your apartment:

- Stay low to the floor under the smoke.
- Call the Fire Emergency Number which should be pasted near your telephone along with police and other emergency services and let them know that you are trapped by smoke.
- If you have a balcony and there is no fire below it, go out.
- If there is fire below, go out to the window. **DO NOT OPEN THE WINDOW** but stay near the window.
- If there is no fire below, go to the window and open it. Stay near the open window.
- Hang a bed sheet, towel or blanket out of the window to let people know that you are there and need help.
- Be calm and wait for someone to rescue you.

(*b*) **Kitchen Fires:** It is important to know what kind of stove or cooking oven you have in your home gas, electric, kerosene or where firewood is used. The stove is the No. 1 cause of fire hazards in your kitchen and can cause fires, which may destroy the entire house, especially in rural areas where there are thatched roof or other inflammable materials like straw kept near the kitchen. For electric and gas stoves **ensure that the switch or the gas valve is switched off/turned off immediately after the cooking is over.** An electric burner remains hot and **until it cools off, it can be very dangerous**. The oven using wood can be dangerous because burning embers remain. When lighting the fire on a wooden fuel oven, **keep a cover on the top** while lighting the oven so that sparks do not fly to the thatched roof. After the cooking is over, ensure that the remaining **fire is extinguished off by sprinkling water** if no adult remains in the kitchen after the cooking. **Do not keep any inflammable article like kerosene near the kitchen fire.**

Important Do's in the Kitchen:

- **Do** have an adult always present when cooking is going on the kitchen. Children should not be allowed alone.
- **Do** keep hair tied back and do not wear synthetic clothes when you are cooking.
- **Do** make sure that the curtains on the window near the stove are tied back and will not blow on to the flame or burner.
- **Do** check to make sure that the gas burner is turned off immediately if the fire is not ignited and also switched off immediately after cooking.
- **Do** turn panhandles to the centre of the stove and put them out of touch of the children in the house.
- **Do** ensure that the floor is always dry so that you do not slip and fall on the fire.
- **Do** keep matches out of the reach of children.

Important Don'ts

- **Don't** put towels, near a stove burner.
- **Don't** wear loose fitting clothes when you cook, and **don't** reach across the top of the stove when you are cooking.
- **Don't** put things in the cabinets or shelves above the stove. Young children may try to reach them and accidentally start the burners, start a fire, catch on fire.
- **Don't** store spray cans or cans carrying inflammable items near the stove.
- **Don't** let small children near an open oven door. They can be burnt by the heat or by falling onto the door or into the oven.
- **Don't** lean against the stove to keep warm.
- **Don't** use towels as potholders. They may catch on fire.
- **Don't** overload an electrical outlet with several appliances or extension cords. The cords or plugs may overheat and cause a fire.
- **Don't** use water to put out a grease fire. ONLY use baking soda, salt, or a tight lid. Always keep a box of baking soda near the stove.
- **Don't** use radios or other small appliances (mixers, blenders) near the sink.

Fire at Indian Oil Corporation, Jaipur: Latest Example

The devastating fire at the Indian Oil Corporation (IOC) depot near Jaipur that began on October 29, 2009 was finally extinguished on November 9,2009 evening, though district officials maintained that smoke was still visible from one container. IOC officials were advised on Monday to remove any remaining flammable materials from the depot to avoid further untoward incidents.

The blaze, which began 11 days ago following a leak in the fuel pipeline and then a blast, has claimed 11 lives, including six IOC employees who were stationed in the depot, and left more than 100 injured. Following an explosion on October 29, the fire spread to all 11 containers in the IOC depot, which became uncontrollable, forcing fire fighting experts to hold back and let the fuel burn off.

The fire at Indian Oil Corporation's bulk storage depot at Jaipur was likely to have destroyed petroleum products worth Rs 140–150 crore.

10. Environmental Hazards

Environmental hazard' is a generic term for any situation or state of events which causes a threat to the surrounding environment. This term incorporates topics like pollution and natural hazards such as storms and earthquakes.

There are five types of environmental hazards:

1. Chemical
2. Physical
3. Mechanical
4. Biological
5. Psychosocial.

Anthropogenic environmental disasters

1. Bhopal: the Union Carbide gas leak
2. Chernobyl: Russian nuclear power plant explosion
3. Seveso: Italian dioxin crisis
4. The 1952 London smog disaster
5. Major oil spills of the 20th and 21st century
6. The Love Canal chemical waste dump
7. The Baia Mare cyanide spill
8. The European BSE crisis
9. Spanish wastewater spill
10. The Three Mile Island near nuclear disaster

Natural environmental disasters

1. Global epidemics (the Plague, Spanish Flu and AIDS)
2. The Bangladesh arsenic crisis
3. The 2004 Asian earthquake and tsunami
4. Hurricane Mitch
5. Izmit: the 1999 Turkish earthquakes
6. The 1953 Dutch flood disaster
7. Brazil: the 1998 Roraima wildfires
8. Phillipines: the Mount Pinatubo volcanic eruption
9. The 1925 Ellington, Missouri tornado
10. The Beijing and Queensland dust storms

Some unfortunate Environmental Disasters:

1. **Bhopal: the Union Carbide gas leak**

   December 3, 1984 has become a memorable day for the city of Bhopal in Madya Pradesh country, India. Shortly after midnight, a poisonous gas cloud escaped from the Union Carbide India Limited (UCIL) pesticide factory. The cloud contained 15 metric tons of methyl isocyanate (MIC), covering an area of more than 30 square miles. The gas leak killed at least

4.000 local residents instantly and caused health problems such as oedema for at least 50.000 to perhaps 500.00 peoples. These health problems killed around 15.000 more victims in the years that followed. Approximately 100.000 people still suffer from chronic disease consequential to gas exposure, today. Research conducted by the BBC in 2004 pointed out that this pollution still causes people to fall ill, and ten more die every year. This event is now known as the worst industrial environmental disaster to ever have occurred.

The cause of the accident has been researched after the disaster. Apparently water ended up in MIC storage tanks, causing an exothermal reaction that released an amount of poisonous gas large enough to open the safety valves. Normally scrubbers would intercept escaping gas, but these were temporarily out of order for repair.

Research showed that factory personnel neglected a number of safety procedures. There were no valves to prevent water from entering the storage tanks. The cooling installation of the tanks and the flaring installation that might have flared the gas that was released were out of order. Safety was very low in this factory of Union Carbide, compared to its other locations. The safety procedures were neglected because of budget cuts.

Union Carbide was accused of deliberate evasion of regular safety procedures. During lawsuits where victims demanded compensation, documents were revealed which proved that Union Carbide regularly used untested technology in the Bhopal factory. When the gas leak occurred doctors were not informed of the nature of the gas. This caused the correct treatment and emergency measures to be held off. The manager of Union Carbide, Warren Anderson, was accused of culpable homicide. However, he did not occur in court and both the Indian and American government did not take adequate measures to make sure this man was tried. This led to a series of protests organized by environmental organizations, such as Greenpeace.

Union Carbide denied responsibility for the accident on their website, stating that: "The Bhopal plant was owned and operated by Union Carbide India, Limited (UCIL), an Indian company in which Union Carbide Corporation held just over half the stock. The other stockholders included Indian financial institutions and thousands of private investors in India. The plant was designed, built, and managed by UCIL using Indian consultants and workers."

About the cause of the accident, they claimed that: "A thorough investigation was conducted by the engineering consulting firm Arthur D. Little. Its conclusion: the gas leak could only have been caused by deliberate sabotage. Someone purposely put water in the gas storage tank, causing a massive chemical reaction. Process safety systems had been put in place that would have kept the water from entering into the tank by accident."

After a long procedure in February 1989 eventually a settlement was achieved. Union Carbide promised to pay 470 million dollars compensation. Only a very small part of this compensation was paid to survivors of the environmental disaster. Union Carbide states on its website that it paid the full settlement to the Indian government within 10 days time. In 2004 the Supreme Court forced the Indian government to pay the remaining 330 million dollars compensation to the victims and their families.

Union Carbide sold the Indian factory to a battery producer. In 2001 Dow Chemical Company took control of Union Carbide. These take-overs led to a discussion on responsibility for cleaning up the tons of poisonous waste that are still present in the environment consequential to the 1984 disaster. Environmental activists are trying to convince Dow Chemical Company to clean up this potential minefield of toxic chemicals. These could cause nervous system failure, liver and kidney disease and possibly cancer for many years to come.

Today, the location is still polluted with thousands of tons of toxic chemicals, such as hexachlorobenzene and mercury. These chemicals are stored in open barrels. Rainfall causes rinsing out of pollution to local drinking water sources. According to BBC research, some wells even contain up to 500 times the legal limit of these toxins. Local residents still suffer from a number of diseases, which appear to be very uncommon among people that do not live in the disaster area.

2. The 1952 London smog disaster

December 1952 brought an episode of heavy smog to London, which lasted until March 1953. Light winds and a high moisture content created ideal conditions for smog formation. The unusual cold in London in the winter of 1952-1953 caused additional coal combustion and many people travelled only by car, which caused the occurrence of a combination of black soot, sticky particles of tar and gaseous sulphur dioxide. This resulted in the heaviest winter smog episode known to men.

Measurements suggested that the concentration of particulate matter in the air had reached 56 times its normal level. Sulphur dioxide concentrations increased to seven times its peak level. The smoke particles trapped in the fog gave it a yellow-black colour. Sulphur dioxide reacted with substances in foggy droplets to form sulphuric acid, adding an intense form of acid rain to the process.

By night of December 5 the smog was so dense that visibility dropped to only a few meters. Smog easily entered buildings, causing cinemas, theatres and stores to be closed. Transport became largely impossible. Motor vehicles were abandoned, trains were disrupted and airports were also closed.

The smog episode killed approximately 12.000 people, mainly children, elderly people and people suffering from chronic respiratory or cardiac disease. The number of deaths during the smog disaster was three or four times that on a normal day. They could be attributed to lung disease, tuberculosis and heart failure. Mortality from bronchitis and pneumonia increased more than sevenfold.

Peaks of smoke and sulphur dioxide were in line with peaks in deaths. However, most deaths occurred because of breathing in acid aerosols, which irritates or inflames the bronchial tubes. Acidity was not measured, but estimates show that the pH probably fell to 2 during the peaks in the smog episode.

The highest death rate during the smog episode occurred on December 8 and December 9, at 900 deaths per day. In some parts of the city death rates even increase to nine times the normal number. Until spring the death rate remained high at almost a thousand more deaths per week than expected in a normal winter.

This heavy pollution and its resulting death toll made people aware of the seriousness of air pollution. The London smog disaster resulted in the introduction of the first Clean Air Acts in 1956.

3. The Bangladesh arsenic crisis

Arsenic is a poisonous metalloid that can be found in three forms; yellow, grey and black arsenic. Arsenic compounds are used as pesticides and in various alloys. It is not only toxic to insects and some plants, but also to humans. Its toxicity stems from the similarity of the chemical structure to phosphorus, causing it to partly substitute phosphorus in chemical reactions.

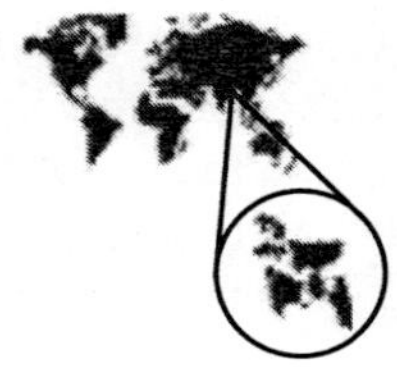

Bangladesh, Asia (see picture) has had major drinking water problems for many decades. Most people used to drink surface water, which led to the spread of pathogens such as cholera and dysentery. International organizations started promoting groundwater welling for drinking water production. It was however not known that groundwater in Bangladesh contained significant amounts of arsenic. The arsenic present in the groundwater is of natural origin, being released from subsurface sediment layers under anoxic conditions. Many other Asian countries, such as Vietnam, Cambodia and Tibet are thought to have similar geological environments as Bangladesh. These countries may also have high-arsenic groundwater.

When drinking water wells were installed in Bangladesh, approximately 57 million Bangladeshi people started drinking groundwater with arsenic concentrations far above the legal limit of 0,05 mg/L. After several years of applying groundwater as drinking water over a quarter of the Bangladeshi population exhibited symptoms of arsenic poisoning (arsenicosis).

Arsenic poisoning kills people by disrupting the digestive system. Symptoms include changes in skin colour, formation of skin patches stomach pains, vomiting, delirium and gangrene. Chronic low level arsenic poisoning in Bangladesh also results in cancers, such as lung cancer, skin cancer, kidney cancer and bladder cancer.

The arsenic problem was first discovered in the early 1980s, but public awareness of the arsenic crisis did not emerge until the mid-1990s. The World Health Organisation has described the naturally occurring arsenic as the largest mass poisoning of a population in history. Today, more than 85 million Bangladeshi people are drinking the arsenic-rich groundwater. It is very probable at least 80 million people now suffer from arsenic poisoning. The exact number is uncertain because it may sometimes take up to 10 years before arsenic poisoning can be diagnosed.

Legal proceedings began in London in 2003 to determine whether the British Geological Survey was negligent in failing to detect arsenic in Bangladeshi water supplies. The organization conducted research on behalf of the Bangladesh government in 1992, but did not test the groundwater for arsenic. The organization pleas 'not guilty' and argues that at the time of its report little was known about the geological origins of arsenic poisoning.

4. **The 2004 Asian earthquake and tsunami**

We have very recently experienced a natural environmental disaster that made worldwide news. In December 2004 an oceanic earthquake of 9-9,3 on a Richter scale caused devastation in Asian countries. The earthquake was among the 10 deadliest in history. Scientists reported that it had lasted nearly ten minutes when most major earthquakes last no more than a few seconds. Since 1900 the only earthquakes recorded with a greater magnitude were the 1960 Great Chilean Earthquake (9,5), the 1964 earthquake in Prince William Sound (9,2), and the 1957 earthquake near the Andreanof Islands (9,1). But all these areas were less densely populated than the Asian earthquake area in 2004 and therefore had a much smaller death toll.

The earthquake centred in the Indian Ocean off the coast of northern Sumatra, Indonesia. It caused the entire planet to vibrate at least a few centimetres. It also triggered earthquakes in entirely different regions, such as Alaska.

The earthquake resulted in a tsunami a series of harbour waves of more than 30 meters high, which washed ashore. It caused in serious damage and killed people in Sri Lanka, South India, Thailand, Indonesia, Somalia, Myanmar, Malaysia, the Maldives and other countries. It took officials months to establish an eventual death toll, which is now estimated at between 230.000 and 310.000 people. Tens of thousands of people are still missing. The death toll of the Asian earthquake is surpassed only by an earthquake in China in 1557, which killed over 830.000 people.

In February 2005, every day 500 bodies were found. These were collected and identified, a process that lasted a very long time and caused a terrible stench in the afflicted regions. Bodies were buried as soon as possible for fear of disease. Among the dead were at least 9.000 foreign tourist that stayed in the region. Sweden was the hardest hit European country with 500 people dead or missing. The true total death toll is not known as many bodies were swept out to sea. Additional deaths may be caused by epidemics and starvation. Millions of people have lost their houses and are packed together in refugee camps on very small space.

The Tsunami not only impacted humans, but also the environment. Severe damage was inflicted on ecosystems such as coral reefs, sand dunes and rock formations, animal and plant biodiversity and groundwater. In addition, the spread of solid and liquid waste and industrial chemicals polluted the water. The destruction of sewage collectors and treatment plants still form a serious threat to the environment. The United Nations Environment Program (UNEP) works with local governments to determine ecological damage and appropriate policy measures.

The Tsunami caused by the earthquake in the Asian region was the worst in human history. The United Nations declared that the relief operation would be the costliest ever. A worldwide collection for tsunami funds resulted in an unbelievable amount of money for emergency aid and restoration of the area. It is estimated this will take between 5 and 10 years.

This year (2006) a report was released by the Tsunami Evaluation Coalition (TEC) on the efficiency of international aid after the earthquake and tsunami in Asia. TEC is a cooperation between UN-organisations such as Unicef and WHO, consultancies, donors and organisations for humanitarian aid such as Oxfam Novib.

The main positive conclusions were:

- High efficiency of the first aid
- Rapid reconstruction of schools and health centres
- Rapid restoring of the fishery sector

The main areas of improvement were:

- The quality of international aid, and support of large international organisations for small local organisations
- A general lack of cooperation between organisations because of large funding
- Unequal distribution of funding; first aid was more adequately funded, whereas more money was required for restoration
- International organisations focus too much on promoting their own name, causing information transfer to be inefficient
- Costs for housing and personnel were far too high
- Disaster response quality should be improved, and this was not attended to so far.

Disaster Management Cycle

It is the illustration of the ongoing process by which governments, businessman and society plan to reduce the impact, during and immediately following a disaster and takes steps to recover after a disaster occurred.

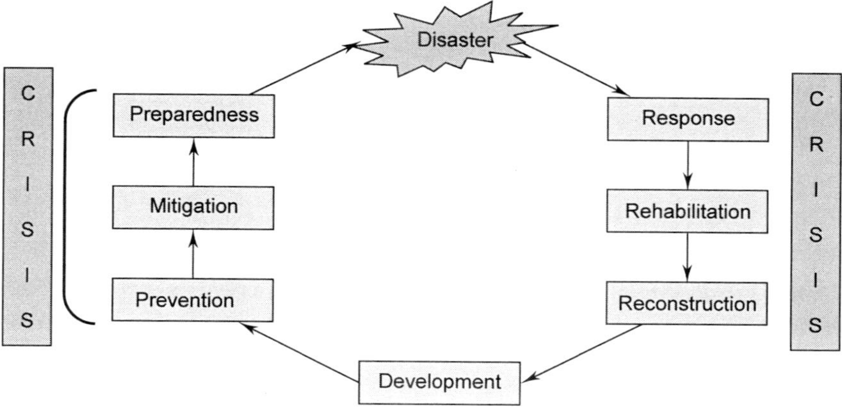

The fundamental aspects of Disaster Management Program.

- Disaster Prevention
- Disaster preparedness
- Disaster response
- Disaster mitigation
- Rehabilitation
- Reconstruction

Three Fundamental Aspects of Disaster Management:

1. Disaster response
2. Disaster Preparedness
3. Disaster Mitigation.

These 3 aspects of Disaster Management corresponds to the 2 phases in the Disaster Cycle, *i.e.*,

1. Risk Reduction Phase, before a Disaster
2. Recovery Phase, after a Disaster.

1. DISASTER RESPONSE

- Appropriate application of current technology can prevent much of the death, injury, and economic disruption resulting from disasters
- Morbidity and mortality resulting from disasters differ according to the type and location of the event.
- In any disaster, prevention should be directed towards reducing:
  1. Losses due to the disaster event itself
  2. Losses resulting from the Mismanagement of disaster relief.

Therefore, the public health objectives of disaster management can be stated as follows:

1. Prevent unnecessary morbidity, mortality, and economic loss resulting directly from the disaster.
2. Eliminate morbidity, mortality, and economic loss directly attributable to Mismanagement of disaster relief efforts.

Nature and Extent of the Problem

Morbidity and mortality, which result from a disaster situation, can be classified into four types:

1. Injuries,
2. Emotional stress,
3. Epidemics of diseases,
4. Increase in indigenous diseases.

The relative numbers of deaths and injuries differ on the type of disaster:

Injuries usually exceed deaths in explosions, typhoons, hurricanes, fires, famines, tornadoes, and epidemics.

Deaths frequently exceed injuries in landslides, avalanches, volcanic eruptions, tidal waves, floods, and earthquakes.

Disaster victims often exhibit emotional stress or the **"disaster shock" syndrome**. The syndrome consists of successive stages of shock, suggestibility, euphoria and frustration.

Each of these stages may vary in extent and duration depending on other factors.

Epidemics are included in the definition of disaster; however, they can also be the result of other disaster situations.

Diseases, which may be associated with disasters, include:

- specific food and/or water bone illnesses (*e.g.*, typhoid, gastroenteritis and cholera),
- vector bone illnesses(*e.g.*, plague and malaria),
- diseases spread by person-to-person contact (*e.g.*, hepatitis A and shigellosis)
- Diseases spread by the respiratory route (*e.g.*, measles and influenza).
- The current status of environmental sanitation, disease surveillance, and preventive medicine has led to a significant reduction in the threat of epidemics following disaster.
- Immunization programs are rarely indicated as a specific post disaster measure.
- A disaster is often followed by an increase in the prevalence of diseases indigenous to the area due to the disruption of medical and other health facilities and programs.

Morbidity and Mortality from Mismanagement of Relief

Ideally, attempts to mitigate the results of a disaster would not add to the negative consequences;

However, there have been many instances in which inappropriate and/or incomplete management actions taken after a disaster contributed to unnecessary morbidity, mortality, and a waste of resources.

Many of the Causalities and much more of the Destruction occurring to natural disaster are due to ignorance and neglect on the part of the individuals and public authorities.

There is a plethora of literature describing the inappropriate actions taken to manage past disasters. Many of the same mismanagement problems tend to recur.

- Physicians and nurses have been sent into disaster areas in numbers far in excess of actual need.
- Medical and paramedical personnel have often been hampered by the lack of the specific supplies they need to apply their skills to the disaster situation.
- In some disasters, available supplies have not been inventoried until well after the disaster, resulting in the importation of material which is used or needed.

In a study of past disaster mismanagement problems and their causes, these problems were categorized as follows:

1. Inadequate appraisal of damages
2. Inadequate problem ranking
3. Inadequate identification of resources
4. Inadequate location of resources
5. Inadequate transportation of resources
6. Inadequate utilization of resources.

2. DISASTER RELIEF (Rehabilitation)

An effective plan for public health and other personnel during a disaster would outline activities designed to minimize the effects of the catastrophe.

These efforts can be summarized as closely **situation analysis and response;** the two types of activities are interrelated.

Although many relief workers may be needed to obtain surveillance information, analyze the data, provide relief services, evaluate results, and provide information to the public, it is essential that a single person with managerial experience be placed in absolute charge of the entire disaster relief operation.

Following a disaster, the desire to provide immediate relief may lead to hasty decisions which are not based on the actual needs of the affected population.

The disaster relief managers can determine the actual needs of the population and make responsible relief decisions.

Reliable information must be obtained on problems occurring in the disaster stricken area, relief resources available and relief activities already in progress.For This, a Surveillance systems must be set up immediately.

The objective of Surveillance in a disaster situation is to obtain information required for making relief decisions.

The specific information required would vary from disaster to disaster, but a basic, three -step processes includes:

1. Collect data
2. Analyze data
3. Respond to data.

The analysis involves collecting and interpreting the data and can include asking questions as the following:

- What problems are occurring? Why are they occurring?
- Where are problems occurring?
- Who is affected?
- What problems are causing the greatest morbidity and mortality?
- What problems are increasing or decreasing?
- What problems will subside on their own?
- What problems will increase if unattended?

- What relief resources are available?
- Where are relief resources available?
- How can relief resources be used most efficiently?
- What relief activities are in progress?
- Are relief activities meeting relief needs?
- What additional information is needed for decision making?

After answering such questions one can carry out the third part, *i.e.*, planning an appropriate Response to the situation described in the surveillance data.

In developing this plan one will decide what types of relief responses are appropriate and what the relative priorities are among the relief activities.

This 3-step process of Data Collection, Analysis and Response can be described as a closed feedback system involving re-evaluation of relief needs and their effects.

Surveillance following a disaster evolves in phases:

1. Immediate Assessment
2. Short term assessment
3. Ongoing Surveillance.

Immediate Assessment

The object of this phase of surveillance is to obtain as much general information as possible and as quickly as possible.

The most basic information needed at this point is the following:

1. The geographical extent of the disaster-stricken area
2. The major problems occurring in the area
3. The number of people effected.

This information can be obtained by whatever means seems most efficient. Listening carefully and asking questions is the best way to begin.

An Arial survey may be useful in defining the geographical extent of the disaster-stricken area and in observing major damage and destruction.

Census data can be examined to determine how many people previously lived in the disaster-stricken area and thus were at risk.

Hospitals, clinics, and morgues, which were in operation, may be able to obtain numbers of known deaths and injuries.

It is useful to determine the most frequent causes of deaths and types of injuries in order to predict whether demands for medical care will be increasing or decreasing.

Some problems likely to occur after a disaster can be predicted according to past experience with that particular type of disaster.

For example, experience has shown that disruption of water supplies has often been a problem following earthquakes.

New types of disasters, such as chemical emergencies and nuclear accidents, still present many unknown problems.

Short-term Assessment

The short-term assessment involves more systematic methods of collecting data and is likely to result in more detailed reliable information on problems, relief resources, and relief information on problems, relief resources and relief activities in progress.

One way to organize data collection during this phase of assessment is to divide the disaster-stricken area into smaller areas or "blocks" to be surveyed simultaneously by different workers or teams of workers.

Simple reporting forms can be developed and workers sent out to survey the different areas and report at a specified time.

The following is a list of Information, which may be needed in order to make relief decisions:

- The geographical extent of the affected area as defined by streets and other clear boundaries.
- The number of persons known to be dead, possibly according to age groups and sex.
- The estimated number of persons severely injured and/requiring medical care, possibly according to age group, sex, and type of injury or medical problem.
- Estimated number of homes destroyed, homes uninhabitable, and homes, which are still habitable.
- Condition of schools, churches, temples and other public buildings etc.
- Condition and extent of water supply.
- Condition and extent of food supply.
- Condition of roads, bridges, communication facilities and public utilities.
- Location and condition of health facilities
- Estimates of medical personnel, equipment's and supplies available
- Description of relief activities already in progress (*e.g.*, search and rescue, first aid, food relief etc).

Ongoing Surveillance

Depending on the factors above, short-term assessment may take as little as 5-6 hours or up to 3-4 days. As early as possible, relief priorities should be determined, resources ordered and full scale relief activities initiated.

Once the short-term assessment is complete and appropriate relief is in progress, surveillance becomes an ongoing system.

When information obtained by the ongoing surveillance is analyzed, new problems may become apparent, requiring investigation.

The surveillance report is one way of coordinating different agencies and preventing duplication of relief efforts.

A relief plan developed during any of the surveillance cycle may include some or all of the following activities:

- Rescue of victims
- Provision of emergency medical care
- Elimination of physical dangers (fire, gas leak etc)
- Evacuation of the population (nuclear and chemical emergencies)
- Provision of preventive and routine medical care

- Provision of water
- Provision of food
- Provision of clothing
- Provision of shelter
- Disposal of human waste
- Control of vector born diseases
- Disposal of human bodies
- Disposal of solid waste.

3. MASS CASUALTY MANAGEMENT

Management of mass casualties is divided into three main areas:

1. Pre-Hospital Emergency Care:
    - Search and Rescue
    - First Aid
    - Field Care
    - Stabilization of the victims
    - Triage
    - Tagging.
2. Hospital Reception and Treatment:
    - Organizational structure in the hospital with a disaster management team consists of senior officers in the medical, nursing and administrative fields.
    - Standardized simple therapeutic procedures followed.
3. Re-distribution of Patients between Hospitals.

4. DISASTER PREPAREDNESS

The objectives of the disaster preparedness is to ensure that appropriate systems, procedures and resources are in place to provide prompt, effective assistance to disaster victims, thus facilitating relief measures and rehabilitation services.

Disaster preparedness is an ongoing, multi-sectoral activity to carry out the following activities;

- Evaluate the risk of the country or particular region to disasters.
- Adopt standards and regulations.
- Organize communication, information and warning systems.
- Ensure coordination and response mechanisms.
- Adopt measures to ensure that financial and other resources are available for increased readiness and can be mobilized in disaster situations.
- Develop public education programs.
- Coordinate information sessions with news media.
- Organize disaster simulation exercises that test response mechanisms.

  For the Health Sectors Disaster Preparedness plan to be successful, clear mechanisms for coordinating with other sectors and internationally must be in place.

The Health Disaster Coordinator is in charge of preparedness activities and coordinating plans with.

- Govt. Agencies.
- Foreign Relations- UN,UNICEF.WHO & other international agencies.
- NGO's- Red Cross etc.,
- Those responsible for power, communication, Housing, water services etc.,
- Civil Protection agencies-Police, armed forces.

5. DISASTER MITIGATION

It is virtually impossible to prevent occurrence of most Natural Disasters, but it is possible to minimize or mitigate their damage effects.

Mitigation measures aim to reduce the Vulnerability of the System [*e.g.*, By improving & enforcing building codes etc].

Disaster prevention implies complete elimination of damages from a hazard, but it is not realistic in most hazards. [*e.g.*, Relocating a population from a flood plain or from beach front].

Medical Casualty could be drastically reduced by improving the Structural Quality of Houses, Schools, Public or Private buildings.

Also ensuring the Safety of Health facilities, Public Health Services, Water Supply, Sewerage System etc.

Mitigation complements the Disaster Preparedness and Disaster Response activities.

A Specialised Unit within the National Health Disaster Management Program should coordinate the works of experts in the field of:

- Health, Public Policy & Public Health
- Hospital Administration
- Water Systems
- Engineering & Architecture
- Planning, Education etc

The Mitigation Program will direct the following activities:

1. Identify areas exposed to Natural Hazards and determine the vulnerability of key health facilities and water systems.
2. Coordinate the work of Multi Disciplinary teams in designing.
   and developing building codes and protect the water distribution from damages.
3. Hospitals must remain operational to attend to disaster victims.
4. Include Disaster Mitigation Measures in the planning and development of New facilities.
5. Identify priority hospitals and critical health facilities that complies with current building codes and standards.
6. Ensure that mitigation measures are taken into account in a facility's maintenance plans.
7. Inform, sensitize and train those personnel's who are involved in planning, administration, operation, maintenance and use of facilities about disaster mitigation.
8. Promote the inclusion of Disaster Mitigation in the curricula of Professional training institutes.

TECHNICAL HEALTH PROGRAMS

- Treatment of casualties
- Identification and disposal of bodies
- Epidemiological surveillance and disease control
- Basic sanitation and sanitary engineering
- Health management in shelters or temporary settlements
- Training health personnel and the public
- Logistical resources and support
- Simulation exercises/Mock Exercises
    1. Desktop simulation exercises [war games]
    2. Field exercises
    3. Drills designed to impart skills.

EPIDEMIOLOGIC SURVEILLANCE AND DISEASE CONTROL

Natural disasters may increase the risk of preventable diseases due to adverse changes in the following areas:

- Population density
- Population displacement
- Disruption and contamination of water supply and sanitation services
- Disruption of public health programs
- Ecological changes that favor breeding of vectors
- Displacement of domestic and wild animals
- Provision of emergency food, water and shelter in disaster situation

The principles of preventing and controlling communicable diseases after a disaster are to;

- Implement as soon as possible all public health measures to reduce the risk of disease transmission
- Organize a reliable disease reporting system to identify outbreaks and to promptly initiate control measures
- Investigate all reports of disease outbreaks rapidly. Early clarification of the situation may prevent unnecessary dispersion of scarce resources and disruption of normal progress.

ENVIRONMENTAL HEALTH MANAGEMENT

Post disaster environmental health measures can be divided into two priorities

1. Ensuring that there are adequate amounts of safe drinking water, basic sanitation facilities, disposal of excreta, waste water and solid wastes and adequate shelter.
2. Providing food protection measures, establishing or continuing vector control measures, and promoting personal hygiene
    - Water Supply

        Alternate water sources

        Mass distribution of Disinfectants

- Food Safety
- Basic Sanitation and Personal Hygiene
- Solid Waste Management
- Vector Control
- Burial of the Dead
- Public information and the Media

9.3 EVALUATION

In the case of disaster management, the Evaluator will be looking at the **"actual" verses the "desired"** on two levels, *i.e.,* the overall outcome of disaster management efforts and the impact of each discrete category of relief efforts. (Provision of food, shelter, management of communications etc)

A critical step in the management of any disaster relief is the setting of objectives, which specify the intended outcome of the relief.

The general objectives of the disaster management will be the elimination of unnecessary morbidity, mortality and economic loss directly and indirectly attributable to mismanagement of disaster relief.

The comparison of the "actual" with "desired" is the first critical step of evaluation. If the objectives were met, those who have participated in the relief have demonstrated that they have accomplished what they set out to do.

On the other hand, if the objectives were not met, it is desirable for those conducting the evaluation to continue with the evaluation process, identify the reasons for the discrepancy and suggest corrective action.

Simulated Disaster Preparedness Operations should be undertaken to test the various components before actual need arise.

Evaluation of the health disaster management program:

- Evaluation of the preparedness program
- Evaluation of the mitigation measures
- Evaluation of the training.

9.4 PREVENTION OF DISASTERS

Existing knowledge that might reduce the undesirable effects of disasters is often not applied.

- Hurricane/Tornado/Cyclone warning systems
- Legislation preventing building in the flood prone areas
- Requirement of protective cellars/shelters in disaster prone areas
- A Seismic housing code for earthquake-prone area
- Strict procedural code followed to prevent Nuclear, Toxicological and Chemical disasters
- Early warning systems, and Disaster preparedness which will help to minimize morbidity, mortality and economic loss.

OVERVIEW OF THE DISASTER RISK MANAGEMENT PROGRAMME

Government of India [GoI], Ministry of Home Affairs [MHA] and United Nations Development Programme [UNDP] have signed an agreement on August 2002 for implementation of ***"Disaster Risk Management"*** Programme to reduce the vulnerability of the communities to natural disasters, in identified multi–hazard disaster prone areas.

Goal: *"Sustainable Reduction in Natural Disaster Risk in some of the most hazard prone districts in selected states of India".*

The four main ***objectives*** of this programme are:

1. National capacity building support to the Ministry of Home Affairs
2. Environment building, education, awareness programme and strengthening the capacity at all levels in natural disaster risk management and sustainable recovery
3. Multi-hazard preparedness, response and mitigation plans for the programme at state, district, block and village/ward levels in select programme states and districts
4. Networking knowledge on effective approaches, methods and tools for natural disaster risk management, developing and promoting policy frameworks.

Programme Phases:

The programme has been divided into two phases over a period of six years. Phase I [2002-2004] would provide support to carry out the activities in 28 select districts in the states of Bihar, Gujarat and Orissa. In phase II [2003-2007], the Programme would cover 141 districts in the states of Assam, Meghalaya, Sikkim, West Bengal, Uttaranchal, Uttar Pradesh, Delhi, Maharashtra, Tamilnadu, Manipur, Mizoram, Tripura, Arunachal Pradesh and Nagaland.

Special Focus: 38 Earthquake prone cities having more than half a million population.

VULNERABILITY OF INDIAN CONTINENT TO DIFFERENT TYPES OF DISASTERS

(India's natural disasters proneness):

On the basis of geographic and climatic considerations, India can be divided into 5 Zones according to its disaster proneness to natural disasters;

1. **Northern mountain region including foot hills**; this region is prone to strong Snow Storms leading to Land slides and strong Cold waves and also is Earthquake prone belt with violent subterranean Volcanic activity.
2. **Indo-gangetic plains**; heavy rains during monsoon make these plains vulnerable to Floods.
3. **Deccan plateau**; a Drought prone area.
4. **The western desert**; a Drought prone area.
5. **Coastal areas**; they are prone to Sea erosion, Cyclones and Tidal waves

India's Disaster Ridden History

- About 60% of India's land mass is prone to Earth Quakes
- Over 40 million Hectares are prone to Floods
- Nearly 3 lakh sq. km are at risk of Cyclones
- The Earth quake in Bhuj killed 14,000 people
- Cyclone in Orissa took away 10,000 lives.

- Between 1990 and 2000 an average of about 3400 people lost their lives annually.
- About 3 crore people were affected by Disasters every year.
- About 17,000 people perished by the Tsunami on 26 Dec.04.

9.5 POPULATION GROWTH AND DISASTERS

Hardly a day passes without news about a major or complex emergency happening in some part of the World.

Disasters continue to strike and cause destruction in developing and developed countries alike, raising peoples concern about their vulnerability to occurrences that can gravely affect their day to day life and their future.

Major disasters have had a big impact on the migration of populations and related health problems, and many millions are struggling for minimum vital health and sanitation needs and suffer from malnutrition.

Vulnerable Populations

Emergencies, especially those that occur in Nature, only become catastrophic events when they combine with vulnerability factors such as human settlements and population density.

An earthquake occurring in a deserted area would be considered a natural hazard; but if it occurred in a mega city it would be recognized as a major disaster.

Man made emergencies and another type of emergency that has to do with population vulnerability concerns technological disasters such as those of a Chemical or Radiological or Nuclear in nature. *e.g.*, Bhopal Gas Tragedy and Chernobyl nuclear disaster.

These examples demonstrate that major and complex emergencies are closely linked to anarchic population growth, leading to unplanned population settlement, environmental degradation and poverty.

The lack of minimum health services and basic health education are aggravating factors which could make a disaster out of an emergency and a complex emergency out of social tension

EXERCISE

1. What do you mean by the term disaster? Mention the types of disasters and explain Earthquake hazards in India.
2. Explain the extent of disaster and detailed account at disaster management steps for any of the national disaster.
3. Explain disaster management programme in detail.
4. Define a natural hazard. Discuss the problems of earthquakes with special reference of India.
5. Write short notes on:
   (*i*) Flood
   (*ii*) Drought
   (*iii*) Natural disaster management
   (*iv*) Migration and preparedness

6. Write short notes on:
   (*i*) Earthquake
   (*ii*) Cyclone
   (*iii*) Epidemic
   (*iv*) Disaster and violence
7. Explain in brief the causes, prevention and management of flood and drought.
8. What is a cyclone? What hazards are associated with a cyclone. Briefly discuss the steps you will take in preparing for a cyclone.
9. With suitable example, discuss the impact of development process on environmental degradation and human health.
10. What is an earthquake? Enumerate its effect. What measures should be taken to mitigate this disaster.
11. Mention the steps which should be taken for prevention of life and property due to earthquake.
    (*i*) When the earthquake shaking is over.
    (*ii*) During the period of earthquake.
12. Enumerate the various sources of nuclear hazards? Briefly discuss the effects and control of nuclear hazards.
13. Define flood. What are its causes. What measures should be taken to mitigate flood disaster.
14. Write short notes on:
    (*i*) London smog disaster 1552
    (*ii*) Bhopal disaster 1984.

10

Seismology and Earthquakes

10.1 INTRODUCTION: WHAT IS SEISMOLOGY?

Seismology is the study of earthquakes and seismic waves that move through and around the earth. This is relatively young science, developed later than most of the physical sciences and is more related to geophysics, geology and geography. Seismology is a combination of two words *seismos* (earthquake) and *logos* (study). Therefore a **seismologist** is a scientist who studies earthquakes and seismic waves.

Seismic Waves are the waves of energy caused by the sudden breaking of rock within the earth or an explosion. They are the energy that travels through the earth and is recorded on seismographs.

Types of Seismic Waves

There are several different kinds of seismic waves, and they all move in different ways. The two main types of waves are **body waves** and **surface waves.** Body waves can travel through the earth's inner layers, but surface waves can only move along the surface of the planet like ripples on water. Earthquakes radiate seismic energy as both body and surface waves.

1. **Body Waves:** Traveling through the interior of the earth, **body waves** arrive before the surface waves emitted by an earthquake. These waves are of a higher frequency than surface waves. It can be further of two types, **Primary Waves (P-waves)** and **Secondray Waves (S-waves).**

   (*i*) **Primary Waves (P-waves):** The first kind of body wave is the **P wave** or **primary wave.** This is the fastest kind of seismic wave, and, consequently, the first to 'arrive' at a seismic station. The P wave can move through solid rock and fluids, like water or the liquid layers of the earth. It pushes and pulls the rock it moves through just like sound waves push and pull the air. Have you ever heard a big clap of thunder and heard the windows rattle at the same time? The windows rattle because the sound waves were pushing and pulling on the window glass much like P waves push and pull on rock. Sometimes animals can hear the P waves of an earthquake. Dogs, for instance, commonly begin barking hysterically just before an earthquake 'hits' (or more specifically, before the surface waves arrive). Usually people can only feel the bump and rattle of these waves.

   P waves are also known as **compressional waves,** because of the pushing and pulling they do. Subjected to a P wave, particles move in the same direction that the wave is moving in, which is the direction that the energy is traveling in, and is sometimes called the 'direction of wave propagation'.

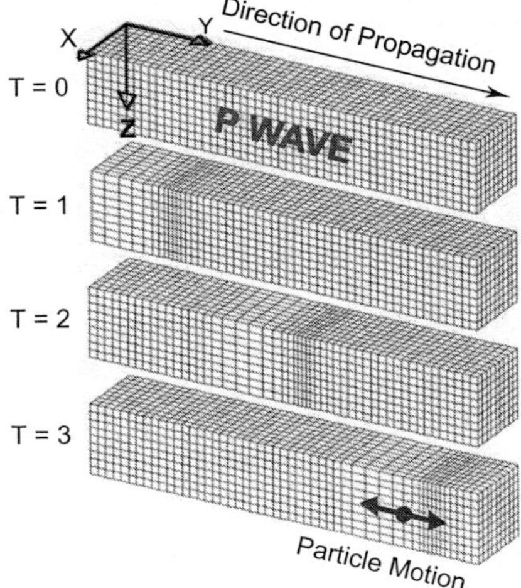

Fig. 10.1. A p wave travels through a medium by means of compression and dilation. Particles are represented by cubes in this model.

(*ii*) **Secondray Waves (S-waves):** The second type of body wave is the **S wave** or **secondary wave**, which is the second wave you feel in an earthquake. An S wave is slower than a P wave and can only move through solid rock, not through any liquid medium. It is this property of S waves that led seismologists to conclude that the Earth's **outer core** is a liquid. S waves move rock particles up and down, or side-to-side--perpindicular to the direction that the wave is traveling in (the direction of wave propagation).

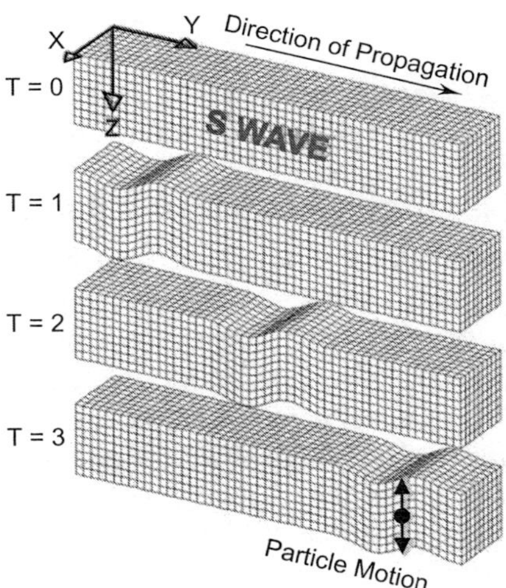

Fig. 10.2. An s wave travels through a medium. Particles are represented by cubes in this model.

2. **Surface Waves:** Travelling only through the crust, **surface waves** are of a lower frequency than body waves, and are easily distinguished on a seismogram as a result. Though they arrive after body waves, it is surface waves that are almost enitrely responsible for the damage and destruction associated with earthquakes. This damage and the strength of the

surface waves are reduced in deeper earthquakes. It can be further of two types, **Love waves** and **Rayleigh Waves.**

(*i*) **Love Waves:** The first kind of surface wave is called a **Love wave,** named after A.E.H. Love, a British mathematician who worked out the mathematical model for this kind of wave in 1911. It's the fastest surface wave and moves the ground from side-to-side. Love waves are confined to the surface of the crust and produce entirely horizontal motion.

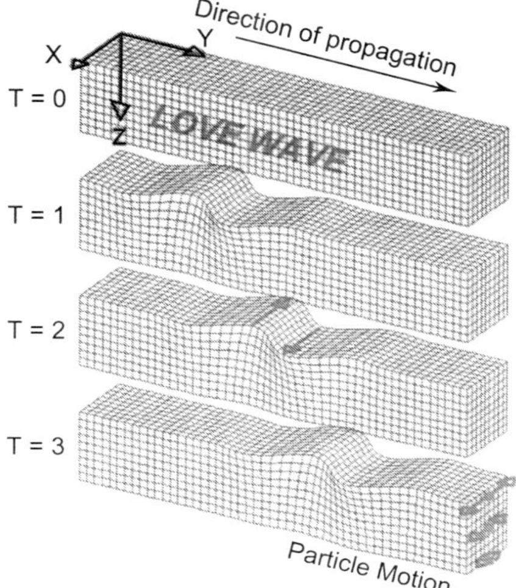

Fig. 10.3. A love wave travels through a medium. Particles are represented by cubes in this model.

(*ii*) **Rayleigh Waves:** The other kind of surface wave is the **Rayleigh wave,** named for John William Strutt, Lord Rayleigh, who mathematically predicted the existence of this kind of wave in 1885. A Rayleigh wave rolls along the ground just like a wave rolls across a lake or an ocean. Because it rolls, it moves the ground up and down, and side-to-side in the same direction that the wave is moving. Most of the shaking felt from an earthquake is due to the Rayleigh wave, which can be much larger than the other waves.

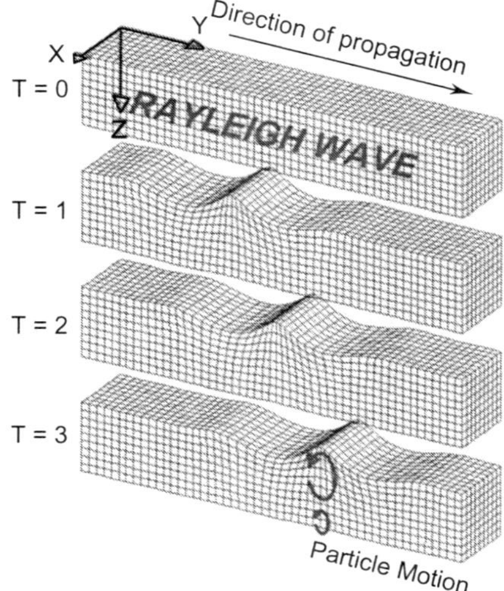

Fig. 10.4 A rayleigh wave travels through a medium. Particles are represented by cubes in this model.

What Are Earthquake Hazards?

"An earthquake is a sudden motion or trembling of the ground produced by the abrupt displacement of rock masses".

Most earthquakes result from the movement of one rock mass past another in response to tectonic forces.

Earthquakes really pose little direct danger to a person. People can't be shaken to death by an earthquake. Some movies show scenes with the ground suddenly opening up and people falling into fiery pits, but this just doesn't happen in real life.

1. **Ground Shaking.** The first main earthquake hazard (danger) is the **effect of ground shaking.** Buildings can be damaged by the shaking itself or by the ground beneath them settling to a different level than it was before the earthquake **(subsidence).**

   Buildings can even sink into the ground if soil liquefaction occurs. **Liquefaction** is the mixing of sand or soil and **groundwater** (water underground) during the shaking of a moderate or strong earthquake. When the water and soil are mixed, the ground becomes very soft and acts similar to quicksand(daldal). If liquefaction occurs under a building, it may start to lean, tip over, or sink several feet. The ground firms up again after the earthquake has past and the water has settled back down to its usual place deeper in the ground. Liquifaction is a hazard in areas that have groundwater near the surface and sandy soil.Buildings can also be damaged by strong surface waves making the ground heave and lurch. Any buildings in the path of these surface waves can lean or tip over from all the movement. The ground shaking may also cause landslides, mudslides, and avalanches on steeper hills or mountains, all of which can damage buildings and hurt people.
2. **Ground Displacement.** The second main earthquake hazard is **ground displacement** (ground movement) along a fault. If a structure (a building, road, etc.) is built across a fault, the ground displacement during an earthquake could seriously damage or rip apart that structure.
3. **Flooding.** The third main hazard is **flooding**. An earthquake can **rupture** (break) dams or levees along a river. The water from the river or the reservoir would then flood the area, damaging buildings and maybe sweeping away or drowning people.

   Tsunamis and seiches can also cause a great deal of damage. A **tsunami** is what most people call a tidal wave, but it has nothing to do with the tides on the ocean. It is a huge wave caused by an earthquake under the ocean. Tsunamis can be tens of feet high when they hit the shore and can do enormous damage to the coastline. **Seiches** are like small tsunamis. They occur on lakes that are shaken by the earthquake and are usually only a few feet high, but they can still flood or knock down houses, and tip over trees.
4. **Fire.** The fourth main earthquake hazard is **fire.** These fires can be started by broken gas lines and power lines, or tipped over wood or coal stoves. They can be a serious problem, especially if the water lines that feed the fire hydrants are broken, too. For example, after the Great San Francisco Earthquake in 1906, the city burned for three days. Most of the city was destroyed and 250,000 people were left homeless. Most of the hazards to people come from man-made structures themselves and the shaking they receive from the earthquake. The real dangers to people are being crushed in a collapsing building, drowning in a flood caused by a broken dam or levee, getting buried under a landslide, or being burned in a fire.

Plate Tectonics Theory

Plate Tectonic theory is based on an earth model characterized by a small number of lithospheric plates, 70 to 250 km (40 to 150 mi) thick, that float on a viscous underlayer called the asthenosphere. These plates, which cover the entire surface of the earth and contain both the continents and seafloor, move relative to each other at rates of up to ten cm/year (several inches/year). The region where two plates come in contact is called a plate boundary, and the way in which one plate moves relative to another determines the type of boundary: **spreading,** where the two plates move away from each other; **subduction,** where the two plates move toward each other and one slides beneath the other; and **transform,** where the two plates slide horizontally past each other. Subduction zones are characterized by deep ocean trenches, and the volcanic islands or volcanic mountain chains associated with the many subduction zones around the Pacific rim are sometimes called the Ring of Fire.

According to the plate tectonic model, the surface of the Earth consists of a series of relatively thin, but rigid, plates which are in constant motion. The surface layer of each plate is composed of oceanic crust, continental crust or a combination of both. The lower part consists of the rigid upper layer of the Earth's mantle. The rigid plates pass gradually downwards into the plastic (soft) layer of the mantle, the astenosphere. The plates may be up to 70 km thick if composed of oceanic crust or 150 km incorporating continental crust. Plates move at different velocities, The African plate moves about 25 mm per year, whereas the Australian plate moves about 60 mm per year.

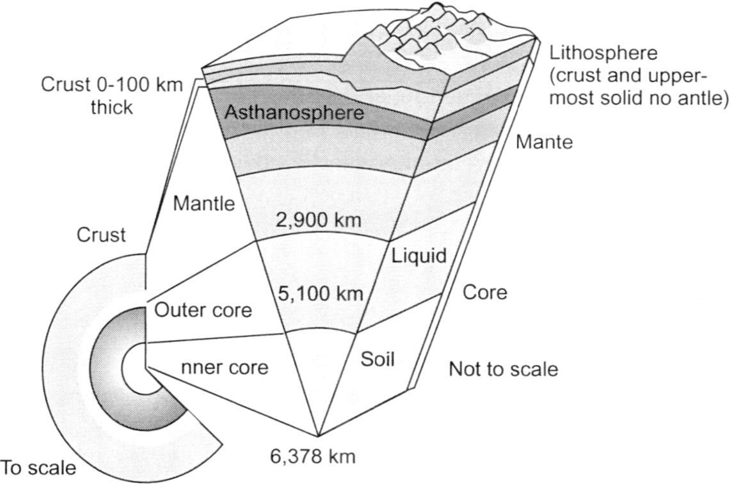

Fig. 10.5. Earth's Layers

Most of the Earth's tectonic, seismic and volcanic activity occurs at the boundaries of neighbouring plates. There are three type of plate boundaries: divergent, convergent and transform boundaries.

1. **Divergent plate margins.** At this type of boundary new oceanic crust is formed in the gap between two diverging plates. Plate area is increased as the plates move apart. Plate movement takes place laterally away from the plate boundary, which is normally marked by a rise or a ridge. The ridge or rise may be offset by a transform fault. Presently, most divergent margins occur along the central zone of the world's major ocean basins. The process by which the plates move apart is referred to as **sea floor spreading.**

   The Mid-Atlantic Ridge and East Pacific Rise provide good examples of this type of plate margin.

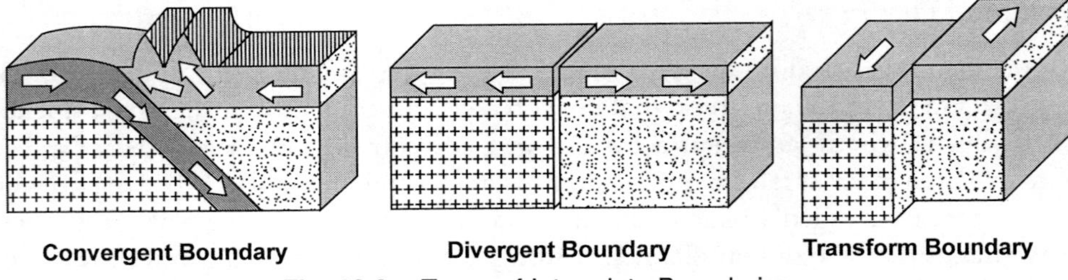

Fig. 10.6. Types of Inter-plate Boundaries

The rate at which each plate moves apart from a divergent margin varies from less than 50 mm per year to over 90 mm per year and can be determined from the pattern of magnetic anomalies either side of a spreading ridge. Either side of a spreading centre, weak magnetic anomalies 5-50 km wide and hundreds of kilometres long can be identified. Molten rock cools between diverging plates the magnetic minerals present align themselves with the orientation of the Earth's magnetic field at that time. The polarity of the Earth has changed at regular intervals throughout geological time. Magnetic north has alternated between the Arctic (normal polarity) and the Antarctic (reversed polarity). As a result of this, sections of crust formed during a period of normal polarity have a **paleomagnetic remnance** which is oriented towards today's magnetic north, while a section of crust formed during a period of reversed polarity does not. These long linear strips of magnetic anomalies form a symmerical pattern either side of a spreading centre. A record of the changes in the Earth's magnetic polarity has been established and dated for the Cenozoic and is the basis for **magnetostratigraphy**. This record, in conjunction with the magnetic stripes found either side of a spreading ridge, allows the rate and pattern of sea floor spreading to be examined.

Magnetostratigraphy uses records of changes in polarity of the geomagnetic field preserved in sedimentary sequences to correlate between wells and to date the sediment. Individual normal and reverse polarity intervals ("Chrons") typicaly range from ~10 thousand to 10 million years in duration. For example:

Correlating and dating non-marine sequences by Magnetostratigraphy

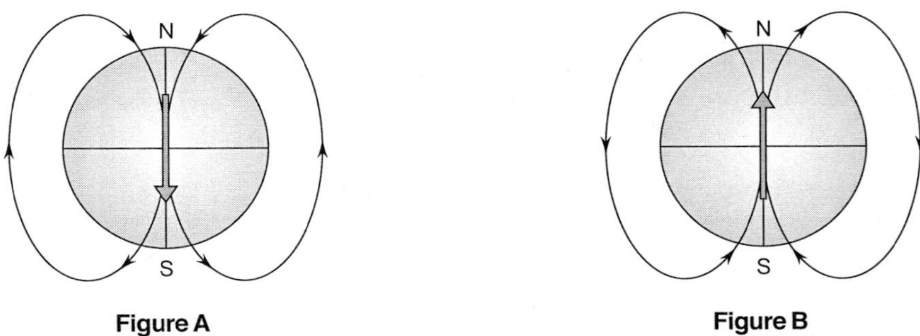

Fig. 10.7. Configuration of the axial dipole geomagnetic field during (A) a normal polarity interval and (B) a reverse polarity interval.

Since geomagnetic polarity reversals are globally synchronous, their records represent "absolute" time planes in sedimentary sequences which can provide a robust stratigraphic correlation framework. This is especialy useful in biostratigraphically-barren sequences. Furthermore, sets of reversals often carry distinctive "fingerprints", which can be matched with appropriate parts of the standard.

2. **Convergent plate boundaries.** At a convergent boundary two plates are in relative motion towards each other. One of the two plates slides down below the other at an angle of around 45 degrees and is incorporated into the Earth's mantle along a **subduction zone.** The path of this descending plate can be found from analysis of deep earthquakes and the initial point of descent is marked on the surface by a deep **ocean trench.** Plate area is reduced along the subduction zone. When two plates of oceanic crust collide a **volcanic island arc** may form. As one of the plates is subducted beneath the other it begins to melt at a depth of between 90 and 150 km and the resulting magma rises to the surface above the subduction zone to form a chain or arc of volcanoes. The edge of the plate which is not descending is therefore marked by a chain of volcanic islands.
3. **Conservative or transform margins.** Transform plate boundaries, where plates slide horizontally against each other, neither create nor destroy lithosphere. However, at these boundaries, or transform faults, powerful earthquakes can occur. The San Andreas fault system is the most famous example of this type of boundary. Here two plates move laterally past each other and oceanic crust is neither created nor destroyed.

What causes plates to move ?

This question has yet to be fully resolved. Four main hypotheses have been put forward to explain this.

1. **Convection currents.** This hypothesis suggests that flow in the mantle is induced by **convection currents** which drag and move the lithospheric plates above the astenosphere. Convection currents rise and spread below divergent plate boundaries and converge and descend along convergent. Three sources of heat produce the convection currents:
    1. cooling of the Earth's core
    2. radioactivity within the mantle and crust
    3. cooling of the mantle

    The convection hypothesis has been proposed in several different forms throughout the last 60 years. Convective models of plate evolution clearly show how important convective heat transport is to the modern Earth, over length scales as small as 100 km and times of 60 million years. Earth is a spendthrift, living on its inherited capital of primaeval heat, not on its radiogenic modern income.
2. **Magma injection.** This hypothesis invokes the injection of magma at a spreading centre pushing plates apart and thereby causing plate movement.
3. **Gravity** Oceanic lithosphere thickens as it moves away from a spreading centre and cools, a configuration which might tend to induce plates to slide under the force of gravity, from a divergent margin towards a convergent margin.
4. **Descending plates.** This hypothesis suggests that a cold dense plate descending into the mantle at a subduction zone may pull the rest of the plate with it and thus cause plate motion.

10.2 TYPES OF EARTHQUAKES

There are many different types of earthquakes: tectonic, volcanic, and explosion. The type of earthquake depends on the region of occurence and the geological make-up of that region. The most common are **tectonic earthquakes.** These occur when rocks in the earth's crust break due to geological forces created by movement of tectonic plates. Another type, **volcanic earthquakes,** occur in conjunction with volcanic activity. **Collapse earthquakes** are small earthquakes in underground caverns and mines, and **explosion earthquakes** result from the explosion of nuclear and chemical devices.

1. **Tectonic Earthquake:** Earthquake that occurs when the earth's crust breaks due to geological forces on rocks and adjoining plates that cause physical and chemical changes.
2. **Volcanic Earthquakes:** Earthquakes that result from tectonic forces which occur in conjunction with volcanic activity.
3. **Collapse Earthquakes:** Small earthquakes in underground caverns and mines that are caused by seismic waves produced from the explosion of rock on the surface.
4. **Explosion Earthquakes:** Earthquakes which are the result of the detonation of nuclear and chemical devices.

Why Do Earthquakes Happen?

Earthquakes are usually caused when rock underground suddenly breaks along a fault. This sudden release of energy causes the seismic waves that make the ground shake. When two blocks of rock or two plates are rubbing against each other, they stick a little. They don't just slide smoothly; the rocks catch on each other. The rocks are still pushing against each other, but not moving. After a while, the rocks break because of all the pressure that's built up. When the rocks break, the earthquake occurs. During the earthquake and afterward, the plates or blocks of rock start moving, and they continue to move until they get stuck again. The spot underground where the rock breaks is called the **focus** of the earthquake. The place right above the focus (on top of the ground) is called the **epicenter** of the earthquake.

Try this little experiment

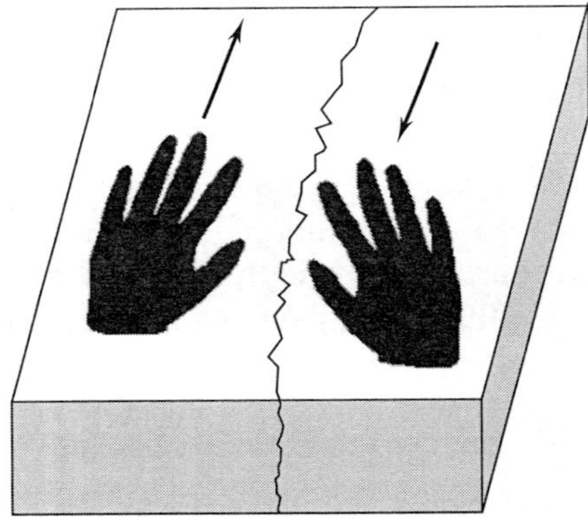

Fig. 10.8.

1. Break a block of foam rubber in half.
2. Put the pieces on a smooth table.
3. Put the rough edges of the foam rubber pieces together.
4. While pushing the two pieces together lightly, push one piece away from you along the table top while pulling the other piece toward you. See how they stick?
5. Keep pushing and pulling smoothly. Soon a little bit of foam rubber along the crack (the fault) will break and the two pieces will suddenly slip past each other. That sudden breaking of the foam rubber is the earthquake. That's just what happens along a strike-slip fault.

Earthquake-like seismic waves can also be caused by explosions underground. These explosions may be set off to break rock while making tunnels for roads, railroads, subways, or mines. These explosions, however, don't cause very strong seismic waves. You may not even feel them.

Sometimes seismic waves occur when the roof or walls of a mine collapse. These can sometimes be felt by people near the mine. The largest underground explosions, from tests of nuclear warheads (bombs), can create seismic waves very much like large earthquakes. This fact has been exploited as a means to enforce the global nuclear test ban, because no nuclear warhead can be detonated on earth without producing such seismic waves.

Where Do Earthquakes Happen?

Earthquakes occur all the time all over the world, both along plate edges and along faults.

Along Plate Edges

Most earthquakes occur along the edge of the **oceanic** and **continental plates.** The earth's **crust** (the outer layer of the planet) is made up of several pieces, called **plates**. The plates under the oceans are called oceanic plates and the rest are continental plates. The plates are moved around by the motion of a deeper part of the earth (the **mantle**) that lies underneath the crust. These plates are always bumping into each other, pulling away from each other, or past each other. The plates usually move at about the same speed that your fingernails grow. Earthquakes usually occur where two plates are running into each other or sliding past each other.

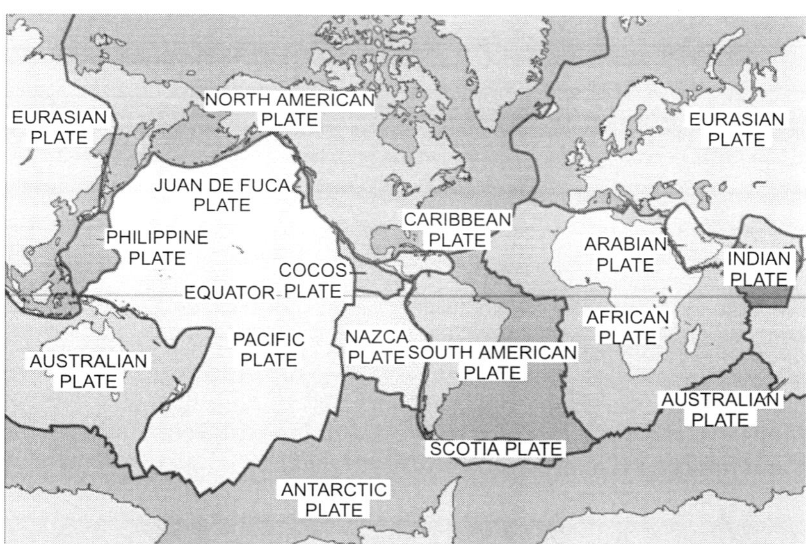

Fig. 10.9. An image of the world's plates and their boundaries. Notice that many plate boundaries do not coincide with coastlines.

Along Faults

Earthquakes can also occur far from the edges of plates, along faults. **Faults** are cracks in the earth where sections of a plate (or two plates) are moving in different directions. Faults are caused by all that bumping and sliding the plates do. They are more common near the edges of the plates.

Types of Faults

Normal faults are the cracks where one block of rock is sliding downward and away from another block of rock. These faults usually occur in areas where a plate is very slowly splitting apart or where two plates are pulling away from each other. A normal fault is defined by the hanging wall moving down relative to the footwall, which is moving up.

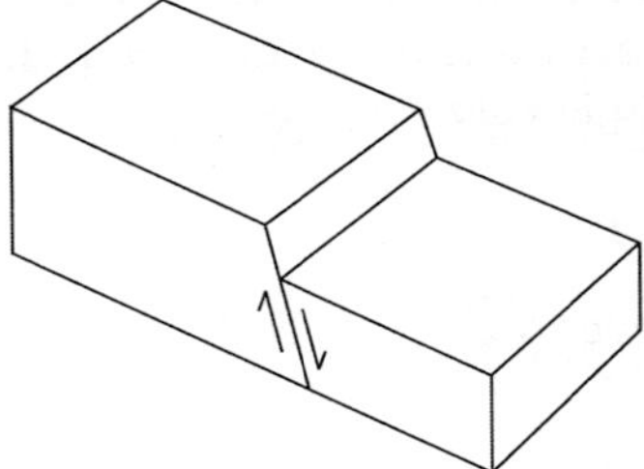

Fig. 10.10. A normal fault. the 'footwall' is on the 'upthrown' side of the fault, moving upwards. the 'hanging wall' is on the 'downthrown' side of the fault, moving downwards.

Reverse faults are cracks formed where one plate is pushing into another plate. They also occur where a plate is folding up because it's being compressed by another plate pushing against it. At these faults, one block of rock is sliding underneath another block or one block is being pushed up over the other. A reverse fault is defined by the hanging wall moving up relative to the footwall, which is moving down.

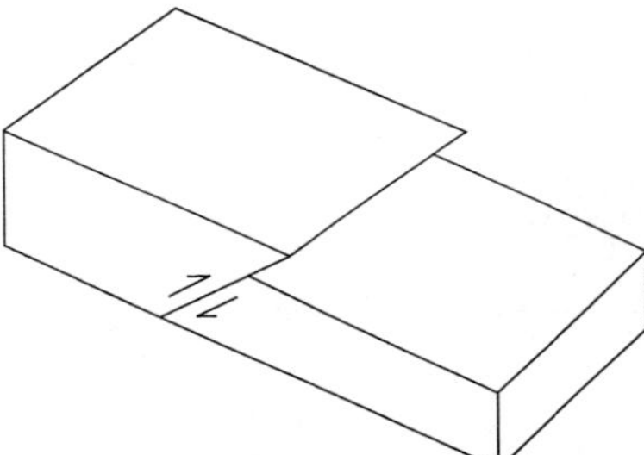

Fig. 10.11. A reverse fault. this time, the 'footwall' is on the 'downthrown' side of the fault, moving downwards, and the 'hanging wall' is on the 'upthrown' side of the fault, moving upwards. when the hanging wall is on the upthrown side, it 'hangs' over the footwall.

Strike-slip faults are the cracks between two plates that are sliding past each other. You can find these kinds of faults in California. The San Andreas fault is a strike-slip fault. It's the most famous California fault and has caused a lot of powerful earthquakes.

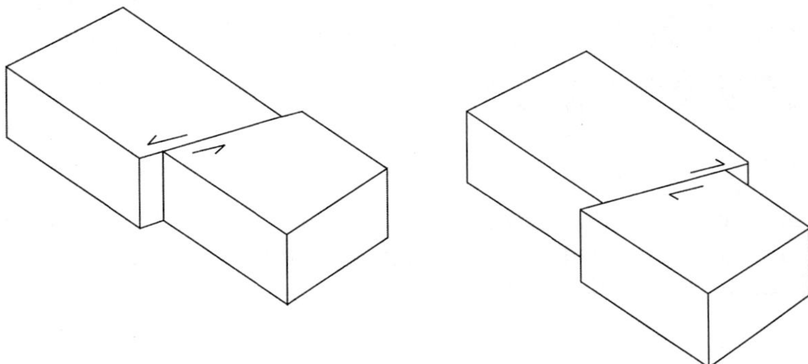

Fig. 10.12. Two strike-slip faults. (left), a left-lateral strike-slip fault. No matter which side of the fault you are on, the other side is moving to the left. (right), a right-lateral strike-slip fault. No matter which side of the fault you are on, the other side is moving to the right.

How Are Earthquakes Studied?

Seismologists study earthquakes by going out and looking at the damage caused by the earthquakes and by using seismographs. A **seismograph** is an instrument that records the shaking of the earth's surface caused by seismic waves. The term **seismometer** is also used to refer to the same device, and the two terms are often used interchangably.

The First Seismograph

The first seismograph was invented in 132 A.D. by the Chinese astronomer and mathematician Chang Heng. He called it an "earthquake weathercock."

Each of the eight dragons had a bronze ball in its mouth. Whenever there was even a slight earth tremor, a mechanism inside the seismograph would open the mouth of one dragon. The bronze ball would fall into the open mouth of one of the toads, making enough noise to alert someone that an earthquake had just happened. Imperial watchman could tell which direction the earthquake came from by seeing which dragon's mouth was empty.

In 136 A.D. a Chinese scientist named Choke updated this meter and called it a "seismoscope." Columns of a viscous liquid were used in place of metal balls. The height to which the liquid was washed up the side of the vessel indicated the intensity and a line joining the points of maximum motion also denoted the direction of the tremor.

Fig. 10.13. A Large-scale Model of Cheng Heng's Original Earthquake Weathercock.

Modern Seismographs

Most seismographs today are electronic, but a basic seismograph is made of a drum with paper on it, a bar or spring with a hinge at one or both ends, a weight, and a pen. The one end of the bar or spring is bolted to a pole or metal box that is bolted to the ground. The weight is put on the other end of the bar and the pen is stuck to the weight. The drum with paper on it presses against the pen and turns constantly. When there is an earthquake, everything in the seismograph moves except the weight with the pen on it. As the drum and paper shake next to the pen, the pen makes squiggly lines on the paper, creating a record of the earthquake. This record made by the seismograph is called a **seismogram**.

By studying the seismogram, the seismologist can tell how far away the earthquake was and how strong it was. This record doesn't tell the seismologist exactly where the epicenter was, just that the earthquake happened so many miles or kilometers away from that seismograph. To find the exact epicenter, you need to know what at least two other seismographs in other parts of the country or world recorded. We'll get to that in a minute. First, you have to learn how to read a seismogram.

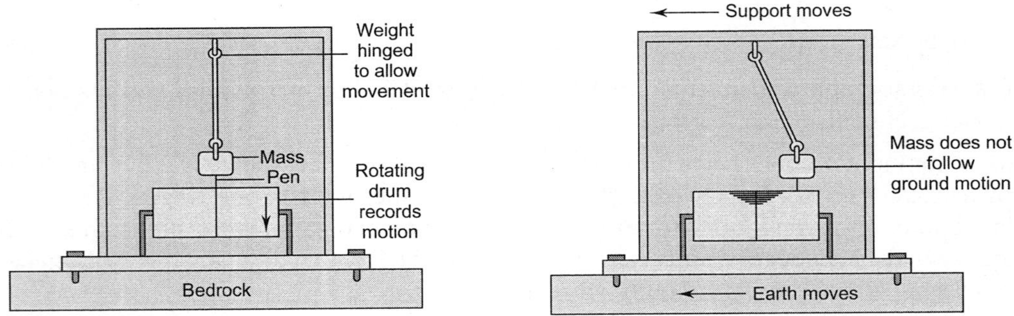

Fig. 10.14. Two illustrations of a Modern Seismograph in action (from Lutgens & Tarbuck, 1989).

Some Important Definitions

Earthquake Magnitude is a measure of the strength of an earthquake as calculated from records of the event made on a calibrated seismograph. In 1935, Charles Richter first defined local magnitude, and the Richter scale is commonly used today to describe an earthquake's magnitude.

Intensity of Earthquakes and their Scales

Earthquake intensity is a measure of the effects of an earthquake at a particular place. It is determined from observations of the earthquake's effects on people, structures and the earth's surface. Among the many existing scales, the Modified Mercalli Intensity Scale of 12 degrees, symbolized as MM, is frequently used.

The intensity of an earthquake becomes weaker outward from the epicenter. However, various types of ground respond differently to earthquake vibrations. Buildings on filled ground are damaged more than structures built on solid rock even though both may be at the same distance from the epicenter.

The magnitude of a particular earthquake is a single number which does not vary from place to place. Magnitude is the total energy released by an earthquake at its focus. Earthquakes of large magnitude are stronger and generally more destructive than those of small magnitude. The amount of destruction depends not only on the magnitude but on the kind of ground and types of buildings thereon, and on the location of the focus in relation to heavily populated areas.

Large earthquakes are preceded by many aftershocks, which may persist for days or weeks. The first shock is the most damaging. However, sometimes an aftershock may be even more powerful than the original shock.

The intensity of an earthquake is measured in terms of its geological effects and the overall damage it brings. There are two major scales in which earthquakes are measured. These two scales are the Mercalli Scale and the Richter Scale.

1. **Mercalli Scale.** The Mercalli scale was introduced at the turn of the 20th century by the Italian seismologist Giuseppe Mercalli. This scale measures the intensity of shaking with numbers from I to XII. Intensity I on this scale is defined as an event felt by very few people,

whereas intensity XII is assigned to a catastrophic event that causes total destruction. Events of intensities II to III are roughly equivalent to quakes of magnitude 3 to 4 on the Richter scale, and XI to XII on the Mercalli scale correspond with magnitudes 8 to 9 on the Richter scale.

(I) Hardly felt

(II) Felt only by a few persons at rest, especially on upper floors of buildings.

(III) Can be felt by persons indoors, especially on upper floors of buildings. Many people do not recognize it as an earthquake.

(IV) Felt indoors by many, outdoors by few during the day. At night, some awakened. Dishes, windows, doors disturbed. Standing motor cars rocked noticeably.

(V) Felt by nearly everyone; many awakened. Some dishes, windows broken. Unstable objects overturned.

(VI) Felt by all

(VII) Considerable damage in poorly built or badly designed structures.

(VIII) Damage slight in specially designed structures. Damage great in poorly built structures. Heavy furniture overturned.

(IX) Damage considerable in specially designed structures. Damage great in substantial buildings, with partial collapse. Buildings shifted off foundations.

(X) Many objects destroyed, buildings collapse.

(XI) Few structures remain standing. Bridges destroyed. Rails bent greatly.

(XII) Total Damage.

2. **Richter Scale .** The Richter scale was named after the American seismologist Charles Francis Richter. This scale measures the motion of the land surface 60 mi from the epicenter, or focus, of the earthquake. An estimated 800 quakes of magnitudes 5 to 6 occur worldwide each year. About 50,000 quakes of magnitudes 3 to 4 occur each year, and only about one of magnitude 8 to 9 each year.

Between to 0-4.3 on the Richter scale:

People at rest upstairs notice shaking.

Shaking felt indoors; hanging objects swing.

Between 4.3-4.8:

Sleeping people are awakened.

Dishes, doors and trees shake and rock.

Between 4.8-6.2:

Difficult to stand; people walk unsteadily.

Windows break; plaster,bricks, and tiles fall.

Between 6.2-7.3:

General panic.

Damage to foundations; buildings destoyed.

Water thrown out of river.

Between 7.3-8.9:

Total destruction; roads break up, rocks fall.

Large cracks appear in ground.

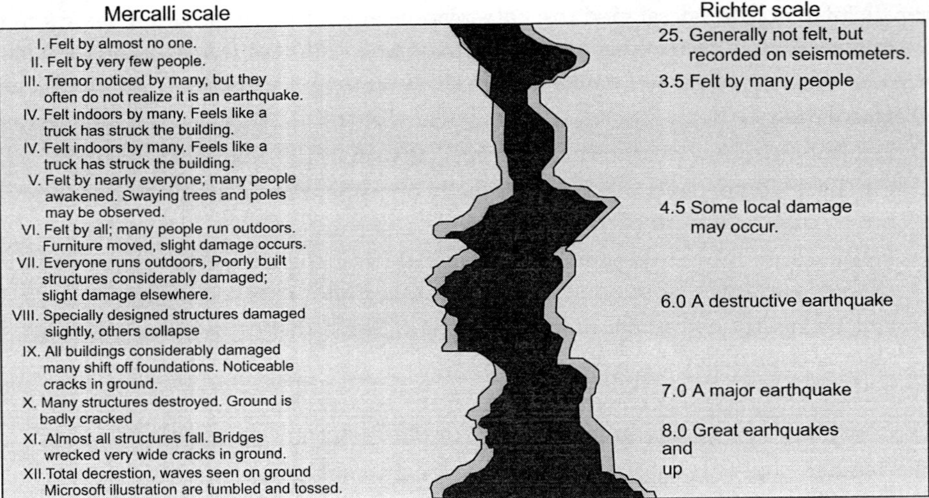

Focus of Earthquake

Earthquakes mainly occur when the different blocks or plates that makes up the Earth's surface move relative to each other causing distortion in the rock. The distortion builds up very slowly, over tens or hundreds of years. Eventually the distortion is enough to cause the rock to break and move, releasing energy in the form of an earthquake. The real place of origin of an earthquake is where the initial break propagates in the plates under the earth which releases energy along the fault. The earthquake originates from the entire fault – which may be 1000 km long. However, it is useful to refer to one part of the fault as the place of origin of an earthquake, and this is taken as the initial break. It is called the *Focus or Hypocenter* of the earthquake. This is usually below the Earth's surface.

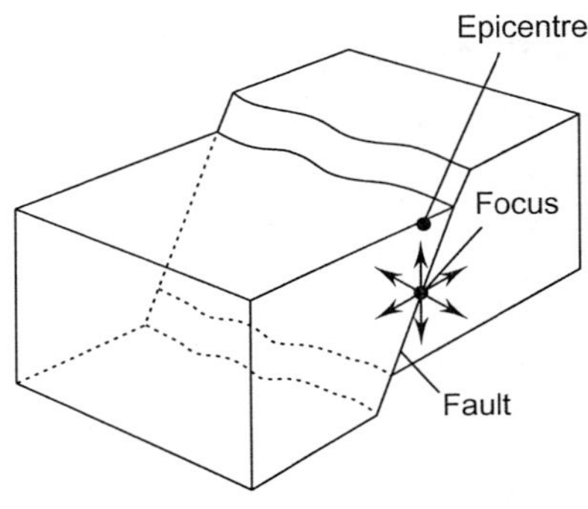

Fig. 10.15.

Most earthquake foci are within a few tens of kilometers of the surface. Earthquakes less than 70 km deep are classified as shallow-focus. Earthquakes with foci 70–300 km deep are classified as intermediate-focus and those below 300 km are deep-focus. Shallow-focus earthquakes occur more often than deeper ones; about 75% of the total energy released from earthquakes is from shallow-focus ones. Unfortunately, the shallower an earthquake, the more damage it can produce at the surface; intermediate and deep-focus earthquakes are rarely destructive. No earthquakes are known to have occurred below about 700 km. at greater depths the rocks are very hot and under high pressure so they deform by flowing rather than breaking and faulting.

Focal Depth of Earth

Focal depth of an earthquake is the distance from the surface of earth to the focus point of earthquake. The most obvious indication on a seismogram that a large earthquake has a deep focus is the small amplitude, or height, of the recorded surface waves and the uncomplicated character of the Primary and Secondary waves. Although the surface-wave pattern does generally indicate that an earthquake is either shallow or may have some depth, the most accurate method of determining the focal depth of an earthquake is to read a depth phase recorded on the seismogram. The focal depth can be calculated from measurements based on seismic wave phenomena. As with all wave phenomena in physics, these is an uncertainty in such measurements that grows with the wavelength so the focal depth of the source of these long-wavelength (low frequency) waves is difficult to determine exactly. Very strong earthquakes radiate a large fraction of their released energy in seismic waves with very long wavelengths and therefore a stronger earthquake involves the release of energy from a larger mass of rock.

Epicenter

It is the point on the earth's surface that is directly above the focus.Its the location on the surface of the Earth that is above the focal point of an earthquake. From this point the waves of the earthquake radiate outward like the ripples in a pond when a stone is thrown in.

Epicentral Distance

During an earthquake, Seismic waves propagate spherically out from the hypocenter. Seismic shadowing occurs on the opposite side of the Earth from the earthquake epicenter because the liquid outer core refracts the longitudinal or compressional (P-waves) while it absorbs the transverse or shear waves (S-waves). Outside of the seismic shadow zone both types of wave can be detected, but due to their different paths through the Earth, they arrive at different times. By measuring the time difference on any seismograph as well as the distance on a travel-time graph at which the P-wave and S-wave have the same separation, geologists can calculate the distance to the earthquake's epicenter. This distance is called the *epicentral distance*, commonly measured in ° (degrees) and denoted as Δ (delta) in seismology.

Once epicentral distances have been calculated from at least three seismographic measuring stations, it is a simple matter to find out where the epicenter was located using trilateration.

Earthquake Energy

The amplitude of a seismic signal is proportional to the energy (E) released in an earthquake. There is no consensus on the actual equation. This is based on the Richter magnitude but also depends on the frequency of the wave, distance from the source, etc., and is of the form $\log E = aM + b$. The annual energy dissipation by this means is estimated to be more than 10^{11} watts, of which 75% is released in shallow earthquakes and 3% in deep.

Surprisingly, considering the forces that we see at the surface, not all the energy involved in an earthquake is transformed into ground motion; much of it is consumed by frictional heating and rock pulverization. An analysis of pulverized rock, or gouge, collected from the San Andreas Fault in California and from a rupture zone at 2 km depth in a South African gold mine contradicts the commonly held view that gouge surface energy is a negligible component of earthquake energy balance. The analysis reveals unexpected diminutive particles in gouge samples, indicating that pulverization, not frictional heating, is the major consumer of earthquake energy at depth. This might explain the heat flow anomaly seen in the San Andreas fault system and perhaps some other long-standing puzzles in earthquake energetics.

Basic Concepts of Earthquake Resistant Houses and Construction Practices

Step 1: Design ground beams for earthquake-resistant housing. Join steel beams together by the feet and allow them to go outside the line of the building to decrease the chance of the structure overturning.

Step 2: Build floors with a light material similar to the roof. Use plywood, chipboard or plain timber with joists bolted firmly to the floor for earthquake-resistant housing.

Step 3: Ensure that buildings resist sideways pressure. Build the roof with a lightweight material. Brace the building diagonally to resist sideways loads of debris.

Step 4: Construct wood-frame housing. Although it's lightweight, wood provides a strong resistance to earthquakes. Tie wall wells to the house foundation when building and nail wood firmly to studs to increase the earthquake resistance of the structure. Use a shear wall construction with studs forming the wall's shell and horizontal joists making up the floors and rafters supporting the roof.

Step 5: Establish a large panel system for residences. Set up concrete panels vertically and horizontally to surround the rooms in a apartment building. These panels uphold the building's structural integrity along with vertical and horizontal floor panels.

Step 6: Use a modular building system. Stack ready-made boxes and connect them together. The walls that form resist shaking caused by earthquakes from both directions. Pipes as well as wall and floor finishes fit in an earthquake-resistant concrete box.

Concept of Seismic Zoning

A seismic zone is a region in which the rate of seismic activity remains fairly consistent. This may mean that seismic activity is incredibly rare, or that it is extremely common. Some people often use the term "seismic zone" to talk about an area with an increased risk of seismic activity, while others prefer to talk about "seismic hazard zones" when discussing areas where seismic activity is more frequent.

Many nations have government agencies concerned with seismic activity. These agencies use the data they collect about seismic activity to divide the nation into various seismic zones. A number of different zoning systems are used, from numerical zones to colored zones, with each number or color representing a different level of seismic activity. In the United States, for example, the seismic zones are divided between one and five, with zone five being the most at risk of seismic activity.

Most high-activity seismic zones are located along what are known as fault zones, regions of the Earth's crust which are prone to seismic activity. Fault zones often occur where continental plates meet, but they can also be found around volcanoes. A major fault zone in North America far from any plate boundaries is caused by a huge bubble of magma under the Earth's crust which periodically bubbles up into an explosive volcanic eruption.

By breaking a country up into different seismic zones, a nation can identify areas which are at increased risk. These areas may have more stringent building codes which are designed to make them safer in the event of an earthquake, and emergency services in a high risk seismic zone may be required to have special earthquake training and frequent drills to practice responding to an earthquake. Insurance companies usually also increase their rates in a high-activity seismic zone.

One of the biggest hazards beyond the basic shaking of an earthquake for people in a high-activity seismic zone is liquefaction. Soil liquefaction occurs when loose sediments become suspended in water as a result of seismic activity which pushes the water table upward. When soil liquefaction occurs, the ground is no longer able to support the weight of buildings, highways, and other structures, causing collapses to occur. Underground utility lines may be severed in the process, resulting in potentially explosive leaks of gas along with widespread contamination with sewage. In high-activity zones which are also subject to liquefaction, an earthquake can be extremely dangerous.

Seismic Zoning Map of India

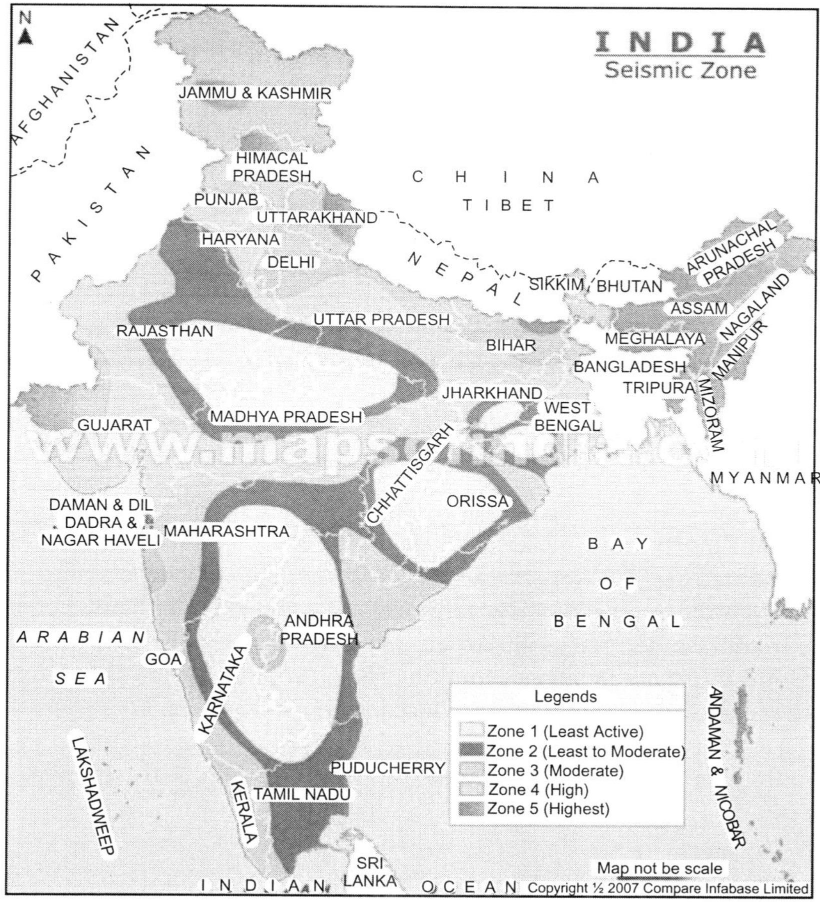

Fig. 10.16.

Important Terms

Aftershocks

When one earthquake happens, others usually take place at nearby locations. The earthquake with the largest magnitude is called the mainshock; anything before it is called a foreshock and anything after it is called an aftershock.

Bigger earthquakes have more and larger aftershocks. As the magnitude of the mainshock increases, the magnitude of the largest aftershock increases as well. In general, an earthquake large enough to cause damage will produce several felt aftershocks within the first hour. The rate of aftershocks dies off quickly with time.

Bolt

One of the most common causes of damage to houses in an earthquake is when they slide off their foundation. Bolts are used to secure the house to its foundation which can prevent this major cause of damage.

Crust

The outermost layer of the Earth is called the crust. The difference between the crust and mantle is chemical, with different, denser rocks in the mantle.

Epicenter

The epicenter of an earthquake is the point on a fault where an earthquake starts. Earthquakes happen because of sudden slip on a fault. The stress in the Earth's crust pushes on the fault holding it together and causing frictional resistance between the rocks on each side. Eventually enough slip is built up and the rocks slip suddenly, releasing energy in the form of sound waves and shear waves that travel through the rock to cause the shaking that we feel as the earthquakes. Your distance from the fault plane is the most important factor in determining how much shaking you feel from a particular size earthquake.

Fault

Earthquakes always occur on faults. Faults are places in the earth where the rocks are broken and the rocks on one side have moved in some direction relative to the other. There are three types of faults --- thrust, normal and strike-slip. A thrust occurs on a plane that is at an angle to the surface such that one side slides up over the other. Normal faults are also on angled fault planes, but one side moves down and away from the other. Strike-slip events occur on vertical faults and one side slides by the other. Strike-slip and thrust are the most common types in southern California. Every point on the fault plane releases energy so bigger faults produce bigger earthquakes. We can thus estimate how big an earthquake to expect by mapping the length of the fault. The longest fault in southern California is the San Andreas fault that is long enough to produce a magnitude 8 earthquake.

Intensity

Magnitude represents the energy released in the earthquake and is not what you feel in the event. Intensity is judged by the size of the area affected by the earthquake and the damage that it does on that area.

Magnitude

Magnitude is the most commonly reported measure of an earthquake's size. They defined it in terms of the amplitude of ground velocity recorded on a particular seismograph, scaled by the distance from the instrument to the earthquake.

P-waves

Earthquakes produce three general types of waves to radiate energy. Two are body waves, which means that they travel through the body of the Earth and the other is surface waves, which means that they travel along the surface of the Earth. The two body waves are called P waves and S waves. P-waves are compressional waves while S waves are shear waves. Shear waves cannot travel through a fluid so P-waves are the only ones that travel through the Earth's core.

P waves travel faster, but S waves are usually 2–3 times larger than the P wave. The time between the P and S wave can tell you how far away the earthquake is.

Richter Scale

The Richter scale is used to describe the magnitude scale, or the power of the earthquake.

Seismograph

We record earthquakes through the use of seismographs. Seismographs are instruments that create an electrical signal when the ground moves. By comparing the signal from many seismographs at different locations we can determine if an earthquake occurred and by timing the waves that travel from the earthquake, determine where and when it occurred.

Shaking

The shaking one feels during an earthquake is not the direct movement of the fault but rather the movement carried by the waves generated because of that slip. This shaking, and not direct fault motion, is the primary cause of damage in an earthquake. Both thrust and strike-slip earthquakes produce both vertical and horizontal shaking.

Strike-slip

Strike-slip faults are fault planes where the plane of the fault is vertical and the rock blocks on either side move horizontally during an earthquake. The San Andreas is an example of a strike-slip fault in southern California.

Tsunamis

When thrust-faulting earthquakes happen under the ocean, the earthquake can push large blocks of ocean floor up. When the ocean floor moves up, the water that was in that spot has to go somewhere else. That somewhere else is into a large wave called a tsunami.

Uplift

Uplift describes how the ground is lifted up during earthquakes. Many of southern California's mountains have formed because of repeated earthquakes. How tall the mountain depends both on how fast the faults are pushing it up and how long that fault has been active. How steep a mountain is tells you how fast its fault is moving.

Waves

A wave travels through a material when a force pushes on that material and the material resists being pushed. Earthquakes produce two types of body waves, called P- and S-waves. P waves are sound waves. The particle motion in all sound waves is compressional which means that the particles move in the same direction as the wave as described in the example above. S waves are shear waves. The particle motion in shear waves is perpendicular to the direction of the wave. Shear waves are only seen in solids and cannot transmit through a liquid or a gas.

EXERCISE

1. What is seismic wave? Discuss its different types.

2. What is Earthquake? Why earthquake happens? Give its reasons.

3. What causes plates to move?

4. Write short notes on -

(*i*) Types of Earthquakes

(*ii*) Types of faults

(*iii*) Mircalli scale

(*iv*) Richter scale

(*v*) Earthquake Energy

(*vi*) Seismic zoning

5. Define Focus and focal depth. Also define epicenter and epicentral distance.

INDEX